THE Finite Difference Time Domain Method for Electromagnetics

KARL S. KUNZ
RAYMOND J. LUEBBERS

CRC Press
Taylor & Francis Group
Boca Raton London New York

CRC Press is an imprint of the
Taylor & Francis Group, an **informa** business

CRC Press
Taylor & Francis Group
6000 Broken Sound Parkway NW, Suite 300
Boca Raton, FL 33487-2742

First issued in paperback 2019

ISBN-13: 978-0-8493-8657-2 (hbk)
ISBN-13: 978-0-367-40237-2 (pbk)
Library of Congress Card Number 92-33535

Library of Congress Cataloging-in-Publication Data

Kunz, Karl S.
The finite difference time domain method for electromagnetics / by Karl S. Kunz and Raymond J. Luebbers.
p. cm.
Includes bibliographical references and index.
ISBN 0-8493-8657-8
1. Electromagnetic fields—Mathematics. 2. Time-domain analysis I. Luebbers, Raymond J. II. Title.
QC665.E4K82 1993
530.1′41—dc20 92-33535
CIP

Visit the Taylor & Francis Web site at
http://www.taylorandfrancis.com

and the CRC Press Web site at
http://www.crcpress.com

AUTHORS

Karl S. Kunz, Ph.D., has worked in industry, government, and academia since receiving his Ph.D. in physics at NMSU in 1971. Now at Penn State, as Professor, he specializes in computational engineering, in particular FDTD applications and developments for modeling aerospace systems, antennas, biological systems, and plasmas to name just a few areas.

His past associations include Lawrence Livermore National Laboratory (LLNL), TRW, Mission Research Corporation, BDM, and White Sands Missile Range. He was in charge of the Microwave and EMP Group at LLNL with responsibility for their indoor transient range. At TRW he was a Technology Section Leader and an EMP Group Leader at MRC. While at BDM he was the ARES EMP Simulator Scientist. At White Sands he worked as a Systems Analyst spending part of that time at the Madison NJ Bell Telephone facilities.

Dr. Kunz is an active consultant having worked with LLNL, Chemring LTD, Lockheed, EG&G, and Dynaeletron. He performs some of this work out of the consulting firm he founded in the late 70s, Kunz Associates, Inc., with his father, Kaiser S. Kunz. While at Penn State he served as Department Head from 1986 to 1989 and has enjoyed an active research career throughout his tenure there.

Raymond Luebbers, Ph.D., is Professor of Electrical Engineering at the Pennsylvania State University. He received the B.S.E.E. degree from the University of Cincinnati, and M.S.E.E. and Ph.D. degrees from the Ohio State University, where he was a member of the ElectroScience Laboratory. He has been an electrical engineering faculty member at Ohio University, and a Research Scientist at the Lockheed Palo Alto research laboratory. He was also a Visiting Professor at Tohoku University, Sendai, under the National Science Foundation program for U.S./Japan Cooperative Research. Dr. Luebbers has written series of technical papers in several areas, including frequency selective surfaces, geometrical theory of diffraction, and FDTD. He is a Senior Member of the Antennas and Propagation Society of the IEEE, and is on the Board of Directors of the Applied Computational Electromagnetics Society. He has chaired a number of IEEE and URSI conference sessions, including an URSI General Assembly session, and reviews articles for several technical journals.

AUTHORS

PREFACE

The finite difference time domain (FDTD) approach is rapidly becoming one of the most widely used computational methods in electromagnetics. There are several reasons for this, including the increased availability of low cost but powerful computers, and increasing interest in electromagnetic interactions with complicated geometries, which include penetrable dielectric and/or magnetic materials. Just as important perhaps is the extreme simplicity of the method. The fundamentals of FDTD can be grasped easily by undergraduate students, more easily than traditional frequency-domain approaches to electromagnetics. Yet FDTD is capable of computing electromagnetic interactions for problem geometries that are extremely difficult to analyze by other methods. It is this combination of simplicity and power that makes FDTD such a popular method.

This also allows this book to serve a wide range of potential readers. It can be used to introduce undergraduate students to time domain electromagnetics, in which case only the first two parts of the book need be covered. It can be used in a graduate level course, in which case the mathematical basis of FDTD would be emphasized with topics selected from the sections on special capabilities and advanced applications at the discretion of the instructor. Finally, someone who wants to use FDTD to solve a particular problem in electromagnetics can use the book to learn FDTD basics and special capabilities necessary for their application. In addition, it can be seen, through the examples in this book, how FDTD has been applied to a variety of problems.

The goals of the book are to provide the basic information necessary to apply FDTD to problems in electromagnetics, and to illustrate some of the types of problems that can be analyzed using it. The theoretical and mathematical basis for much of FDTD is included, but the emphasis is on the practical aspects of applying FDTD.

While equations are given in the text as needed to develop and understand the method and applications, complete detail is not included. For example, equations usually are not given for all vector components but only for one representative component. There are two reasons for this. One is to avoid a book filled with pages of almost but not quite identical equations. The other is that the book includes FORTRAN listings of a complete 3-D FDTD program based on the concepts presented in Part 1. This provides better documentation of the actual details of the method than equations included in the text.

While a joint effort of the two authors, the primary responsibilities were divided between the authors by chapter. Karl Kunz wrote the majority of Chapters 1 and 2, all chapters in Part 2, the last section of Chapter 10, Chapter 12, all chapters in Part 5, and Appendix A. Raymond Luebbers wrote the majority of the remaining chapters, i.e., Chapter 3, Chapters 7 through 11 (except for the last section of Chapter 10), all chapters in Part 4, and Appendix B. Despite this division of labor we have attempted to make the book consistent throughout in both notation and philosophy.

In addition to the authors themselves, many people contributed to this book in many ways, both directly and indirectly. Acknowledging all of them is not possible, but some contributions are too important to remain unmentioned. First, let me acknowledge the tremendous effect Karl Kunz has had, not just on this book, but on my career. He introduced me to FDTD and initiated the project of writing this book. Several others made direct contributions. Chapter 15 is closely based on the work of Forrest P. Hunsberger, Jr. John Beggs is responsible for much of the content of Chapter 9. Both were graduate students at Penn State when this work was done. Other students who contributed results which are incorporated in the book include Deirdre Ryan and David Steich of Penn State, and Li Chen and Ken Kumagai of Tohoku University, Sendai, Japan. Results provided by Chris Trueman and Stan Kubina of Concordia University in collaboration with Shantnu Mishra and C. Larose of David Florida Laboratory are also greatly appreciated. Kent Chamberlin of the University of New Hampshire provided a critical review of the original manuscript and his numerous suggestions for improvement have been incorporated.

While many people provided encouragement, support, or technical advice, a few provided these in greater measure and deserve acknowledgment: Vaughn Cable, Randy Jost, Doug Havens, Tom Campbell, Fred Beck, Alex Woo, Scott Ray, Ken Demarest, Kane Yee, Ron Standler, Richard Holland, Kultegin Aydin, Michael Schneider, Saburo Adachi, Kunio Sawaya, Toru Uno, Linda Kellerman, and Julie Corl. Without the support of these people this book would not have its present form.

Raymond Luebbers
University Park, PA
August 1992

FOREWORD

Any book on the finite difference time domain (FDTD) technique can expect to be out of date as soon as it is published. We hope that this is the first book devoted to FDTD and will therefore gladly suffer this fate. We must apologize to the reader, however, as the reader would undoubtedly prefer to have a timeless text. We have tried to provide some measure of timelessness combined with some measure of currency.

The timeless elements are the applications of FDTD which are as broad as the Maxwell equations they embody in discretized form, the basic formulation using centered finite difference expressions which can treat virtually any material type and geometry and the basic "housekeeping" needed to implement the technique. Our introduction and first section covers these matters and should be of lasting value.

We choose next to address some basic modeling issues well suited to FDTD — exterior and interior coupling, waveguide propagation and coupling and scattering from lossy dielectrics. These cases illustrate the operation of the basic code and also provide, for the most part, the historical origins of the particular formulation of FDTD we emphasize, the scattered field formulation. This formulation is easy to implement as a code, is only slightly more complex in the equations employed than its alternative, the total field formulation, and offers some pleasing physical insights. Part 2, with these basic applications, should also remain useful as a pedagogical necessity for introducing the reader to the actual use of FDTD. One can argue which "history" is the best one to use. We took the direct approach and used what we have been personally involved with. *Chacun á son goût.*

One of the reasons for stating how short an expiration date comes with any version of FDTD arises from the work we did for Part 3. Here, we have placed some of our best work, extensions to the basic FDTD capabilities. Some are very basic extensions, near to far zone capabilities and frequency-dependent material modeling capabilities to name two. The need for these capabilities will not diminish, but we can expect other researchers to improve on what we offer here. This is all to the good and we trust the reader will make themselves aware of any evolutions in these areas after reading our book which we would consider current as of 1992.

A fourth section trades timelessness for currency. It is our section on advanced applications and it makes use of the extensions of the prior section. We know that in time this will become old news and perhaps only hold a little bit of historic interest. It is today "hot off the press" and tells better than anything else where some of the research frontiers are and how hard a straightforward technique such as FDTD has to be pushed to get there.

Finally, we close with some of the mathematical foundations and some of the alternatives available to FDTD — a mix of currency and timeliness. A FORTRAN listing of a simple version of the code completes our offering.

My colleague, Professor Raymond Luebbers, and I have had a stimulating time putting this book together. I wish to recognize and say that I much admire the pioneering work he has contributed to this effort, much of which appears in Parts 3 and 4. We both remain active in FDTD research and are working to make further contributions. I expect that we will need to do this in collaboration with the many students and friends Professor Luebbers so ably acknowledges in the Preface; their contributions have been invaluable.

To our friend, Dr. Kane Yee, who started all this in 1966, we send our regards and our thanks.

Karl S. Kunz

TABLE OF CONTENTS

Part 3: Special Capabilities

Part 5: Mathematical Basis of FDTD and Alternate Methods

Chapter 1

INTRODUCTION

Of the four forces in nature — strong, weak, electromagnetic, and gravitational — the electromagnetic force is the most technologically pervasive. Of the three methods of predicting electromagnetic effects — experiment, analysis, and computation — computation is the newest and fastest-growing approach. Of the many approaches to electromagnetic computation, including method of moments, finite difference time domain, finite element, geometric theory of diffraction, and physical optics, the finite difference time domain (FDTD) technique is applicable to the widest range of problems.

This text covers the FDTD technique. Emphasis is placed on the separate field formalism, in which the incident field is specified analytically and only the scattered field is determined computationally. This approach is only slightly more complex in its basic implementation than the total field approach and more readily allows for the absorption of scattered fields at the limits of the problem space. The total field can be easily obtained from the combination of the scattered and incident fields. Though slight advantages may be found in either approach, they are very similar in concept and capabilities. The FDTD technique treats transients (e.g., pulses) in the time domain, and it is applicable over the computationally difficult-to-predict resonance region in which a wavelength is comparable to the interaction object size.

As a form of computational engineering, FDTD is part of a three-tier hierarchy consisting of:

Computer Science
- Stresses the mathematics underlying algorithms as well as the structure and development of the algorithm

Computer Engineering
- Hardware based on and concerned with hardware architecture and capabilities including parallelism and fault tolerance

Computational Engineering
- Explores various engineering problems via numerical solutions to systems of equations describing the phenomenon or process in question

Computational engineering relies on computer science and engineering, but is not hardware, language, or operating system specific. It requires a computer powerful enough to accomodate the problem in question, running within acceptable times and costs while producing the desired accuracy.

Electromagnetic computational engineering encompasses the electromagnetic modeling, simulation, and analysis of the electromagnetic responses of complex systems to various electromagnetic stimuli. It provides an understanding of the system response that allows for the better design or modification of the system.

The FDTD technique offers many advantages as an electromagnetic modeling, simulation, and analysis tool. Its capabilities include:

- Broadband response predictions centered about the system resonances
- Arbitrary three-dimensional (3-D) model geometries
- Interaction with an object of any conductivity from that of a perfect conductor, to that of a real metal, to that of low or zero conductivity
- Frequency-dependent constitutive parameters for modeling most materials
  - Lossy dielectrics
  - Magnetic materials
  - Unconventional materials, including anisotropic plasmas and magnetized ferrites
- Any type of response, including far fields derived from near fields, such as
  - Scattered fields
  - Antenna patterns
  - Radar cross-section (RCS)
  - Surface response
  - Currents, power density
  - Penetration/interior coupling

These capabilities are available for a variety of diverse electromagnetic stimuli covering a broad range of frequencies. Typical stimuli include:

- Lightning
- EMP (electromagnetic pulse)
- HPM (high power microwave)
- Radar
- Lasers

The systems responding to these stimuli are equally diverse. They can be small to large, inorganic or organic, in an exoatmospheric environment to a subterranean one. Samples of the diverse types of systems that can be treated are

- Aerosols
- Shelters
- Aircraft
- Humans
- Satellites
- Buried antennas

What ties the above stimuli and systems together is that typically the wavelengths of interest and the characteristic system dimensions are usually within an order of magnitude of each other. Thus, the broadband response predictions will typically encompass at least a few system resonances.

Hybrid techniques employing the geometrical theory of diffraction (GTD) or physical optics (PO) along with FDTD can in principle provide predictions from below resonance to extremely high frequencies. Numerical techniques, such as Prony's method, allow arbitrarily long-time response prediction extensions of the FDTD generated time response predictions. Thus, resonance and below-resonance predictions can be extended to extremely low frequencies. Alternately, low frequency versions of FDTD, in which the displacement current is ignored and the equations become diffusive, can be used to extend low frequency capabilities. In short, FDTD can span the critical resonance region over more than four orders of magnitude in frequency, and with low and high frequency extensions this range can exceed six orders of magnitude.

FDTD has run on a diverse set of host computers, ranging from PCs to supercomputers, and is extremely well suited to implementation on parallel computers because only nearest-neighbor interactions are involved. The important variables are problem space size in cells required to model the system and the number of time steps needed. These determine the computer run time and computation cost. Less important are the material types modeled and the number of response locations monitored. Of little or no impact on computational capability requirements are the type of stimuli and the type of response, except in cases of far fields, which require a modest amount of postprocessing.

Over 1 million cells can be accommodated on personal workstations for a 3-D problem space $100 \times 100 \times 100$ cells large. At a typical 10 FDTD cells per wavelength, this space is a 10-wavelength cube. The limits of today's supercomputers are reached at roughly 100 million cells with computation times on the order of hours.

The advantages of FDTD can be summarized as its ability to work with a wide range of frequencies, stimuli, objects, environments, response locations, and computers. To this list can be added the advantage of computational efficiency for large problems in comparison with other techniques such as the method of moments, especially when broadband results are required. Further, the FDTD code, while inherently volumetric, has successfully treated thin plates and thin wire antennas. Its accuracy, using a sufficiency of cells, can be made as high as desired. Conversely, engineering estimates of a few decibels' accuracy can be made with surprisingly few cells. Finally, powerful visualization tools are being developed to enhance the user's understanding of the essential physics underlying the various processes that FDTD can model, simulate, and analyze.

The basis of the FDTD code is the two Maxwell curl equations in derivative form in the time domain. These equations are expressed in a linearized form by means of central finite differencing. Only nearest-neighbor interactions need be considered as the fields are advanced temporally in discrete time steps over spatial cells of rectangular shape (as emphasized here, other cell shapes are possible, as are reduced 2- and 1-D treatments).

It should also be noted that at least six kinds of electromagnetic computational problems exist:

- Generation (power, devices, klystrons, etc.)
- Transmission (transmission lines, waveguides, etc.)
- Reception/detection/radiation (antennas)
- Coupling/shielding/penetration
- Scattering
- Switching/nonlinearities

All but the first can be treated using FDTD. The first area either requires, as for the klystron, the addition of charged particles or a 60-Hz calculation for power frequencies which require inordinately many time steps (on the order of 1 billion) for analysis.

While FDTD is most suited to computing transient responses, FDTD may be the computational approach of choice even when a single frequency or continuous wave (CW) response is sought. This is especially the case when complex geometries or difficult environments, such as an antenna that is buried or dielectrically clad, are considered. Interestingly, interior coupling into metallic enclosures is also a situation wherein FDTD is the method of choice. A CW analysis, using the method of moments, for example, will most likely fail to capture the highly resonant behavior of a metallic enclosure, even when made at many frequency points. The highly resonant nature of interior coupling was verified first experimentally and then computationally with FDTD. Indeed, low frequency resonances may be poorly characterized experimentally because of their extremely resonant behavior, but are revealed by FDTD runs of 1 million time steps.

This book is organized into an introduction followed by five sections:

1. Fundamental concepts
2. Basic applications
3. Special capabilities
4. Advanced applications
5. Mathematical basis of FDTD and alternate methods

The first section treats the most basic aspects of FDTD, providing the reader with the formalism and the basic procedures for FDTD operation. Along with the introduction describing FDTD's utility and areas of application, the two chapters of the first section allow the reader to apply FDTD to a host of problems using the nondispersive lossy dielectric FDTD code listed in Appendix B:

- Chapter 2: Scattered Field FDTD Formulation. Discusses the discretized central differenced or "leap frog" Maxwell curl equations upon which the separate field formalism is based. The equations are formulated for lossy dielectrics, which in the limit of infinite conductivity become perfect conductors. The rudimentary computer code requirements and architecture needed to support the formalism in an operational code are presented as well. Much more detail about these issues is given in later chapters.

- Chapter 3: FDTD Basics. Provides guidance for applying the FDTD formulations of Chapter 2. This includes limitations on cell and time step sizes, specifying the incident field and the object to be analyzed, estimating the computer resources required, and applying an outer boundary condition at the extremities of the FDTD computation space.

The second section treats the basic applications of the basic formulation of FDTD given in the first section:

- Chapter 4: Coupling Effects. The scattered field formulation of FDTD was first applied to an F-111 aircraft to calculate the induced surface currents and charges from a simulated EMP field. This brought together all the elements representative of FDTD modeling, and for this reason and for some sense of history the modeling effort is discussed in detail. Only exterior coupling is treated with this example and to complete the discussion of coupling, interior coupling modeling of a simple shield, penetrated by an aperture and containing an interior wire is presented. The response is examined above and below aperture cutoff with resonant and highly resonant behavior noted in the two regimes. A strongly stressed point is the large number of time steps needed to accurately characterize the highly resonant behavior of the currents induced on the interior wire.
- Chapter 5: Waveguide Aperture Coupling. A transient wave propagating in a waveguide and coupled to a second waveguide via circular aperture(s) is examined in this section. At issue is whether FDTD can be employed successfully to model waveguide behavior. It is shown to work well and FDTD modeling is being rapidly extended into this modeling regime.
- Chapter 6: Lossy Dielectric Scattering. Chapter 2 develops the scattered field FDTD formulation for perfect electric conductors and lossy dielectrics without frequency-dependent constitutive parameters. Here, the capability for modeling lossy dielectrics is applied to a simple sphere geometry where there are known analytic surface responses and to a complex human body geometry where there are experimental results available for comparison. Excellent to good agreement is obtained in these early modeling efforts. As will be seen in later chapters, the agreement has only gotten better. Where FDTD once had to "prove itself" it has become something of a standard. Results for the models of complex geometries are treated as nearly exact, with the limits established mainly by the skill of the practitioner in defining the geometry and in properly setting the constitutive parameters.

The third section treats extensions to the basic formulation that provide special capabilities. These extensions and special capabilities are quite varied. They represent in most cases the fruition of past research efforts to extend the basic capabilities of FDTD.

- Chapter 7: Near to Far Field Transformation. Many scattering problems, in particular RCS problems, require the far fields. In addition, radiation from antennas and inadvertent antennas or radiators such as transmission lines require far fields for the radiation pattern. While these problems typically involve perfect conductors/metals they can also involve more complex materials as for a stripline antenna. This chapter develops a broadband time domain near to far field transformation and shows how it may be applied to 2- or 3-D problems.
- Chapter 8: Frequency Dependent Materials. An advantage of the FDTD method is its capability to produce wide frequency band results from one computation with pulse excitation. If materials with constitutive parameters that vary with frequency are involved, the FDTD formulation of Chapter 2 must be extended to include this variation if more than an approximate result is desired.
- Chapter 9: Surface Impedance. In frequency domain calculations, finding the fields interior to a volumetric scatterer can be avoided by specifying the impedance relating the electric and magnetic fields at the surface of the scatterer. This concept is extended to the time domain for materials with both constant and frequency-dependent constitutive parameters.
- Chapter 10: Subcellular Extensions. Often an FDTD calculation includes a structural element much smaller than one cell size for calculation. Using a finer mesh throughout the problem space is typically too computationally "expensive". In this case, a subcellular reduction of the mesh on and (possibly) about the element is needed. Examples treated in this chapter include "thin" wires for antennas with diameters well below a cell size, lumped circuit elements, and the expansion technique for regridding a subvolume more finely, as in the interior of an aircraft where important structural details would otherwise be lost.
- Chapter 11: Nonlinear Loads and Materials. Nonlinear materials are more easily accomodated in FDTD than in frequency domain methods. Some examples, including transients in antennas with nonlinear loads, are given.
- Chapter 12: Visualization. Immense amounts of data can be generated with FDTD, in the terabyte range in some instances. Only by applying visualization techniques to such data can it be rendered readily comprehensible. This chapter discusses the progress made in the visualization of electromagnetic fields. It is noted that this area of visualization is in itself very computationally intensive and demanding.

The three chapters of the fourth section treat advanced applications made possible by the extensions developed in Part 3:

- Chapter 13: Far Zone Scattering. The scattered field FDTD formulation given in Chapter 2 is extremely well suited to scattering calculations. Several of the special capabilities of Part 3 are combined to provide scattering cross-section results, including the far zone transformation and frequency-dependent materials. Scattering examples are also given in Chapters 7 and 9 in association with the development of far zone transformation and surface impedance.
- Chapter 14: Antennas. A basic approach to using FDTD to determine antenna self and mutual impedance, efficiency, and gain are presented for wire antennas, followed by results for more challenging geometries. Total fields are directly computed, and far zone transformation and sub-cell methods from previous chapters are utilized.
- Chapter 13: Gyrotropic Media. Gyrotropic media possess both strong frequency dependence of their constitutive parameters and anisotropy. In this chapter the frequency-dependent FDTD methods of Chapter 8 are extended to include these materials. Both magnetized plasmas and ferrites are considered.

The final part of the book provides the detailed mathematical foundations for FDTD and related techniques, and includes:

- Chapter 16: Difference Equations in General. The curl equations, from which a wave equation may be derived and vice versa, are a set of hyperbolic equations for which a number of differencing schemes are possible. In the interest of mathematical completeness, the different types of possible differencing schemes are presented. The advantages, even necessity in a practical sense, of the "leapfrog" method is stressed. Higher order formulations of the "leapfrog" method are also discussed.
- Chapter 17: Stability, Dispersion, Accuracy. The stability requirement for the "leapfrog" method (and for several other methods), the Courant stability condition, is discussed in detail here along with general stability considerations. Numerical dispersion, a source of error that exists in FDTD computations except under special conditions, spreads or disperses the scattered field leading to time domain envelope errors and frequency domain phase errors.
- Chapter 18: Outer Radiation Boundary Conditions. A finite problem space is subject to reflections of scattered fields at the faces of the problem space. These can be minimized by the selection of an outer radiation boundary condition that absorbs much of the scattered wave, simulating energy scattering into infinite space. A number of approaches exist, and a general discussion of the different approaches is given.

Emphasis is placed on the widely employed Mur absorbing boundary condition.

- Chapter 19: Alternate Formulations. While the scattered field FDTD formulation is extremely flexible and effective, other formulations have their place and are treated here. Most commonly encountered is the total field formalism, an approach that can be obtained (quite simply in many cases) from the scattered field formalism. Less commonly encountered are implicit, as opposed to the explicit, FDTD formulations given in the prior discussions. The sought-after advantage in implicit schemes is arbitrarily long time steps. Another variation is the vector potential formulation. Both low and high frequency extensions of the FDTD technique are also possible.

Appendix A gives FDTD equations in other coordinate systems and reduced dimensions. While less useful than the general 3-D results given in the previous text, they have application in special situations.

Appendix B contains FORTRAN listings of basic FDTD codes and associated computer codes including a fast Fourier transformation (FFT) code. These listings provide precise documentation of the basic FDTD equations given in the text.

While providing a mathematical basis for FDTD, this book is intended more as a practical guide for students and researchers interested in applying FDTD to actual problems in electromagnetics. This intention is the basis for the organization of the book, with practical fundamentals preceding the mathematical basis.

Because FDTD is an area of ongoing research and development, this book can provide only the fundamentals and some example applications, with more advanced topics available only from the current literature.

PART 1: FUNDAMENTAL CONCEPTS

Chapter 2

SCATTERED FIELD FDTD FORMULATION

2.1 MAXWELL CURL EQUATIONS

We begin by examining the differential time domain Maxwell equations in a linear medium:

$$\nabla \times E = -\partial B / \partial t \tag{2.1}$$

$$\nabla \times H = \partial D/\partial t + J \tag{2.2}$$

$$\nabla \cdot D = \rho \tag{2.3}$$

$$\nabla \cdot B = 0 \tag{2.4}$$

where

$$D = \varepsilon E \tag{2.5}$$

$$B = \mu H \tag{2.6}$$

This is all the information needed for linear isotropic materials to completely specify the field behavior over time so long as the initial field distribution is specified and satisfies the Maxwell equations. Conveniently, the field and sources are set to zero at the initial time, often taken as time zero. The two divergence equations are in fact redundant as they are contained within the curl equations and the initial boundary conditions.

Thus, the starting point for the FDTD formulations is the curl equations. They can be recast into the form used for FDTD:

$$\partial H/\partial t = -\frac{1}{\mu}(\nabla \times E) - \frac{\sigma^*}{\mu}H \tag{2.7}$$

$$\partial E / \partial t = \frac{-\sigma}{\varepsilon}E + \frac{1}{\varepsilon}(\nabla \times H) \tag{2.8}$$

where we have let $J = \sigma E$ to allow for lossy dielectric material and have included the possibility of magnetic loss by adding a magnetic conductivity term σ^*.

The formulation only treats the electromagnetic fields, E and H, and not the fluxes, D and B. All four constitutive parameters, ε, μ, σ, and σ^*, are present so that any linear isotropic material property can be specified. In the formulation developed here we do not attempt to simplify the calculations by normalizing to a unity speed of light, or letting the permittivity and permeability of free space to be normalized to 1, as proposed by some practitioners. We feel that this removes the intuitive and physical basis of the calculations for very little, if any, gain in computational accuracy or speed.

It is easily shown that we need only consider the curl equations as the divergence equations are contained in them. To do this simply take the divergence of the curl equations (2.1 and 2.2) to obtain

$$\nabla\cdot(\nabla\times E = -\partial B/\partial t) \rightarrow 0 = -\partial(\nabla\cdot B)/\partial t \rightarrow \nabla\cdot B = \text{constant}$$

$$\nabla\cdot(\nabla\times H = \partial D/\partial t + J) \rightarrow 0 = \partial(\nabla\cdot D)/\partial t + \nabla\cdot J$$

$$\rightarrow \partial(\nabla\cdot D)/\partial t - \partial\rho/\partial t \ (\text{from continuity } \nabla\cdot J + \partial\rho/\partial t = 0)$$

$$\rightarrow \frac{\partial}{\partial t}\left[(\nabla\cdot D) - \rho\right] = 0 \rightarrow \nabla\cdot D - \rho = \text{constant}$$

where we have used the vector identity $\nabla\cdot\nabla\times A = 0$. Because the fields and sources are initially set to zero in FDTD calculations, at that initial time

$$\nabla\cdot B = 0$$

$$\nabla\cdot D - \rho = 0$$

Therefore, $\nabla\cdot B$ and $(\nabla\cdot D - \rho)$ must be zero for all times and the curl equations are sufficient for FDTD calculations.

Note that while the divergence equations are not part of the FDTD formalism they can be used as a test on the predicted field response, so that after forming $D = \varepsilon E$ and $B = \mu H$ from the predicted fields, the resulting D and B must satisfy the divergence equations.

2.2 SEPARATE FIELD FORMALISM

The two curl equations (2.7 and 2.8) can be discretized to obtain a total field FDTD technique. Alternately the fields can be expressed as

$$E = E^{total} \equiv E^{incident} + E^{scattered} \tag{2.9}$$

$$H = H^{total} \equiv H^{incident} + H^{scattered} \tag{2.10}$$

The rationale for the separate field approach is that the incident field components can be specified analytically throughout the problem space while the scattered fields are found computationally and only the scattered fields need to be absorbed at the problem space outer boundaries. This last feature is an important one. The scattered fields, emanating from a scattering or interaction object, can be more readily absorbed than a total field by an outer radiation boundary condition applied at the problem space extremities or faces. This is especially important in situations in which the scattered fields are desired and are of much lower amplitude than the total fields.

On a more philosophical level, this separation allows further insight into the interaction process.

The scattered wave arises on and within the interaction object in response to the incident field so as to satisfy the appropriate boundary conditions on or within the interaction object. These boundary conditions are the Maxwell equations themselves, which in the limit of a perfect conductor require $E^{scattered} = -E^{incident}$ in the scatterer. For anything other than a perfect conductor the scattered fields depend on the constitutive parameters of the material. The scattered fields are subject to the Maxwell equations for this media when in the media, while outside the media they satisfy the free space Maxwell equations. The incident field always propagates in free space (even when passing through the interaction object or scatterer material) and is defined as the field that would be present in the absence of the scatterer.

It is always possible to combine the scattered and incident field to obtain the total field and with it all the insight the total field behavior provides. Also, if total field FDTD equations are desired, they can be obtained from the scattered field equations by setting the incident field to zero and applying initial conditions to the scattered (now also total) field. This is discussed in Section 3.6 and applied in Chapter 14, when total radiated fields from antennas are directly calculated.

Incident and scattered fields must satisfy the Maxwell equations independently (we assume here linear materials). The incident field is specified to be propagating in free space. Free space could be generalized to a uniform media (soil, for example) if necessary. However, we shall assume free space for simplicity. While the incident field travels through free space throughout the problem space, the total field propagates in free space outside the scatterer and in the media of the scatterer when it is propagating within the scatterer.

In the media of the scatterer, the total fields satisfy

$$\nabla \times E^{total} = -\mu \partial H^{total} / \partial t - \sigma^* H^{total} \tag{2.11}$$

$$\nabla \times H^{total} = \varepsilon \partial E^{total} / \partial t + \sigma E^{total} \tag{2.12}$$

while the incident fields traversing the media satisfy free space conditions

$$\nabla \times E^{inc} = -\mu_o \partial H^{inc} / \partial t \tag{2.13}$$

$$\nabla \times H^{inc} = \varepsilon_o \partial E^{inc} / \partial t \tag{2.14}$$

Rewriting the total field behavior as

$$\begin{aligned} \nabla \times \left(E^{inc} + E^{scat}\right) = & -\mu \partial \left(H^{inc} + H^{scat}\right) / \partial t \\ & - \sigma^* \left(H^{inc} + H^{scat}\right) \end{aligned} \tag{2.15}$$

$$\begin{aligned} \nabla \times \left(H^{inc} + H^{scat}\right) = & \varepsilon \partial \left(E^{inc} + E^{scat}\right) / \partial t \\ & + \sigma \left(E^{inc} + E^{scat}\right) \end{aligned} \tag{2.16}$$

we can subtract the incident fields above to obtain the equations governing the scattered fields in the media

$$\begin{aligned} & \nabla \times E^{scat} = -\mu \partial H^{scat} / \partial t - \sigma^* H^{scat} - \\ & \left[(\mu - \mu_o) \partial H^{inc} / \partial t + \sigma^* H^{inc} \right] \end{aligned} \tag{2.17}$$

$$\begin{aligned} & \nabla \times H^{scat} = \varepsilon \partial E^{scat} / \partial t + \sigma E^{scat} + \\ & \left[(\varepsilon - \varepsilon_o) \partial E^{inc} / \partial t + \sigma E^{inc} \right] \end{aligned} \tag{2.18}$$

Outside the scatterer in free space the total fields satisfy

$$\nabla \times E^{total} = -\mu_o \partial H^{total} / \partial t \tag{2.19}$$

$$\nabla \times H^{total} = \varepsilon_o \partial E^{total} / \partial t \tag{2.20}$$

which can be rewritten as

$$\nabla \times \left(E^{inc} + E^{scat}\right) = -\mu_o \partial \left(H^{inc} + H^{scat}\right) / \partial t \tag{2.21}$$

$$\nabla \times \left(H^{inc} + H^{scat}\right) = \varepsilon_o \partial\left(E^{inc} + E^{scat}\right)/\partial t \tag{2.22}$$

Now, subtracting the incident fields we obtain the equations governing the scattered fields in free space:

$$\nabla \times E^{scat} = -\mu_o \partial H^{scat}/\partial t \tag{2.23}$$

$$\nabla \times H^{scat} = \varepsilon_o \partial E^{scat}/\partial t \tag{2.24}$$

as expected. Note that these equations could have been found from the equations for the scattered fields in a media by letting the media become free space; i.e., Equations 2.17 and 2.18 become Equations 2.23 and 2.24 when

$$\begin{aligned} \mu &\rightarrow \mu_o \\ \varepsilon &\rightarrow \varepsilon_o \\ \sigma &\rightarrow 0 \\ \sigma^* &\rightarrow 0 \end{aligned}$$

In summary, only one set of equations is needed for the separate field formalism. The equations for the incident field, 2.13 and 2.14,

$$\nabla \times E^{inc} = -\mu_o \partial H^{inc}/\partial t$$

$$\nabla \times H^{inc} = \varepsilon_o \partial E^{inc}/\partial t$$

merely remind us that the analytically specified incident field must be Maxwellian. Only the scattered field equations, 2.17 and 2.18

$$\nabla \times E^{scat} = -\mu \partial H^{scat}/\partial t - \sigma^* H^{scat} - \left[(\mu - \mu_o)\partial H^{inc}/\partial t + \sigma^* H^{inc}\right]$$

$$\nabla \times H^{scat} = \varepsilon \partial E^{scat}/\partial t + \sigma E^{scat} + \left[(\varepsilon - \varepsilon_o)\partial E^{inc}/\partial t + \sigma E^{inc}\right]$$

with μ, ε, σ^*, and σ inside the scatterer, and with $\sigma^* = \sigma = 0$, $\mu = \mu_0$, and $\varepsilon = \varepsilon_0$ outside the scatterer, are determined computationally.

These scattered field equations can now be rearranged so that the time derivative of the field is expressed as a function of the remaining terms for ease in generating the appropriate difference equations.

$$\frac{\partial H^{scat}}{\partial t} = -\frac{\sigma *}{\mu} H^{scat} - \frac{\sigma *}{\mu} H^{inc}$$
$$- \frac{(\mu - \mu_o)}{\mu} \frac{\partial H^{inc}}{\partial t} - \frac{1}{\mu}(\nabla \times E^{scat}) \tag{2.25}$$

$$\frac{\partial E^{scat}}{\partial t} = -\frac{\sigma}{\varepsilon} E^{scat} - \frac{\sigma}{\varepsilon} E^{inc}$$
$$- \frac{(\varepsilon - \varepsilon_o)}{\varepsilon} \frac{\partial E^{inc}}{\partial t} + \frac{1}{\varepsilon}(\nabla \times H^{scat}) \tag{2.26}$$

We could difference this set of scattered field equations, but it is more instructive to difference these equations in the limit of a perfect conductor first and then the equations as presented here. This allows one to see the essentials of the differencing scheme in the perfect conductor case, as this is the most basic formulation possible. We will then return to the more general case of a scatterer with finite ε, μ, σ, and σ^*.

2.3 PERFECT CONDUCTOR FDTD FORMULATION

Outside the scatterer the scattered fields satisfy the free space conditions where $\sigma^* = \sigma = 0$, $\mu = \mu_o$ and $\varepsilon = \varepsilon_o$, so that Equations 2.25 and 2.26 reduce to

$$\frac{\partial H^{scat}}{\partial t} = -\frac{1}{\mu_o}(\nabla \times E^{scat}) \tag{2.27}$$

$$\frac{\partial E^{scat}}{\partial t} = \frac{1}{\varepsilon_o}(\nabla \times H^{scat}) \tag{2.28}$$

In the perfect conductor, Equation 2.26 governing the scattered field may be written as

$$\frac{\varepsilon}{\sigma}\frac{\partial E^{scat}}{\partial t} = -E^{scat} - E^{inc} - \frac{(\varepsilon - \varepsilon_o)}{\sigma} \partial E^{inc}/\partial t$$
$$+ \frac{1}{\sigma}(\nabla \times H^{scat}) \tag{2.29}$$

For a perfect conductor $\sigma = \infty$, and for this situation Equation 2.29 reduces to

$$E^{scat} = -E^{inc} \tag{2.30}$$

Inside the perfect conductor we apply Equation 2.30, rather than 2.26 with $\sigma = \infty$. Thus, if only free space and perfect conductor are present, only a specification of the incident field, the free space equations 2.27 and 2.28 for the scattered field, plus the relation Equation 2.30 are needed to apply FDTD. A further simplification is to note that Equation 2.30 need only be applied at the surface of the perfect conductor. Interior portions of the perfect conductor, if present, are completely isolated from the rest of the problem space.

We now difference the free space scattered field equations. In essence finite differencing replaces derivatives with differences:

$$\frac{\partial f}{\partial t} \equiv \lim_{\Delta t \to 0} \frac{f(x,t_2) - f(x,t_1)}{\Delta t} \approx \frac{f(x,t_2) - f(x,t_1)}{\Delta t} \tag{2.31}$$

$$\frac{\partial f}{\partial x} \equiv \lim_{\Delta x \to 0} \frac{f(x_2,t) - f(x_1,t)}{\Delta x} \approx \frac{f(x_2,t) - f(x_1,t)}{\Delta x} \tag{2.32}$$

where in the above approximation Δt and Δx are finite rather than infinitesimal. In short, calculus becomes algebra.

Some critical issues aside from this algebraic replacement include:

- **What form the differencing takes:**
  We use an explicit central difference scheme here that only retains first order terms. The E and H fields are interleaved spatially and temporally because of the central differencing. What results is often referred as a "leapfrog" scheme.

- **Stability:**
  Only for Δt given by the Courant stability condition, $\Delta t \leq (\Delta x)/c\sqrt{3}$ for cubical cells, is the formulation stable.

We shall return to these issues in later chapters. For now, we will proceed with the algebra. We must further decompose the vector Maxwell curl equations governing the scattered fields into their component scalar parts, obtaining

$$\frac{\partial E_x^{scat}}{\partial t} = \frac{1}{\varepsilon_o}\left(\frac{\partial H_z^{scat}}{\partial y} - \frac{\partial H_y^{scat}}{\partial z}\right) \tag{2.33a}$$

$$\frac{\partial E_y^{scat}}{\partial t} = \frac{1}{\varepsilon_o}\left(\frac{\partial H_x^{scat}}{\partial z} - \frac{\partial H_z^{scat}}{\partial x}\right) \tag{2.33b}$$

$$\frac{\partial E_z^{scat}}{\partial t} = \frac{1}{\varepsilon_o}\left(\frac{\partial H_y^{scat}}{\partial x} - \frac{\partial H_x^{scat}}{\partial y}\right) \tag{2.33c}$$

$$\frac{\partial H_x^{scat}}{\partial t} = \frac{1}{\mu_o}\left(\frac{\partial E_y^{scat}}{\partial z} - \frac{\partial E_z^{scat}}{\partial y}\right) \tag{2.33d}$$

$$\frac{\partial H_y^{scat}}{\partial t} = \frac{1}{\mu_o}\left(\frac{\partial E_z^{scat}}{\partial x} - \frac{\partial E_x^{scat}}{\partial z}\right) \tag{2.33e}$$

$$\frac{\partial H_z^{scat}}{\partial t} = \frac{1}{\mu_o}\left(\frac{\partial E_x^{scat}}{\partial y} - \frac{\partial E_y^{scat}}{\partial x}\right) \tag{2.33f}$$

For simplicity, we will only treat the pair E_x^{scat}; and H_y^{scat}, the other components follow naturally. (Note that E_x^{scat} and H_y^{scat} could be used alone in a 1-D transmission line analysis with propagation in the z direction if only E_x and H_y field components exist.)

Replacing derivatives with differences we find

$$\frac{E_x^{scat,n} - E_x^{scat,n-1}}{\Delta t} = \frac{1}{\varepsilon_o}\left[\frac{\Delta H_z^{scat,n-\frac{1}{2}}}{\Delta y} - \frac{\Delta H_y^{scat,n-\frac{1}{2}}}{\Delta z}\right] \tag{2.34}$$

$$\frac{H_y^{scat,n+\frac{1}{2}} - H_y^{scat,n-\frac{1}{2}}}{\Delta t} = \frac{1}{\mu_o}\left[\frac{\Delta E_z^{scat,n}}{\Delta x} - \frac{\Delta E_x^{scat,n}}{\Delta z}\right] \tag{2.35}$$

This completes the perfect conductor separate field formulation. We next consider how the formulation can be implemented as an operational computer code using the FORTRAN language.

2.4 PERFECT CONDUCTOR FDTD FORTRAN CODE

With a little more work, we can recast the above formula into the form used in a perfectly conducting version of an FDTD code. We quantize space, letting $x = I\,\Delta x$, $y = J\,\Delta y$, and $z = K\,\Delta z$, and time, letting $t = n\,\Delta t$. We can define uniform cells in the problem space and locate them by the I, J, K indices. Within each cell we choose to locate the field components at offsets (Figure 2-1) as given by Yee.[1] This "Yee cell", as it is called, results in spatially centered differencing. In Yee notation E_z^n (I,J,K) represents the z component of the electric field at time $t = n\Delta t$ and at spatial location $x = I\Delta x$, $y = J\Delta y$, and $z = (K+1/2)\Delta z$, as can be seen in Figure 2.1. Other field components will have different offsets as can be seen from the figure.

As a mnemonic aid we shall write $E_x^{scat,\,n}$ in the I,J,Kth Yee cell as the FORTRAN subscripted array variable EXS(I,J,K), with the time step determined by an index (integer variable) N in the code itself. Similarly $H_y^{scat,\,n+\frac{1}{2}}$ in the I,J,K cell is HYS(I,J,K) and the spatial offset, different from the offset for EXS, is determined by the H_y location in the Yee cell. The temporal offset is also understood, so that the time index variable N in the computer code corresponds to time step number n = N for electric fields and n = N + 1/2 for magnetic fields (this order can be reversed with no loss of generality). We write E_x^{inc} as EXI(I,J,K), and for the lossy dielectric version of the code very naturally we shall write $\partial E_x^{inc} / \partial t = \dot{E}_x^{inc}$ as DEXI(I,J,K).

With minimal algebraic manipulation, then, the FDTD equations for scattered fields propagating in free space are recast in the form of FORTRAN statements (remember these are FORTRAN assignments, not equalities)

$$\begin{aligned} \mathrm{EXS(I,J,K)} &= \mathrm{EXS(I,J,K)} \\ &+ \frac{\Delta t}{\varepsilon_o}\left[\frac{\mathrm{HZS(I,J,K)} - \mathrm{HZS(I,J-1,K)}}{\Delta Y} \right. \\ &\left. - \frac{\mathrm{HYS(I,J,K)} - \mathrm{HYS(I,J,K-1)}}{\Delta Z}\right] \end{aligned} \tag{2.36}$$

$$\begin{aligned} \mathrm{HYS(I,J,K)} &= \mathrm{HYS(I,J,K)} \\ &+ \frac{\Delta t}{\mu_o}\left[\frac{\mathrm{EZS(I+1,J,K)} - \mathrm{EZS(I,J,K)}}{\Delta X} \right. \\ &\left. - \frac{\mathrm{EXS(I,J,K+1)} - \mathrm{EXS(I,J,K)}}{\Delta Z}\right] \end{aligned} \tag{2.37}$$

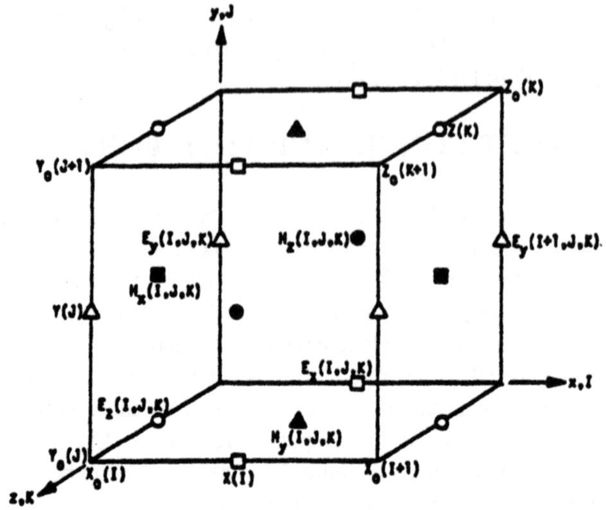

FIGURE 2-1. Convention for imposing the (I,J,K) indices on the (x,y,z) problem space, and location of the six-field evaluation points in a typical cell.

where it is understood that terms such as ΔZ, ΔY, ε_0, etc. would actually be stored as FORTRAN variables. The above notation shows E at time corresponding to n = N updated from its prior value at time n = N – 1 and the curl of H at time n = N – 1/2, where t = Δt. Next, H is evaluated at n = N + 1/2 from its earlier value at n = N – 1/2 and the curl of E at n = N. This interleaves E and H temporally and results in a centered difference or "leapfrog in time" approach. After each update of E and H the index N is increased by 1 and the process is repeated. The spatial indices in the curl calculations are determined by the Yee cell geometry. The curl calculations are also center differenced, and represent nearest-neighbor interactions.

The above equations apply to cells containing free space. In electric field cell locations I,J,K containing perfect electrical conductor the FORTRAN assignment EXS(I,J,K) = –EXI(I,J,K) is made at each time step rather than FORTRAN assignment Equation 2.36.

2.5 LOSSY MATERIAL FORMULATION

Using

$$E^{tot} = E^{inc} + E^{scat} \tag{2.9}$$

$$H^{tot} = H^{inc} + H^{scat} \tag{2.10}$$

we previously derived the equations for scattered field propagation in lossy materials

$$\frac{\partial H^{scat}}{\partial t} = -\frac{\sigma^*}{\mu}H^{scat} - \frac{\sigma^*}{\mu}H^{inc} - \frac{(\mu-\mu_o)}{\mu}\frac{\partial H^{inc}}{\partial t} - \frac{1}{\mu}\left(\nabla \times E^{scat}\right) \tag{2.25}$$

$$\frac{\partial E^{scat}}{\partial t} = -\frac{\sigma}{\varepsilon}E^{scat} - \frac{\sigma}{\varepsilon}E^{inc} - \frac{(\varepsilon-\varepsilon_o)}{\varepsilon}\frac{\partial E^{inc}}{\partial t} + \frac{1}{\varepsilon}\left(\nabla \times H^{scat}\right) \tag{2.26}$$

These are equivalent to the equations in the original scattered field paper 2. We now proceed, however, to apply linear differencing rather than the exponential differencing used in Reference 2, because it is simpler and (as implemented here) will yield a more stable formulation for high values of conductivity.

Considering Equation 2.26 and using i for inc and s for scat, we express it as

$$\frac{\varepsilon \partial E^s}{\partial t} + \sigma E^s = -\sigma E^i - (\varepsilon - \varepsilon_o)\frac{\partial E^i}{\partial t} + \left(\nabla \times H^s\right) \tag{2.38}$$

which is approximated using centered finite differences as

$$\varepsilon\left(E^{s,n} - E^{s,n-1}\right) + \sigma \Delta t E^{s,n} = -\sigma \Delta t E^{i,n} - (\varepsilon - \varepsilon_o)\Delta t \dot{E}^{i,n} + \left(\nabla \times H^{s,n-\frac{1}{2}}\right)\Delta t \tag{2.39}$$

and can be reexpressed as

$$(\varepsilon + \sigma\Delta t)E^{s,n} = \varepsilon E^{s,n-1} - \sigma \Delta t E^{i,n} - (\varepsilon - \varepsilon_o)\Delta t \dot{E}^{i,n} + \left(\nabla \times H^{s,n-\frac{1}{2}}\right)\Delta t \tag{2.40}$$

or finally

$$E^{s,n} = \left(\frac{\varepsilon}{\varepsilon + \sigma\Delta t}\right)E^{s,n-1} - \left(\frac{\sigma\Delta t}{\varepsilon + \sigma\Delta t}\right)E^{i,n}$$
$$-\left(\frac{(\varepsilon - \varepsilon_o)\Delta t}{\varepsilon + \sigma\Delta t}\right)\dot{E}^{i,n} + \left(\nabla \times H^{s,n-\frac{1}{2}}\right)\left(\frac{\Delta t}{\varepsilon + \sigma\Delta t}\right) \tag{2.41}$$

The use of $E^{s,n}$ in the expressions involving σ, using the most recent value of electric field to determine the current density, is the key to obtaining stability for large conductivity values. Note that as σ becomes infinite Equation 2.41 correctly gives $E^{scat} = -E^{inc}$.

We can now write the FORTRAN-like expression for updating the electric field in a lossy dielectric medium as

$$E^s_x(I,J,K)^n = E^s_x(I,J,K)^{n-1}\left(\frac{\varepsilon}{\varepsilon + \sigma\Delta t}\right)$$
$$-\left(\frac{\sigma\Delta t}{\varepsilon + \sigma\Delta t}\right)E^i_x(I,J,K)^n - \left(\frac{(\varepsilon - \varepsilon_o)\Delta t}{\varepsilon + \sigma\Delta t}\right)\dot{E}^i_x(I,J,K)^n$$
$$+ \frac{H^s_z(I,J,K)^{n-\frac{1}{2}} - H^s_z(I,J-1,K)^{n-\frac{1}{2}}}{\Delta y}\left(\frac{\Delta t}{\varepsilon + \sigma\Delta t}\right) \tag{2.42}$$
$$+ \frac{H^s_y(I,J,K)^{n-\frac{1}{2}} - H^s_y(I,J,K-1)^{n-\frac{1}{2}}}{\Delta z}\left(\frac{\Delta t}{\varepsilon + \sigma\Delta t}\right)$$

In a similar way the corresponding equations for updating the other electric field components can be obtained. Should we wish to consider lossy magnetic media the corresponding magnetic field equations can be derived using the same approach.

2.6 LOSSY DIELECTRIC FDTD FORTRAN CODE

We now consider extending the previous perfect conductor FDTD code to lossy dielectric materials. The only addition to the previously described perfect conductor FDTD implementation is that Equation 2.42 must be used to update electric fields at spatial locations where $\varepsilon \neq \varepsilon_0$. Because we are assuming that no magnetic materials are present, Equation 2.37 can be used throughout the FDTD space to update magnetic fields.

Let us now consider implementing Equation 2.42 in FORTRAN. One consideration is that we do not want to repeatedly evaluate the constant multiplying terms, but would prefer to store and reuse them. On the other hand, storing all of the multiplying terms for each field component in each cell will require significant amounts of additional memory. To reduce the storage required significantly for most problems, we recognize that typically only a few different types of materials will exist in a particular FDTD space. The space may have millions of cells, but perhaps only a few different materials with different permittivities, conductivities, and permeabilities will be modeled in the space. The multiplying constants need to be evaluated for each type of material rather than for each cell, with a pointer array used to designate which material is in a given I,J,K location. We let IDONE(I,J,K) be the pointer array designating the type of material interacting with the x component of electric field at Yee cell location I,J,K. We have found it convenient to let IDONE(I,J,K) = 0 mean that in cell I,J,K the x component of electric field is in free space, IDONE(I,J,K) = 1 for a perfect conductor interacting with this electric field, and values > 1 for user-defined values of permittivity and conductivity at this location.

Using this approach, and letting an integer value M denote the Mth type of lossy dielectric contained in the FDTD space, we can express the above constant multipliers for this material as:

$$\mathrm{ECRLY(M)} = \frac{\mathrm{DT}}{(\mathrm{EPS(M)} + \mathrm{SIGMA(M)} * \mathrm{DT}) * \mathrm{DY}} = \frac{\Delta t}{(\varepsilon + \sigma\Delta t)\Delta Y}$$

$$\mathrm{ECRLZ(M)} = \frac{\mathrm{DT}}{(\mathrm{EPS(M)} + \mathrm{SIGMA(M)} * \mathrm{DT}) * \mathrm{DZ}} = \frac{\Delta t}{(\varepsilon + \sigma\Delta t)\Delta Z}$$

$$\mathrm{ESCTC(M)} = \mathrm{EPS(M)} / (\mathrm{EPS(M)} + \mathrm{SIGMA(M)} * \mathrm{DT}) = \frac{\varepsilon}{\varepsilon + \sigma\Delta t}$$

$$\mathrm{EINCC(M)} = \mathrm{SIGMA(M)} * \mathrm{DT} / (\mathrm{EPS(M)} + \mathrm{SIGMA(M)} * \mathrm{DT}) = \frac{\sigma\Delta t}{\varepsilon + \sigma\Delta t}$$

$$\mathrm{EDEVCN(M)} = \mathrm{DT} * (\mathrm{EPS(M)} - \mathrm{EPSO}) / (\mathrm{EPS(M)} + \mathrm{SIGMA(M)} * \mathrm{DT}) = \frac{(\varepsilon - \varepsilon_o)\Delta t}{\varepsilon + \sigma\Delta t}$$

Using these definitions the FORTRAN expression for Equation 2.42 for E_x^s in the presence of a lossy dielectric becomes:

$$\begin{aligned}
\text{EXS(I,J,K)} &= \text{EXS(I,J,K)} * \text{ESCTC}(\text{IDONE(I,J,K)}) \\
&- \text{EINCC}(\text{IDONE(I,J,K)}) * \text{EXI(I,J,K)} \\
&- \text{EDEVCN}(\text{IDONE(I,J,K)}) * \text{DEXI(I,J,K)} \\
&+ (\text{HZS(I,J,K)} - \text{HZS(I,J}-1\text{,K)}) * \text{ECRLY}(\text{IDONE(I,J,K)}) \\
&- (\text{HYS(I,J,K)} - \text{HYS(I,J,K}-1)) * \text{ECRLZ}(\text{IDONE(I,J,K)})
\end{aligned} \tag{2.43}$$

In a lossy dielectric FDTD code the FORTRAN assignment (2.36) is executed for cells containing free space, (2.43) is executed for cells containing lossy dielectric, and in cells containing a perfect conductor, EXS(I,J,K) = –EXI(I,J,K) is executed. The contents of the IDONE(I,J,K) array determines which assignment to execute for EXS in a particular (I,J,K) Yee cell. Arrays IDTWO(I,J,K) and IDTHRE(I,J,K) would similarly determine the material interacting with EYS and EZS, respectively. Because no magnetic materials are present, FORTRAN assignment (2.37) is executed in all cells. If magnetic materials are to be considered, the magnetic version of (2.43) can be derived from (2.25), and arrays IDFOUR(I,J,K), IDFIVE(I,J,K), and IDSIX(I,J,K) can be used to specify the magnetic material interacting with HXS, HYS, and HZS, respectively.

2.7 FDTD CODE REQUIREMENTS AND ARCHITECTURE

We have derived the FDTD algorithm for six field components in a 3-D rectilinear space that is composed of Yee cells. We must still determine what additional computational support is required, and to these code requirements a suitable architecture must be given.

Let us first define the code requirements. There must be a main computer routine that acts as an overseer or driver of the remaining subroutines. This driver steps through time, calling the subroutines in the appropriate order.

Before time stepping can begin there must be a problem space defined, including parameters such as cell size, time step, and incident field. Constant multipliers that need not be computed at each time step may also be evaluated and stored before time stepping begins. There must be a definition of the scatterer or test object provided, which consists of coding the information as to which cell locations contain materials other than free space. Monitor points or test locations at which responses are examined must be specified along with the response type: voltage, current, field, power, etc. If a far zone transformation is being used, the direction(s) of the far zone fields desired must be specified.

After these specifications the fields are then advanced one step at a time. This is, of course, the core of the code. Most of the computer running time will be spent in this section, but it represents a small fraction of the lines of code.

In addition to the time-stepped field subroutines we must also have outer radiation boundary condition subroutines that absorb the scattered field at the outermost portion of the problem space.

When time stepping is completed we must have a way of outputting response data. This data saving routine can store response data at every time step and then at the end of the run "dump" it out as a listing or as a file for postprocessing. This output process may involve transformation of the near zone FDTD fields to the far zone for radiation or scattering calculations.

These code requirements are summarized as:

- Driver
- Problem space setup
- Test object definition
- E,H field algorithms
- Outer radiation boundary condition
- Data saver
- Far zone transformation

These requirements are not changed by the presence of lossy dielectric and/or magnetic materials in addition to perfectly conducting materials. The only change necessary is that the E and/or H field time stepping assignments will need to be further generalized. When attempts are made to treat portions of the problem space in finer detail additional code requirements may be added. Each of the aforementioned code requirements can be implemented as a modular subroutine or set of subroutines.

The required capabilities of each subroutine can now be given:

Driver

- Calls problem space setup subroutine and the test object definition subroutine
- Time steps over an index N
- While looping over N, calls E,H subroutines and the outer radiation boundary condition subroutines
- At each or selected time steps calls data saver or far zone subroutine to store data samples
- At the completion of all the time steps calls appropriate subroutines to write output data

Problem space setup

- Sets the problem space size
    - Sets the number of cells in each dimension
    - Sets the cell size (Δx, Δy, Δz)
- Calculates the Δt time step according to the Courant stability condition using the cell dimensions Δx, Δy, Δz
- Calculates constant multipliers

Test object definition

- Cells or individual field components in the cells are "flagged" (with an integer variable IDONE(I,J,K) for E_x in our example FORTRAN) indicating their composition; this dictates how the E,H algorithm is to process the data: as perfect conductor, lossy dielectric, free space, or other more complicated material; it is usually convenient to set the default material to free space; the array of "flags" can be read for a preprocessing check of the geometry and composition of the object.

E,H field algorithms

- Calculate the response of a component from its own prior time value and that of the nearest-neighbor field quantifies (Es around Hs and Hs around Es) according to the type of material present at that component location:
    - Free space
    - Lossy dielectric
    - Lossy magnetic
    - Perfect conductor

Outer radiation boundary condition

- Absorbs, at least partially, the scattered field at the outermost portion of the problem space

Data saver

- Saves response data such as E and H field components, currents, or other quantities in the FDTD computation space in arrays at chosen time steps

Far zone transformation

- Evaluates tangential electric and magnetic currents on a closed surface surrounding the object and computes the corresponding scattered or radiated fields in the far zone

The architecture of the code is quite straightforward given the modular nature of each subroutine. FORTRAN COMMON blocks can be used to pass data between subroutines, with the main Driver program orchestrating events. A simplified flow chart of the code appears in Figure 2-2. A computer code listing is included in Appendix B.

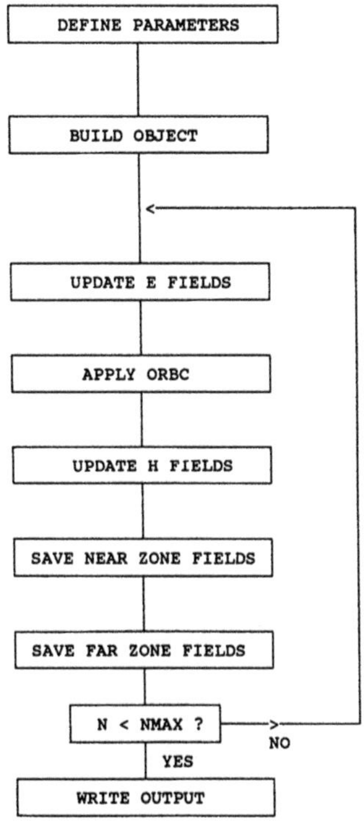

FIGURE 2-2. FDTD flow chart.

REFERENCES

1. **Yee, K. S.,** Numerical solution of initial boundary value problems involving Maxwell's equations in isotropic media, *IEEE Trans. Ant. Prop.*,14(3), 302, 1966.
2. **Holland, R., Simpson, L., and Kunz, K. S.,** Finite-difference analysis of EMP coupling to lossy dielectric structures, *IEEE Trans. EMC*, 22(3), 203, 1980.

Chapter 3

FDTD BASICS

3.1 INTRODUCTION

In this chapter, the practical considerations of FDTD calculations are treated:

- Cell size
- Time step size
- Incident field specification
- Scattering object construction
- Direct calculation of total fields
- Outer radiation boundary condition
- Resource requirements

The choice of cell size is critical in applying FDTD. It must be small enough to permit accurate results at the highest frequency of interest, and yet be large enough to keep resource requirements manageable. Cell size is directly affected by the materials present. The greater the permittivity or conductivity, the shorter the wavelength at a given frequency and the smaller the cell size required.

Once the cell size is selected, the maximum time step is determined by the Courant stability condition. Smaller time steps are permissible, but do not generally result in computational accuracy improvements except in special cases. A larger time step results in instability.

When using the scattered field FDTD formulation the incident field must be analytically specified. An infinite variety of waveforms are possible, but experience has led to the Gaussian pulse as the incident waveform of choice. The exception to this is when frequency-dependent materials are included, in which case a smoothed cosine pulse has advantages, as described in Chapter 8.

FDTD is capable of providing very large dynamic ranges, which can be in excess of 120 dB. Accuracies of 0.1 dB or less can also be obtained depending on the cell size, frequency, and shape of the objects being considered.

The scattering object can be "constructed" for the FDTD calculation using integer arrays for each field component. Different integer values indicate a different material and determine which FDTD field equations are used with what multiplying constants (depending on material) to update the field component. In the scheme described later in this chapter, free space FDTD equations correspond to a value of zero stored in the integer array, while perfectly conducting field equations are used for E field components corresponding to a value of 1. Numbers >1 may be used for lossy dielectric and/or magnetic materials. Any shape and material can be included within the constraints of the cell size chosen and maximum frequency of interest.

For problems which do not have an incident field, antenna radiation calculations, for example, an FDTD computer code based on the scattered field formulation may easily be converted to computing total fields, as discussed in Section 3.6.

Whether total or scattered fields are computed, Mur's first or second order absorbing boundaries provide a relatively reflection-free and easily implemented termination for the FDTD space. Other absorbing boundary conditions are available which provide better absorption with fewer cells required between the object and the outer boundary, but at the expense of added complexity. These are discussed in Chapters 14 and 18, although for most FDTD users, the Mur absorbing boundaries are adequate and relatively simple to apply.

After the user determines the cell size, a problem space large enough to encompass the scattering object, plus space between the object and the absorbing boundary, is determined. Also, a number of time steps sufficiently large to allow a full characterization of the interaction of object and fields, most importantly any resonant behavior, is estimated. From the number of Yee cells needed and the number of time steps required, resource requirements can be closely estimated. These resources include CPU time (determined in part by the speed of the computer being used) and the amount of solid state memory (RAM) and extended storage capacity (megabytes of hard disc memory) needed for the calculation and for storing the results.

Upon completing this chapter, the reader should be able to make FDTD calculations with an appreciation of the general constraints and requirements of the method.

3.2 DETERMINING THE CELL SIZE

The fundamental constraint is that the cell size must be much less than the smallest wavelength for which accurate results are desired. The obvious question is "How much less?", and to this must be added the question, "How accurate do you want the results to be?" An often quoted constraint is "10 cells per wavelength", meaning that the side of each cell should be $1/10\ \lambda$ or less at the highest frequency (shortest wavelength) of interest. For some situations, such as a very accurate determination of radar scattering cross-sections, $1/20\ \lambda$ or smaller cells may be necessary. On the other hand, reasonable results have been obtained with as few as four cells per wavelength. If the cell size is made much smaller than this the Nyquist sampling limit, $\lambda = 2\Delta x$, is approached too closely for reasonable results to be obtained and significant aliasing is possible for signal components above the Nyquist limit.

A word of caution here is that FDTD is a volumetric computational method, so that if some portion of the computational space is filled with penetrable material we must use the wavelength in the material to determine the maximum cell size. For problems containing electrically dense materials this results in cells in the material that are much smaller than if only free space and perfect conductors were being considered. If a uniform cell size is used throughout this forces the cells in all of the problem space to be relatively small, which may greatly increase the number of cells needed. Possible measures to deal with this include nonuniform cells (smaller cells in the dense material, larger cells outside) or surface or sheet "impedance" methods, which are considered in Chapter 9.

To understand why the cell size must be much smaller than one wavelength, consider that at any particular time step the FDTD grid is a discrete spatial sample of the field distribution. From the Nyquist sampling theorem, there must be at least two samples per spatial period (wavelength) in order for the spatial information to be adequately sampled. Because our sampling is not exact, and our smallest wavelength is not precisely determined, more than two samples per wavelength are required. Another related consideration is grid dispersion error. Due to the approximations inherent in FDTD, waves of different frequencies will propagate at slightly different speeds through the grid. This difference in propagation speed also depends on the direction of propagation relative to the grid. For accurate and stable results, the grid dispersion error must be reduced to an acceptable level, which can be readily accomplished by reducing the cell size.

Another cell size consideration is that the important characteristics of the problem geometry must be accurately modeled. Normally this will be met automatically by making the cells smaller than 1/10 λ or so, unless some special geometry features smaller than this are factors in determining the response of interest. An example is thin wire antennas, in which a change in the wire thickness from 1/10 λ to 1/20 λ (and even smaller) will affect the antenna impedance. Another example is computation of low level scattering from smooth targets in which the "staircase" effects of modeling a smooth surface with rectangular cells may cause significant errors. Good results in these and similar situations may require extremely small cells, or alternative measures such as sub-cell modeling or special grids which approximate the actual geometry better than Yee cells of the same size. These situations are also discussed later in this book.

Once the cell size has been determined, the number of cells needed to model the object and a reasonable amount of free space between the object and the outer boundary in each dimension is found, and from this the total size of the FDTD space in cells is determined. Assuming a 3-D problem, a total number of cells from a few hundred thousand up to several million can be accomodated for computers ranging from personal computers (PCs) and work stations to supercomputers. We now consider other basic aspects of FDTD calculations in the following sections of this chapter, with the final section devoted to estimating the computer resources required for an FDTD problem once the number of cells has been determined.

3.3 TIME STEP SIZE FOR STABILITY

Once the cell size is determined, the maximum size of the time step Δt immediately follows from the Courant condition. To understand the basis for the Courant condition, consider a plane wave propagating through an FDTD grid. In one time step any point on this wave must not pass through more than one cell, because during one time step FDTD can propagate the wave only from one cell to its nearest neighbors. To determine this time step constraint we pick a plane wave direction so that the plane wave propagates most rapidly between field point locations. This direction will be perpendicular to the lattice planes of the FDTD grid. For a grid of dimension d (where d = 1, 2, or 3), with all cell sides equal to Δu, we find that with v the maximum velocity of propagation in any medium in the problem, usually the speed of light in free space,

$$v\,\Delta t \le \frac{\Delta u}{\sqrt{d}} \tag{3.1}$$

for stability. More generally for a 3-D rectangular grid[1]

$$v\,\Delta t \le 1\Big/\sqrt{\frac{1}{(\Delta x)^2}+\frac{1}{(\Delta y)^2}+\frac{1}{(\Delta z)^2}} \tag{3.2}$$

Experience has indicated that for actual computations the Δt value given by the equality in (3.1) or (3.2) will provide accurate results, and in most situations more accurate results will not be obtained by using a smaller value of Δt. In fact, when the equality holds, the discretized wave most closely approximates the actual wave propagation, and grid dispersion errors are minimized.

However, exceptions to this occur. One situation in which the time step must be reduced relative to (3.2) is when the conductivity of the material is much greater than zero. For conducting materials ($\sigma > 0$), stable calculations require time steps smaller than the Courant limit. This is usually not a problem, because in most calculations the time step size is set by the speed of light in free space. As the velocity in the conducting material will be smaller than in free space, the time step in an FDTD calculation that includes both free space and conducting materials will be such that the Courant limit will be satisfied everywhere. However, the short wavelength inside highly conducting materials may require much smaller FDTD cells than in surrounding free space regions.

Another situation where the time step must be reduced below the Courant limit occurs for nonlinear materials. Chapter 11 contains discussions of stability concerns for both nonlinear (Sections 11.2 and 11.3) and conducting (Section 11.3) materials.

3.4 SPECIFYING THE INCIDENT FIELD

A key advantage of using the scattered field formulation is that the incident field is specified analytically. In this section the practical considerations for doing this are given for an incident plane wave. Such a plane wave may be needed for a scattering calculation, for example. Other types of incident fields may be specified for other problems and applications, but the procedure should be similar.

The specified incident field will be a Gaussian pulse plane wave, as this provides a smooth roll off in frequency content and is simple to implement. Provision for arbitrary incidence angle is also made.

First, consider specifying a general incident plane wave in the time domain. We assume a spherical coordinate system with the origin coincident with the origin of the FDTD Cartesian system. Following normal usage θ is measured from the z axis and ϕ from the x axis. Following the usual scattering convention, we specify the direction that the incident plane wave is coming *from* by θ and ϕ. Letting a unit vector $\hat{r}$ point from the origin in the θ,ϕ direction, an incident plane wave from this direction can be specified as

$$\overline{E} = \left[E_\theta \hat{\theta} + E_\phi \hat{\phi}\right] f\left(t + (\bar{r}' \cdot \hat{r})/c + R/c\right) \tag{3.3}$$

$$\overline{H} = \left[\frac{E_\phi}{\eta}\hat{\theta} - \frac{E_\theta}{\eta}\hat{\phi}\right] f\left(t + (\bar{r}' \cdot \hat{r})/c + R/c\right) \tag{3.4}$$

where $\hat{\theta}$ and $\hat{\phi}$ are the spherical coordinate system unit vectors, η is the impedance of free space, c is the speed of light, and $\bar{r}'$ is the vector from the origin to the point in the FDTD computation space at which we desire to evaluate the incident field. The function f(t) may be any function of time, a sine wave or a pulse, for instance. R is an arbitrary reference distance. For transient calculations the pulse must propagate into the FDTD space rather than suddenly appear at the scattering object, and the value of R is chosen accordingly.

We can easily obtain the amplitudes of the Cartesian components of the incident fields as

$$\begin{aligned}
E_x &= E_\theta \cos\theta \ \cos\phi - E_\phi \ \sin\phi \\
E_y &= E_\theta \cos\theta \ \sin\phi + E_\phi \ \cos\phi \\
E_z &= -E_\theta \ \sin\theta \\
H_x &= (E_\theta \ \sin\phi + E_\phi \ \cos\theta \ \cos\phi)/\ \eta \\
H_y &= (-E_\theta \ \cos\phi + E_\phi \ \cos\theta \ \sin\phi)/\ \eta \\
H_z &= (-E_\theta \ \sin\phi)/\ \eta
\end{aligned}$$

Next consider specifying a particular field component for a Gaussian pulse. Assuming that the amplitude has been found from the above expressions, we let the function f(t) be a Gaussian pulse and express the x component of the electric field of the plane wave as

$$E_x^i(I,J,K)^n = E_x \exp\left(-\alpha\left((\tau - \beta\Delta t)^2\right)\right) \tag{3.5}$$

where the time delay τ is given by

$$\tau = n\Delta t + \bar{r}' \cdot \hat{r}/c + R/c \tag{3.6}$$

and incorporates the relative time delay for the E_x component in cell I,J,K at time step $n\Delta t$. For our plane wave incident from the θ,ϕ direction we have

$$\begin{aligned} \bar{r}' \cdot \hat{r} = {} & \left((I-1)+0.5\right)\Delta x \cos\phi \sin\theta \\ & + (J-1)\Delta y \sin\phi \sin\theta \\ & + (K-1)\Delta z \cos\theta \end{aligned} \tag{3.7}$$

Note the 1/2 cell offset in (3.7) corresponding to the location in the (I,J,K) Yee cell of the E_x component. Corresponding offsets must be included for each field component.

We now define α and β and constrain τ so as to provide a suitable pulse as a function of the time step size we have chosen. Although an ideal Gaussian pulse extends infinitely in time, ours must be truncated in our calculations, and the effects of this must be considered. We must also specify the time duration of the pulse so that it has a suitable bandwidth.

To start, determine the duration of the Gaussian pulse. We select $\beta = 32$, where β is the number of time steps in the Gaussian pulse from the peak value to the truncation value. The pulse will exist from $\tau = 0$ until $\tau = 2\beta\Delta t$; approximated as zero outside this range, with peak value at $\tau = \beta\Delta t$.

The value at truncation (at $\tau = 0, 2\beta\Delta t$) is determined by α, and as seen from (3.5) the Gaussian pulse at truncation will have a value $\exp(-\alpha(\pm\beta\Delta t)^2)$ down from the maximum value. We now need to determine α so that this truncation does not introduce unwanted high frequencies into our spectrum, and yet does not waste computation time on determining values of the incident field that are essentially zero.

There is no correct answer, but a practical solution is to let α change with β so that at truncation the amplitude of the pulse is always reduced by the same value. We let $\alpha = (4/(\beta\Delta t))^2$. Thus, at truncation the pulse is down by $\exp(-16)$, or almost 140 dB. Because a reasonable goal for a single precision calculation

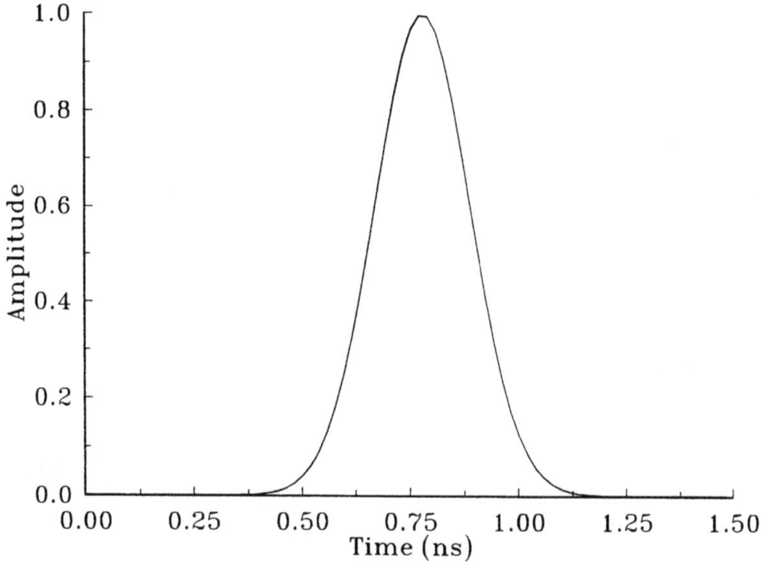

FIGURE 3-1. Gaussian pulse for 1-cm cubical FDTD cells for a time step at Courant stability limit and $\beta = 32$.

is –120 dB (six significant decimal digits) this value of α results in truncation frequencies well below this level. As a maximum a 32-bit world length allows at most a little over 190 dB dynamic range.

To illustrate the results obtained with these values we consider a situation often used for other examples in this book, a 3-D cubic cell with 1-cm sides. Applying the Courant stability condition we obtain a Δt of 1.924E-11 s or 0.01924 ns. Using this time step along with the parameters given above yields the Gaussian pulse of Figure 3-1. The approximation to the pulse is most noticeable near the peak value where the curve of the actual pulse is approximated with straight line segments between the discrete values.

The fast Fourier transformation (FFT) of this pulse, shown in Figure 3-2, makes it clear that the truncation of the pulse did not introduce unwanted high frequencies. At ten cells per wavelength we would hope to get accurate results from our FDTD calculations for frequencies up to 3 GHz, and from Figure 3-2 it is clear that our Gaussian pulse is providing relatively high signal levels out to this frequency. On the other hand, we may be concerned about noise and instability if we have appreciable energy in the incident wave for wavelengths in which our cell size is less than four cells per wavelength. This corresponds to a frequency of 7.5 GHz, and from Figure 3-2 we see that our spectrum is down by approximately 120 dB at this frequency — small enough to provide stability.

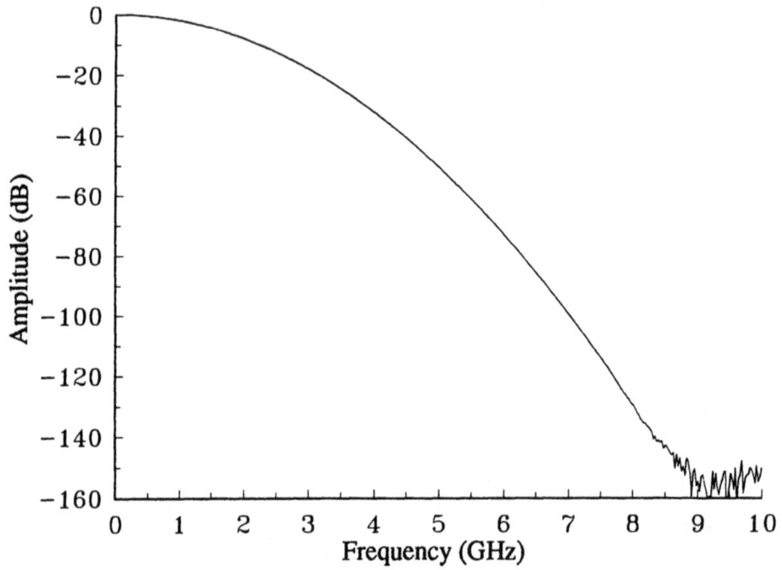

FIGURE 3-2. Fourier transform of the Gaussian pulse of Figure 3-1.

This determination of the Gaussian pulse parameters has assumed that the entire problem space is filled with either free space or perfect conductor. If there is penetrable material we must modify our choice of β. To illustrate this suppose that we still desire to obtain results that are accurate up to 3 GHz, but that our problem space contains penetrable material with a relative permittivity of 4. For the same accuracy at 3 GHz as in the free space/conductor problem, we would need to correspondingly reduce our cell size by a factor of 2, to 0.5 cm on each side, as the wavelength in this material would be one half the free space wavelength.

We next would apply the Courant condition and find that the time step is now one half as large as the previous value. If we let β remain 32, with the time step reduced by one half, our Gaussian pulse would be one half of the previous time duration, and twice the previous spectral band width. Instead, we must double the size of β to 64, so that the pulse width and frequency band remain as shown in Figures 3-1 and 3-2. If we do not, we risk noise and even instability in the volumetric region containing the dielectric material, as the narrower pulse will contain significant energy at wavelengths too small to be adequately sampled inside the dielectric. Thus, the only adjustment required in the above scheme for determining the Gaussian pulse parameters is that the value of β must be increased inversely as the time step size is decreased below the Courant limit.

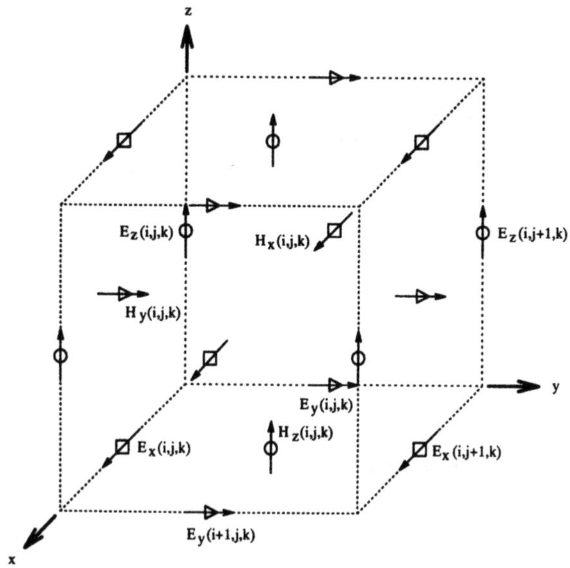

FIGURE 3-3. Yee cell geometry.

3.5 BUILDING AN OBJECT IN YEE CELLS

In Chapter 2, an approach for using integer arrays to specify the materials in FDTD cells was presented. With this approach, the IDONE-IDTHRE arrays are meant for specifying PEC and dielectric materials. IDONE is used to specify which material type is interacting with the E_x field component, IDTWO for the E_y component, etc. The IDFOR-IDSIX arrays are for magnetic materials and correspond to the x, y, and z components of magnetic field, respectively. These correspondences are summarized below.

IDONE	E_X	IDFOUR	H_x
IDTWO	E_y	IDFIVE	H_Y
IDTHRE	E_z	IDSIX	H_z

The I,J,K subscript index identifies in which Yee cell the material is located. The content of the array identifies which material is at this location. For example, a content of 0 might specify free space; 1, perfect conductor; 2, a lossy dielectric with a specific permittivity and conductivity; 3, for a lossy dielectric with different permittivity, etc. The constitutive parameters of each material type that exists in the FDTD space are set once and the corresponding multiplying terms in the FDTD update equations are calculated and stored before time stepping is begun. Note that because the material that

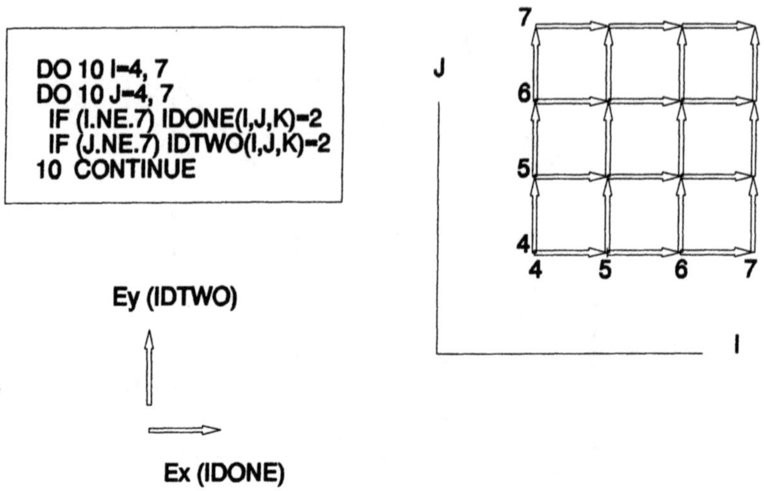

FIGURE 3-4. FORTRAN code and corresponding view of 3 × 3 cell FDTD dielectric plate in xy plane at z = KΔz.

interacts with each field component is specified separately, modeling anisotropic materials with diagonal permittivity or permeability tensors is straightforward in FDTD.

As six separate arrays are used the user can control independently the exact placement of dielectric and magnetic material in the Yee cells. The placement of the field components in a Yee cell is shown in Figure 3-3. For example, setting an element of the IDONE array at some I,J,K location is actually locating dielectric material where it interacts with the x component of electric field located at position I+1/2,J,K in the FDTD space. Setting an element of the IDFOR array at some I,J,K location is actually locating magnetic material where it interacts with an x component of magnetic field located at position I,J+0.5,K+0.5. The spatial difference between the IDONE and IDFOR array locations is a direct result of the field offsets in the Yee cell. This inherent offset in the field locations causes "fuzziness" or "staircasing" approximations when building objects. Smooth surfaces must be approximated by stepped Yee cell locations. Even an object that fits the Cartesian coordinate system (for example a cube) cannot be specified exactly if it is composed of material that has both dielectric and magnetic properties because the electric and magnetic field locations are offset spatially. One remedy for this situation is to use average values of the permittivity or permeability in the "fuzzy" regions where this offset is a factor, but this remedy only reduces rather than removes the effect.

When "building" an object it is necessary to keep in mind that one is not filling Yee cells with material, but rather locating material at the field locations within a Yee cell. To illustrate this let us consider building a flat 3 × 3 cell dielectric plate in the xy plane located at z = KΔz in a 3-D Yee cell space. This

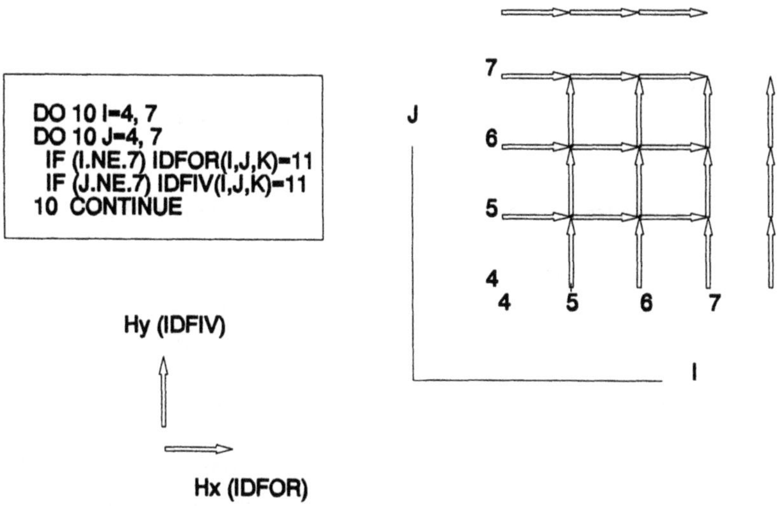
DO 10 I=4, 7
DO 10 J=4, 7
IF (I.NE.7) IDFOR(I,J,K)=11
IF (J.NE.7) IDFIV(I,J,K)=11
10 CONTINUE
J
7
6
5
4
4 5 6 7
I
Hy (IDFIV)
Hx (IDFOR)

FIGURE 3-5. Attempt to build magnetic material plate with same FORTRAN logic as for dielectric plate.

plate will be composed of material type "2", in which the constitutive parameters of this material are set elsewhere in the FDTD computer code. Using the IDONE and IDTWO arrays, this plate can be built with the FORTRAN shown in Figure 3-4. In this example, note the two IF statements that must be included so that the plate is properly formed. Merely setting the ID arrays for nine cells (3×3) will not result in a plate with smooth sides.

To illustrate this further, consider "building" the corresponding magnetic material plate. If the same FORTRAN logic is used to try to generate a magnetic plate, the object generated would actually be unconnected, as illustrated in Figure 3-5. The correct way to build the magnetic plate is with the following FORTRAN code:

```
      DO 10 I=4,7
      DO 10 J=4,7
        IF(I.NE.4)IDFOR(I,J,K)=11
        IF(J.NE.4)IDFIV(I,J,K)=11
10    CONTINUE
```

This complication is a direct consequence of the locations of the field components within the Yee cell.

One can define a correspondence between setting electric and magnetic field locations, so that the code used to generate a dielectric object can be modified to generate a magnetic object. To see this consider that we have set the permittivity at cell locations corresponding to EX(I,J,K), EY(I,J,K), EZ(I,J,K) using the IDONE, IDTWO, and IDTHRE arrays, respectively. This determines one corner of a dielectric cube. If we wish to define the corner of a corresponding magnetic cube, offset one half cell in the x, y, z directions, we would set the locations of the magnetic fields HX(I+1,J,K), HY(I,J+1,K), HZ(I,J,K+1) as magnetic material using the IDFOR, IDFIV, and IDSIX arrays. This example indicates the following correspondence between "building" dielectric and magnetic objects:

Dielectric object		**Magnetic object**
IDONE(I,J,K)	=	IDFOR(I+1,J,K)
IDTWO(I,J,K)	=	IDFIV(I,J+1,K)
IDTHRE(I,J,K)	=	IDSIX(I,J,K+1)

This correspondence agrees with the changes in the FORTRAN code necessary to build the plates in the preceding example.

Before leaving the topic of plates, a pertinent question might be "how thick are the plates that are built in the above examples, i.e., how physically thick is a plate that is one cell thick in FDTD space?" More generally, we can consider where the "surface" of a volumetric conductor "built" of FDTD cells is located. The answers to these questions will give us some insight into how FDTD approximates structures, because no unique answer exists.

First, consider the situation that we are computing the scattering by one of these plates for a plane wave incident from $\theta = 90°$, that is edge-on, and with the electric field polarized in the θ or equivalently the $-z$ direction. As only an E_z component is present, and all IDTHRE and IDSIX array locations are assumed set to 0 (free space; we would not set these to any other values when building the plates), the scattered field will be identically zero, and we conclude that the plates have zero thickness.

However, consider the case in which the plane wave is incident from the $\theta = 0°$ direction, and thus normally incident on the plate. Because the plate is thin and composed of lossy dielectric we can approximate it as a sheet resistance or impedance. This is discussed in Chapter 13, which shows that the correct result is obtained when the FDTD plate is considered to be Δz thick.

We might also consider the location of the real FDTD "surface" for a more general FDTD geometry. For example, consider "building" a perfectly conducting (PEC) cube with FDTD cells. At first we would assume that the surface of the cube corresponds with the outer Yee cell locations, where E fields are set to perfect conductor. Then we also may consider that the FDTD "cube" extends halfway between these outer E field locations and the adjacent E field locations that are in free space. Experience in calculating scattered field cross-sections from perfectly conducting volumetric targets indicates that the surface of the FDTD object extends about one fourth of the Yee cell dimension (Δx, Δy, or Δz) beyond the locations where the E fields are set to perfect conductor.[3]

An explanation for this one fourth cell extension is that when a plane wave is normally incident on a planar perfect conductor, the tangential electric fields are zero, but the tangential magnetic fields are maximum.

In FDTD calculations the zero electric fields are at the Yee locations where perfect conductor is set, but the maximum magnetic fields are at the H field locations just outside the conducting region. Thus, the FDTD "surface" is approximated as being halfway between these electric and magnetic field locations, or one fourth of the Yee cell dimension beyond the outer electric field locations set to perfect conductor.

So just how big is an FDTD object? It is generally a small fraction of a cell larger than the geometry employed. We must keep in mind that FDTD is an approximate method, and our physical reality of the plate is only approximated by FDTD.

Returning to the general topic of "building" objects, a good approach is to generate squares (in 2-D or for thin objects in 3-D) or cubes of material, where setting a cube of material is not the same as setting one Yee cell. Setting a cube of dielectric or magnetic material requires setting 12 Yee cell field locations as shown in the following FORTRAN, which fills a cube of space located in Yee cell I,J,K with dielectric material type "MTYPE":

```
IDONE(I,J,K)=MTYPE
IDONE(I,J,K+1)=MTYPE
IDONE(I,J+1,K+1)=MTYPE
IDONE(I,J+1,K)=MTYPE
IDTWO(I,J,K)=MTYPE
IDTWO(I+1,J,K)=MTYPE
IDTWO(I+1,J,K+1)=MTYPE
IDTWO(I,J,K+1)=MTYPE
IDTHRE(I,J,K)=MTYPE
IDTHRE(I+1,J,K)=MTYPE
IDTHRE(I+1,J+1,K)=MTYPE
IDTHRE(I,J+1,K)=MTYPE
```

This approach to modeling a solid object, a sphere for example, is to determine if a particular Yee cell cube is within the sphere, and if so set all 12 field locations to the material of the sphere as shown above. Setting individual field locations will result in a sphere that does not have a closed surface. This may not be important for penetrable materials, but for a conducting sphere would correspond to having short "wires" sticking out of the surface, where individual field components are set that are not a part of a complete cube.

Another point to keep in mind is that overwriting the value in a particular ID array does not cause problems, but rather may be a useful technique. Writing into the ID arrays is analogous to painting on a canvas, in that the last color painted is the color seen and the last value written into an ID array is the value used in the calculations. Just as an artist may paint a background sky, then partially cover this with grass, and in turn paint a lake over a part of the grass, we can consider that we can fill all the FDTD cells with free space, then fill a section of these with some other material, then fill a portion of this section with some different material by overwriting the ID arrays. This approach is simple and has the advantage that it eliminates the possibility of leaving voids within the material region.

In summary, the point of this section is not that "building" objects in Yee cells is difficult, but rather that it is not trivial and that some care must be taken. Simple objects can be built by writing FORTRAN code specifically, as in the plate example above. For more complicated structures, translation programs may be necessary. These will take geometrical data already in some database, from a computer-aided design (CAD) software package for example, and translate it into Yee cell locations. However, one must always keep in mind the distribution of field components in a Yee cell, and that setting one spatial location to a material type is equivalent to specifying the material which interacts with a specific electromagnetic field component at a specific location.

3.6 DIRECT COMPUTATION OF TOTAL FIELDS

In this book we stress using the scattered field formulation in FDTD problems, an approach that has many advantages. Some have already been discussed in Chapters 1 and 2, and others will be illustrated in later chapters involving applications of FDTD. Mathematically, the scattered and total field equations are equivalent and yield, in the limit of infinitesimal cells, exactly the same results. However, due to the approximations inherent in the finite difference implementation, scattered and total field results can differ, and situations exist in which direct calculation of the total fields may be simpler or yield more accurate results.

One of these involves computing small values of total field, such as in shielding problems where a plane wave is incident in which one is concerned with calculating relatively low level total fields which may penetrate a shielded region of space. This means that the incident and scattered fields are nearly identical in magnitude but opposite in sign. Both total field and scattered field formulations are limited in dynamic range by the precision of the variables used in the computer calculations. However, in the scattered field formulation, the incident field propagates exactly through the FDTD space, while the scattered field propagation is subject to grid dispersion and other errors. This slight error (or more precisely, lack of error in the incident field) may prevent the incident and scattered fields from canceling accurately in the scattered field formulation implementation. The error vanishes as frequency goes toward zero and grows to appreciable size only as the Nyquist frequency is approached.

In the total field formulation, the total field propagates through the FDTD mesh and thus grid dispersion errors exist for both the incident and scattered fields. The hope is that because they are subject to the same error, they may tend to cancel one another better. However, some contributions to the scattered field will travel in directions through the FDTD mesh differently from the propagation direction of the incident field, and therefore will not be subject to the same grid dispersion errors. Thus, one cannot count on total compensation of grid dispersion errors in the total field mode of calculation.

Other situations in which total field computation may be desirable occur due to difficulties in computing the incident field. Consider the problem of applying FDTD to antenna radiation problems. If we are dealing with a wire antenna and wish to excite fields in the gap of the antenna, analytically determining the transient near fields of a pulsed current source radiating in free space would be tedious and time consuming. As shown in Chapters 10 and 14, it is relatively simple to specify a source in this gap and directly compute the total fields that result from this source interacting with the antenna geometry.

When we wish to compute in the total field mode computer codes based on scattered field formulation are simple to use, because total field formulation is a special case of the scattered field formulation. We merely specify a zero amplitude incident field and then insert into the FORTRAN code the necessary source terms. For example, consider the case in which we are modeling a wire antenna and wish to locate a voltage source at the location of E_z (I,J,K) in the FDTD grid. We set the incident field amplitude to zero, and after each update of the E_z field values, we execute a line of FORTRAN specifying the source field at this location. This process is described in Chapter 10, and examples are given in Chapter 14.

In summary, for most situations, and especially for scattering problems, directly computing the scattered field using the scattered field formulation is preferable. However, in some situations direct computation of the total field may be more accurate or simpler, and in these cases the scattered field FDTD code emulates a total field code merely by setting the incident field amplitude to zero.

3.7 RADIATION BOUNDARY CONDITION

An outer radiation boundary condition (ORBC) may not always be necessary when applying FDTD. If the FDTD problem space is bounded by a condition that can be implemented directly into the finite difference equations of Chapter 2, an ORBC is not necessary. For example, if we are modeling electromagnetic phenomena inside a closed waveguide system, the tangential electric field at the walls is zero, and we implement this by setting the ID arrays appropriately.

For many applications, we are attempting to model a structure that is situated in free space, such as a scatterer or radiating antenna, and we would like the scattered or radiated fields to propagate into boundless space, satisfying a radiation condition. Unfortunately, the FDTD computational space is by necessity bounded, and when the scattered or radiated fields arrive at the boundary they will be reflected back into the computational space unless we take preventive measures. The usual measures involve application of an ORBC to absorb the scattered or radiated fields when they arrive at the limits of the FDTD space so that scattering or radiation into boundless free space is at least approximately simulated. The only alternative for these problems is to stop the time marching before reflections from the outer boundary return and corrupt the data, but for most problems this is not a viable alternative, as the large number of cells needed to model an extent of free space surrounding the scatterer or antenna could be greatly reduced with a corresponding savings in computer resources by application of an ORBC.

To understand the need for an ORBC in scattering and radiation problems, consider that field components are found at the boundaries. These cannot be updated using the usual FDTD equations of Chapter 2 because some of the nearest-neighbor field components needed to evaluate the finite-difference curl enclosing it are outside the problem space and not available. The usual basis for ORBCs is to estimate the missing field components just outside the problem space by some means. This typically involves assuming that a locally plane wave is propagating out of the space, and estimating the fields for the outward traveling plane wave on the boundary by looking at the fields just within the boundary. Because in most situations the wave incident on the outer boundary will not be exactly plane, nor will it be normally incident, the absorbing boundary will not absorb the wave perfectly.

There are many different schemes for accomplishing this, and some are discussed in Chapters 14 and 18. However, rather than provide the theoretical basis of these ORBC conditions, a popular and easily applied ORBC is presented. This is commonly called the Mur[2] absorbing boundary, or more particularly first or second order Mur, depending on the order of the approximation used to estimate the field on the boundary. A first order condition looks back one step in time and into the space one cell location; a second order condition looks back two steps in time and inward two cell locations, etc.

Consider that we are at the $x = 0$ limit of our FDTD computational space. We decide that on this plane we will locate E_y and E_z field components. Using these field components we can evaluate the finite difference curl operations needed to update the H_x magnetic field components at $x = 0$, and of course all nearest-neighbor field components will be available for updating the field components located at $x = \Delta x/2$ and beyond (at least to the maximum x dimension included in the problem space, at which location we must again apply an ORBC). However, we cannot update the E_y and E_z field components at $x = 0$ with the usual FDTD equations because the magnetic fields at $x = -\Delta x/2$ are not available.

We may update them, however, with the Mur expressions. Let us consider the E_z component located at $x = 0$, $y = j\Delta y$, and $z = (k+1/2)\Delta z$. The first order Mur estimate of this field component is

$$E_z^{n+1}(0,j,k+1/2) = E_z^n(1,j,k+1/2) + \frac{c\Delta t - \Delta x}{c\Delta t + \Delta x}\left(E_z^{n+1}(1,j,k+1/2) - E_z^n(0,j,k+1/2)\right) \tag{3.8}$$

The second order estimate for E_z at the boundary $x = 0$ is

$$
\begin{aligned}
E_z^{n+1}(0,j,k+1/2) = & -E_z^{n-1}(1,j,k+1/2) \\
& + \frac{c\Delta t - \Delta x}{c\Delta t + \Delta x}\left(E_z^{n+1}(1,j,k+1/2) + E_z^{n-1}(0,j,k+1/2)\right) \\
& + \frac{2\Delta x}{c\Delta t + \Delta x}\left(E_z^{n}(0,j,k+1/2) + E_z^{n}(1,j,k+1/2)\right) \\
& + \frac{\Delta x (c\Delta t)^2}{2(\Delta y)^2 (c\Delta t + \Delta x)} \cdot \\
& \left(E_z^{n}(0,j+1,k+1/2) - 2E_z^{n}(0,j,k+1/2)\right. \\
& + E_z^{n}(0,j-1,k+1/2) + E_z^{n}(1,j+1,k+1/2) \\
& \left. - 2E_z^{n}(1,j+1/2) + E_z^{n}(1,j-1,k+1/2)\right) \\
& + \frac{\Delta x (c\Delta t)^2}{2(\Delta z)^2 (c\Delta t + \Delta x)} \cdot \left(E_z^{n}(0,j,k+3/2)\right. \\
& - 2E_z^{n}(0,j,k+1/2) + E_z^{n}(0,j,k-1/2) + E_z^{n}(1,j,k+3/2) \\
& \left. - 2E_z^{n}(1,j,k+1/2) + E_z^{n}(1,j,k-1/2)\right)
\end{aligned}
$$

where we have extended the expressions given in Reference 1 to include noncubical cells.

Considering the first order Mur approximation, we see that the current value of E_z at $x = 0$ is estimated from the previous and current values at $x = \Delta x$ and the same y and z positions. The second order estimate uses previous values from the preceding two time steps, and values at the adjacent y and z positions. The equations needed to determine other field components at other limiting surfaces of the FDTD space are readily determined by modification of (3.8) and (3.9).

These expressions are straightforward to apply, but there are some practical considerations. First, we must be careful when determining the limits of our FDTD space in terms of the ranges on the DO loop indices that determine which field components in which Yee cells are on the problem space limits. These indices will not be set properly if we merely set all of them to run between the same limits. Instead we must consider the Yee cell geometry, and at each of the six surfaces that limit our FDTD space carefully determine the values of the array subscripts (Yee cell coordinates) of the electric field components on the surface and set the DO loop limits accordingly.

Another consideration is that because the second order ORBC requires field values from adjacent Yee cells it cannot be used for determining electric field values that are adjacent to the intersection of two of the terminating planes. Even if second order Mur is being applied, first order must be used for field components located adjacent to the edges of problem space.

One other consideration is that of determining the distance between the object and the outer boundary. The farther from the object the outer boundary is located the better the absorption of the outward traveling waves. This is due to these waves becoming more like plane waves as they travel farther from the structure that radiates them. However, the number of cells that can be placed between the object and the outer boundary is limited by computer memory. A common criteria is a minimum of ten cells between the object and outer boundary. For some situations more than ten will be required, especially if high accuracy is needed. Some examples are shown in Section 13.5.

Moving the outer boundary too close to the object may cause instabilities in the Mur (and other) outer absorbing boundary implementations. This may be more of a problem for antenna and other calculations, which are excited by a source within the space rather than by a plane wave, as the outer boundary must absorb total fields. Also, some fields that are required for an accurate solution may be absorbed if the outer boundary is too close. For example, sphere scattering includes a "creeping" wave that propagates around the sphere and radiates energy. The radiation from this creeping wave can easily be seen in the transient backscatter results (see Chapter 13). If the absorbing outer boundary is too close to the sphere this wave will be disturbed and the scattered field results will be incorrect.

3.8 RESOURCE REQUIREMENTS

When considering application of FDTD to a particular problem one of the first things that must be considered is (given the computer resources available) will the FDTD method be capable of providing a solution to the problem. While we are primarily working in the time domain, such questions can be considered by first determining how large the problem geometry is when measured in terms of the shortest wavelength for which we desire results. This is because the object size in wavelengths determines the number of cells (a guideline for obtaining good accuracy is that the side of each cell should be one tenth of a wavelength or less at the highest frequency of interest), which in turn determines the amount of computer storage required. The number of cells also provides an indication of the number of time steps needed for the transient fields to dissipate or for the sinusoidal excitation to reach steady state.

This section provides the basic relationships for estimating the computer resources required for a particular problem. It is assumed that based on the smallest wavelength of interest, the cell size has already been determined. From this and the problem geometry the total number of cells in the problem space (here denoted as N) has also been determined. We also assume that the material information is stored in 1 byte (INTEGER*1) ID??? arrays, with both dielectric and magnetic materials considered. Then, to estimate the computer storage in bytes required, and assuming single-precision FORTRAN field variables, we can use the relationship

$$\text{Storage} = \text{N} \times \left(6 \ \frac{\text{Components}}{\text{Cell}} \times 4 \ \frac{\text{Bytes}}{\text{Component}} + 6 \ \frac{\text{IDs}}{\text{Cell}} \times 1 \ \frac{\text{Byte}}{\text{ID}} \right)$$

where we have neglected the relatively small number of auxiliary variables needed to store temporary values, index DO loops, save results for later processing or display, and similar functions, and we are also neglecting the memory needed to store the executable instructions. This overhead is nearly independent of the number of cells in the problem space, so that as the total number of cells increases it will become a smaller fraction of the total memory required.

We can also estimate the computational cost in terms of the number of floating point operations required using

$$\text{Operations} = \text{N} \times 6 \text{ components / cell } 10 \text{ operations / component} \times \text{T}$$

where T is the total number of time steps. The actual number of calculations for each component depends on the material type and existence of the incident field at a particular time step. There are also logical statements which must be executed to determine what type of material (free space, perfect conductor, dielectric) is located in a particular location.

The number of time steps T is typically on the order of ten times the number of cells on one side of the problem space. More precisely, for cubical cells it takes $\sqrt{3}$ time steps to traverse a single cell when the time step is set by the Courant stability condition,

$$\Delta t = \frac{\Delta x}{\sqrt{3} c} \qquad \Delta x = \text{cell side dimension}$$

so an estimate for T is

$$\text{T} \cong 10 \times \sqrt{3} \ \text{N}^{1/3} \approx \text{number cells on a side of the problem space}$$

Combining the above we find that the total number of floating point operations is approximately given by

$$\text{Operations} = 10 \sqrt{3} \ \text{N}^{4/3} \times 6 \frac{\text{Components}}{\text{Cell}} \times 10 \frac{\text{Operations}}{\text{Component}}$$

From this we see that the total number of floating point operations required is proportional to the number of cells in the FDTD space raised to the 4/3 power.

Now, let us consider how the number of floating point operations scales with frequency. The size of the FDTD cell must be scaled proportional to wavelength to maintain a certain number of cells per wavelength, so that the number of cells in each linear dimension will scale in proportion to frequency. This means the number of cells in the (3-D) problem space will be proportional to frequency to the third power, and the number of floating point operations required proportional to frequency raised to the fourth power. This fourth-power scaling with frequency compares favorably to other methods, such as the method of moments. Indications are that as problem sizes become larger in wavelengths, the FDTD method will tend to require fewer operations than approaches which require the solution of a matrix, especially if results over a band of frequencies are required.

Consider a (100 cell)^3 problem space. For this number of cells approximately 30 Mbytes of memory would be required, with the actual amount being somewhat greater due to storage of other variables and instructions. Problems of this size can be run on machines ranging from supercomputers to 32-bit PCs. As available memory is reduced, the maximum number of cells that can be accomodated is correspondingly decreased. For example, with 16 Mbytes of memory, the problem space size would be estimated from the above relationship as (79 cells)^3. In actual experience 16 Mbytes will accomodate approximately (72 cells)^3, indicating a memory overhead for instructions and auxiliary variables for this problem size of about 30% of the memory needed to store the field components. Again, for larger problem spaces with more cells, this overhead percentage would be reduced.

Next, we consider the computer time needed for a typical problem. These estimates are only approximate, as the total number of time steps required depends on the geometry being considered. More time steps will be necessary for resonant geometries; fewer for lossy geometries with highly damped responses.

For a (65 cell)^3 problem with 1024 time steps to be calculated, approximately 16.9×10^9 floating point operations are required, and about 8 Mbytes of memory are required to store the field components and ID??? arrays. Speeds of available machines range from 1000 or more MFLOPS (Million FLoating point Operations Per Second) for a supercomputer through 10 to 50 MFLOPS (or more) for work stations to approximately 2 MFLOPS for a 32-bit PC. The run (CPU) times are then estimated from the above discussion as 17 s, 28 min (for a 10-MFLOPS work station), and 141 min, respectively. In practice, with a 20-cell radius dielectric sphere in this problem space and far zone fields calculated, running times on a 10-MFLOPS work station and a 33-MHz 486-based PC are 38 and 210 min. Obtaining results for problems of this size is feasible on all these machines.

The above estimates of the number of required operations assume basic FDTD calculations and neglect the slight overhead of applying absorbing boundaries, setting up the problem geometry, etc., since for large calculations these will be an extremely small fraction of the total number of operations. However, if the FDTD code being considered requires a significant number of additional computations, such as may occur when calculating far zone fields (Chapter 7), especially for many far zone directions, or evaluating recursive convolutions in large numbers of cells filled with dispersive materials (Chapter 8), the above estimates may be somewhat low and actual resource requirements will depend on the particular application and FDTD computer code. In any case, the above guidelines will provide a basis for estimating the resources required for a particular FDTD calculation.

REFERENCES

1. **Taflove, A. and Brodwin, M. E.,** "Numerical solution of steady-state electromagnetic scattering problems using the time-dependent Maxwell's equations," *IEEE Trans. Microwave Theory Tech.*, 23, 623–630, 1975.
2. **Mur, G.,** "Absorbing boundary conditions for finite-difference approximation of the time-domain electromagnetic-field equations," *IEEE Trans. Electromagn. Compat.*, 23, 1073–1077, 1981.
3. **Trueman C., Kubina, S., Luebbers, R., Kunz, K., Mishra, S., and Larose, C.,** Validation of FDTD RCS computations for PEC targets, presented at IEEE Antennas and Propagation Society Int. Symp., Chicago, July 18 to 25, 1992.

PART 2: BASIC APPLICATIONS

Chapter 4

COUPLING EFFECTS

4.1 INTRODUCTION

While FDTD was introduced to the electromagnetics community by Yee in 1966, it was nearly 1 decade before it was used to any great extent for actual applications. There were several reasons for this, including the lack of the necessary computer hardware capabilities. What really got FDTD started was an application for which it was uniquely well suited: determining the electromagnetic energy which would interact with and penetrate into an "almost" closed conducting object, such as an aircraft, due to an incident pulse of electromagnetic energy. This pulse may be due to natural phenomenon such as lightning, or it may be due to a nuclear detonation. Because the problem dealt with transient fields interacting with (to a good approximation) perfect conductors at relatively low frequencies (but not too low), it was ideally suited to FDTD.

This problem was divided into two parts. The first dealt with determining the exterior surface charges and currents which would be excited on an object (again, perhaps an aircraft) due to the incident pulse. The second part dealt with how electromagnetic pulsed energy might penetrate into a region that was partially enclosed by conducting materials designed to shield it from the incident energy. This problem was often idealized to a representative geometry, since modeling actual electronic equipment, cable bundles, etc. was too challenging for the computer power then available.

In Chapter 2 we developed the governing finite difference equations for lossy material media. A simplified and specialized case was the treatment of a perfect conductor. In this chapter we show how this capability can be used to treat exterior response and interior shielding with pulse excitation. In addition some more recent results for shielding effectiveness at frequencies below the cutoff frequency of the shield aperture are presented. These illustrate real-world applications of FDTD that often require only modest resources, but typically show relatively complex behavior.

Under exterior response we will consider FDTD calculations of aircraft exterior charge and current response to an incident electromagnetic pulse (EMP). This application was an early impetus to FDTD development. A nonoperational F-111 aircraft was EMP tested in one of the EMP simulators in the U.S., the horizontally polarized dipole (HPD) facility, to obtain system responses that could in principle be extrapolated to true threat levels. We show here how well FDTD was able to predict the measured simulated EMP responses for this aircraft test.

In the second broad area of this chapter, interior coupling, we are concerned with shielding of an interior component, in particular a wire, from an incident EMP. We divide this topic into two subsections based on whether the frequencies of interest are above or below the cutoff frequency of the shield aperture. FDTD prediction of shielding involves extremely resonant behavior that requires large expenditures of computer time for accurate characterization. This offers yet another challenge for the FDTD method.

Before proceeding to detailed discussions of FDTD applications in these areas we provide some background on the phenomenon of EMP in general and FDTD attempts to predict and understand it.

4.2 ELECTROMAGNETIC PULSE

The EMP associated with a high-altitude nuclear burst can induce large currents and charges on an aircraft in flight that, in turn, couple into the interior of the aircraft where they can cause failures or temporary outages in mission-critical subsystems, if these systems are not efficiently hardened. Military aircraft in many instances must be hardened in an electronic sense to the nuclear weapons effect known as EMP.[1]

Once called radio flash and later labeled HEMP for high-altitude EMP, it arises from Compton electrons produced by a nuclear detonation in the extreme reaches of the upper atmosphere. These electrons, the direct result of gamma rays colliding with air molecules, are preferentially directed toward the earth, the result of the burst location and air density profile, and spiral about the earth's magnetic field. EMP is the result of collective radiation from these downward spiraling electrons. An area nearly equal to the continental U.S. can be covered with a single pulse. Field strengths on the order of 50 kV/m are possible. The pulse may be approximated by a double exponential of the form $A_0(e^{-\alpha t} - e^{-\beta t})$ with a rise time on the order of 10 ns and with fall time somewhat under 1 μs. There is strong spectral content up to a few megahertz and appreciable content to around 100 MHz.

A related "threat" is naturally occurring lightning. In the older literature rise times were on the order of 2 μs and fall time might be around 50 μs. More recent work with step leaders[2] indicates rise times can be as short as 50 ns or less. Lightning is certainly very energetic with nominal peak currents of 10,000 A and occasionally going as high as 100,000 A and higher. Debate has periodically arisen on which "threat" is more severe, EMP or lightning. Without fully characterizing electronic devices, their failure mechanisms, critical interior coupling geometry considerations, and developing the tool (such as FDTD) to model the interior coupling with fair exactitude, this question cannot be definitely answered. Lighting does occasionally down airplanes and the U.S. military does view EMP as a very serious danger.

In either case much of the early work on assessing potential problems concentrated on external aircraft responses such as skin currents and charges. These are externally generated responses, although just barely, in that normal electric fields could be measured with surface mounted capacitance sensors and may be related to surface charge, just as the magnetic fields linking half-turn solenoidal sensors (also surface mounted) could be used to infer currents, often decomposed into axially and circumferentially flowing, on the aircraft. It was assumed that some transfer function would relate these exterior responses to interior responses and electronic device failure. Awaiting the success of a much more ambitious interior coupling program the transfer function could only be guessed at. The process of characterizing the transfer function may be said, with some debate, to be ongoing. What was taken as a starting point in this earlier work was that the larger the exterior response the more likely electronic damage or upset would occur inside. Certainly much could be learned and was learned in these early efforts in which FDTD played an important role.

4.3 EXTERIOR PULSE RESPONSE

FDTD was first applied to a model of an F-111 aircraft[3] to estimate the external charge and current responses to a simulated EMP. The EMP simulator was the HPD facility located at the Kirtland Air Force Base in Albuquerque, NM. The key issues in this application were the simulator fields illuminating the aircraft, the aircraft model, and the computational resources available for the effort. These issues in slightly altered form are present in all modeling efforts.

What may not always be obvious is the underlying physics. For example, the exterior response of an aircraft is resonant but much more damped than the resonant response of, for example, a thin dipole antenna. As a result a relatively short time record will suffice for an exterior aircraft response prediction as compared to the record required of a thin dipole antenna. If the antenna is placed inside a cavity as in the case of a wire inside a cavity pierced by an aperture, even for a large aperture, the wire/antenna response is now much more resonant than the bare antenna. Extremely long records in terms of number of time steps required are then needed. Rather simple insights provided by the antenna-like features and the distinction between exterior and interior response have proved very useful in anticipating the length of the response and therefore the computer resources required to predict the response using FDTD.

Many other aircraft exterior response studies have been performed by the authors and other investigators. Of particular interest was an F-106 aircraft outfitted by NASA for the assessment of lightning effects on aircraft. It required modeling lightning attachment and detachment channels as wires entering and exiting the aircraft at the nose and tail. The aircraft/lightning model[4] was able to provide reasonable exterior response predictions. Interior response predictions were also attempted with limited success. The problems encountered arose from vastly underestimating the time duration and resources needed to accurately characterize the interior responses.

4.3.1 MEASUREMENT FACILITY

The HDP facility is located in Albuquerque, NM. This is the site of numerous EMP simulators including the vertically polarized dipole (VPD) facility,

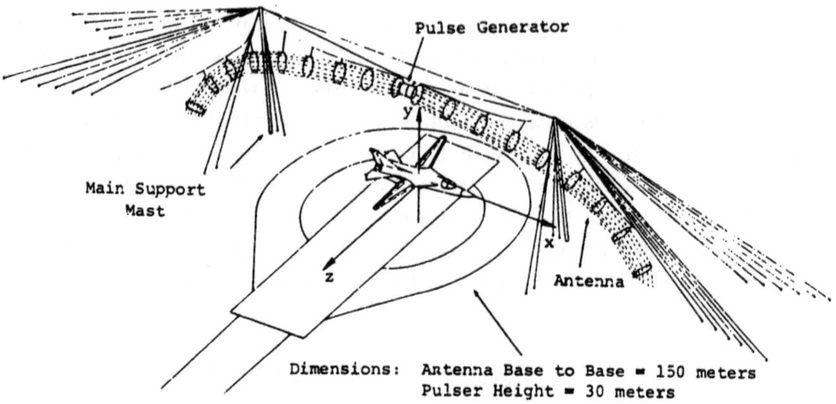

FIGURE 4-1. Horizonitally polarized dipole simulator system (aircraft not to scale).

advanced research EMP simulator (ARES), and TRESTLE, the world's largest EMP simulator, reported to have used 2% of 1 year's lumber production in the U.S. in its construction. The HPD facility (Figure 4-1) is a half loop of wires arrayed on a series of hoops 30 m tall at its center where there is a 1- or 2-MV high voltage pulser, depending on the model employed. The current discharged by the pulser into the antenna-like arms of the loop produces the nominally horizontally polarized E fields desired of the simulator.

The fields radiated by the HPD simulator are quite complex. The electric field is approximately horizontal in the test pad area below the pulser and along the main or z axis of the facility where a test aircraft is typically placed. An approximation of the simulator fields that assumes the electric field to be horizontal was derived from measurements and used in the FDTD model. A free space magnetic field ($E/H = 377\ \Omega$) was associated with the electric field. The ground was modeled as a conducting surface that produced a time varying reflection coefficient, $R = 1 + (0.75)\ e^{-t/200}$ ns. This heuristic field model reproduced reasonably well the measured field data in the volume enclosing the aircraft when placed either directly below the pulser or along the main axis of the facility.

This model did not account for the field variations arising from pulser shot to shot variation, which could be considerable. The pulser exhibited 10 to 90% rise times that varied between 7 to 13 ns, with 9 ns being nominal. Variations of 10% or more in the spectral content between 10 to 50 MHz were common. Comparisons between measurements and predictions do not account for these errors. Above 50 MHz the variations were large enough that the experimental data could not be relied upon.

4.3.2 FDTD MODEL OF THE AIRCRAFT

The aircraft used in the measurements was a nonoperational F-111, with much of its internal components, including the engine, removed. Its exterior surface was, however, intact. The FDTD model used 3-D, rectangular cells

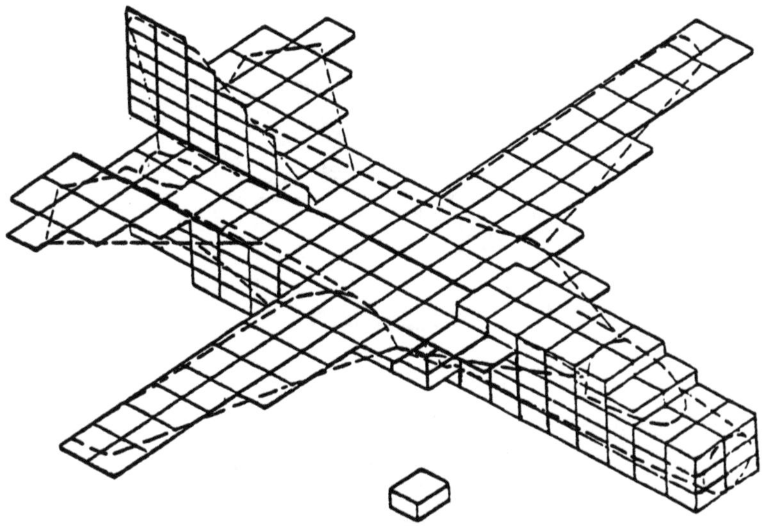

FIGURE 4-2. F-111 FDTD model.

having dimensions of $\Delta x = 1$ m, $\Delta y = 0.5$ m, and $\Delta z = 1$ m that were part of a total cell space of $28 \times 28 \times 28$ cells. This model is shown in Figure 4-2 along with the outline of the actual aircraft. The nonsquare cell allowed increased modeling detail in the y direction, where it was most needed. The problem space provided either four (in the x and y direction) or nine cells (in the y direction) from any extremity of the model, which is centered in the space, to the outer edges of the problem space where the outer radiation boundary condition is imposed. This outer radiation boundary condition was a look back scheme developed prior to Mur's conditions, but was similar in nature.

The aircraft fuselage was defined by zero tangential total fields on the exterior faces of the outermost cells, explicitly $E^{scat}_{tangential} = -E^{inc}_{tangential}$. No specification was made on the interior fields as the surface boundary condition completely decouples the exterior fields from the interior fields. The wings were defined by zero total tangential electric at the interfaces of the appropriate cells, and hence were of zero thickness. By extrapolating and interpolating the fields on either side of the wing to the wing surface to find E_{normal} and $H_{tangential}$ the charges and currents on either side of the wing, which generally will be different, were found for any location on the wing. The nose of the FDTD model of the aircraft was truncated because a dielectric nose cone is used at the end of the actual aircraft. For an FDTD code limited to perfect conductors only, such a dielectric nose cone was best modeled by free space.

The cells inside a box that just encloses the modeled aircraft were kept constant while those outside were made progressively larger as they approached the outer boundary. A scale factor of 1.3 was used in the x and z directions and 1.15 in the y direction. This scaling increased the distance of the outer boundary from the aircraft to lessen reflection. Larger scale factors than

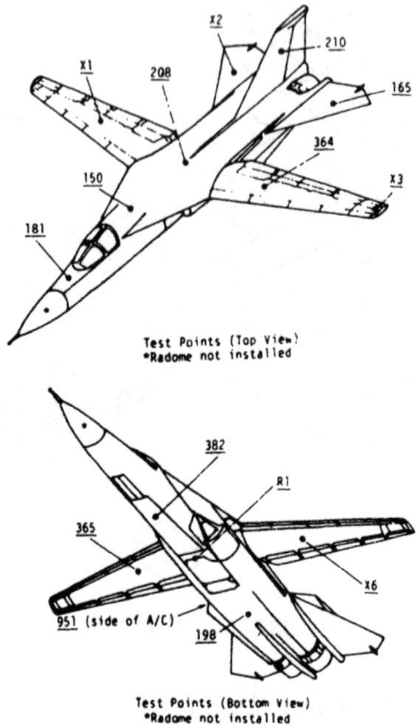

FIGURE 4-3. Model of test aircraft showing sensor locations.

1.3 were observed to lead to numerical instabilities. Using λ/4 equal to a cell size as the upper limit of accurate modeling yields 75 MHz as the highest frequency accurately modeled.

4.3.3 COMPARISON OF PREDICTIONS AND MEASUREMENTS

Let us proceed to compare the FDTD predictions with measurements. A set of comparisons of the data predicted with the field and aircraft models and actual aircraft measurements[3] are presented here for the aircraft located 30 m out along the z axis of the facility, with the fuselage center 2 m above the ground plane. Other test locations were also employed, but the geometries, for example, directly under the pulser, were less demanding when it comes to comparison with the code. The aircraft fuselage is oriented parallel or perpendicular to the HPD loop so that at early times the electric field is parallel to the fuselage or the wings, respectively. Various sensor location about the aircraft surface (Figure 4-3) were employed. The magnetic loop and capacitive gap sensors' derivative output of the charge and current response, namely, $\dot{Q}$, $\dot{J}_A$, and $\dot{J}_C$, where the derivative currents $\dot{J}_A$ and $\dot{J}_C$ are axial and circumferential, respectively, were left unchanged. The predicted responses were differentiated so that a direct comparison was possible.

The comparisons,[3] a sample of which are shown here (Figures 4-4 to 4-11), range in agreement from excellent to poor. It is good to excellent when a strong response is involved and can be poor when a weak response is involved. This characterization of responses rests on treating the aircraft fuselage as a thin half-wavelength dipole when the E-field is parallel to the fuselage, and the

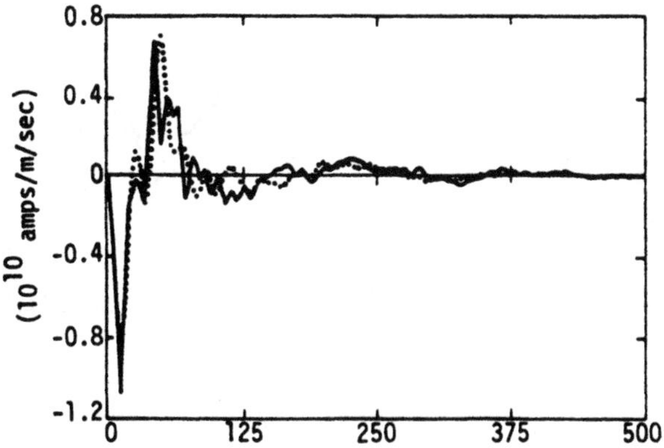

FIGURE 4-4. TP 208, $\dot{J}_A$ (t), E ∥ fuselage.

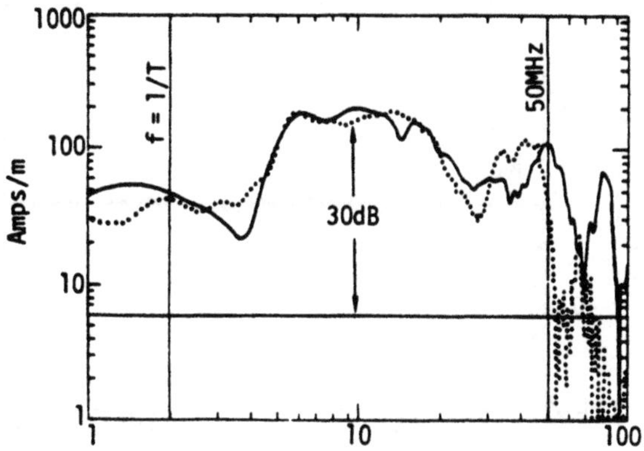

FIGURE 4-5. TP 208, $\dot{J}_A$ (ω), E ∥ fuselage.

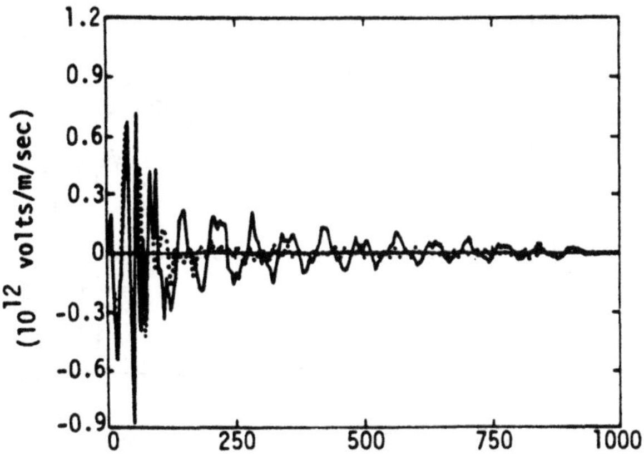

FIGURE 4-6. TP 208, $\dot{Q}$ (t), E ∥ fuselage.

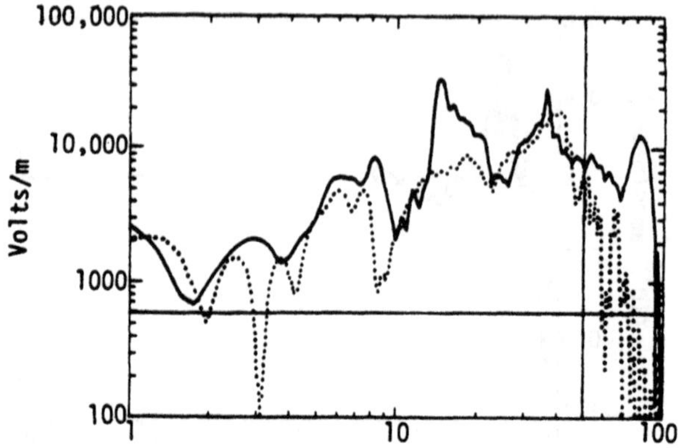

FIGURE 4-7. TP 208, $\dot{Q}$ (ω), E ǁ fuselage.

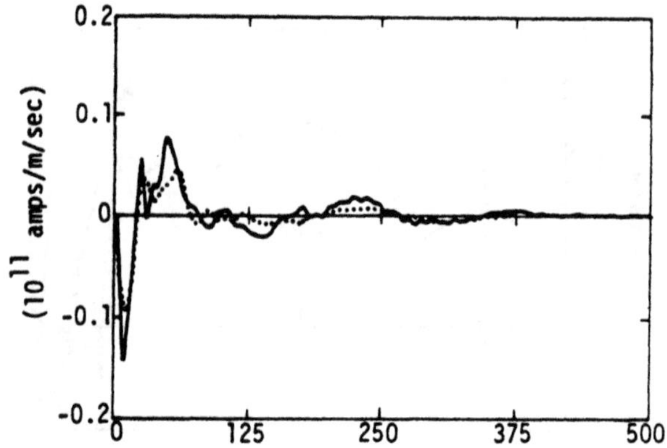

FIGURE 4-8. TP 150, $\dot{J}_A$ (t), E ǁ fuselage.

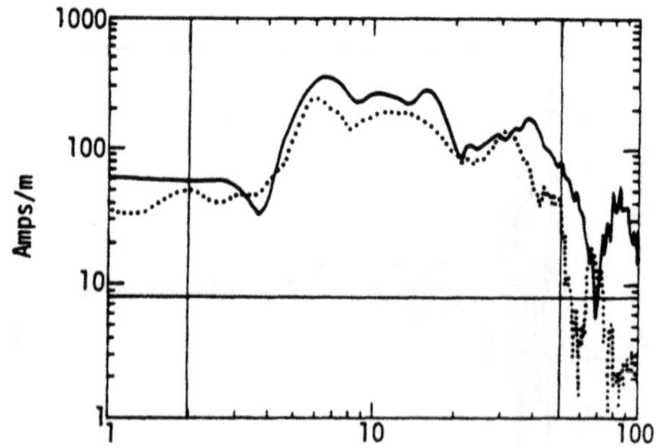

FIGURE 4-9. TP 150, $\dot{J}_A$ (ω), E ǁ fuselage.

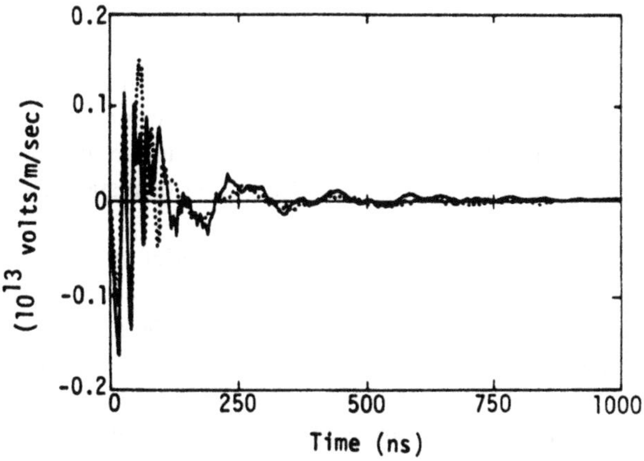

FIGURE 4-10. TP 150, $\dot{Q}$ (t), E ∥ fuselage.

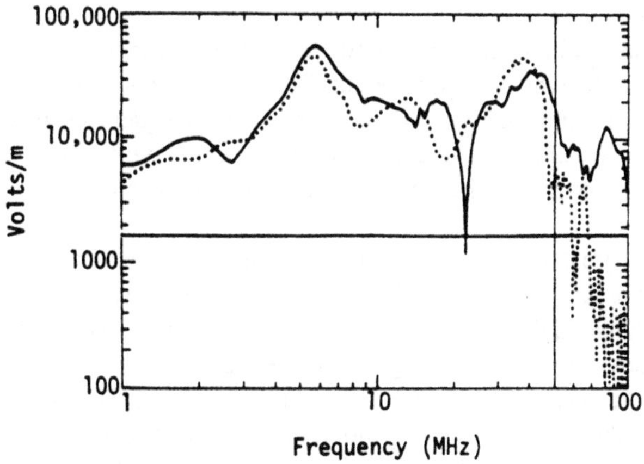

FIGURE 4-11. TP 150, $\dot{Q}$ (ω), E ∥ fuselage.

wings to behave the same when E is parallel to the wings. Note that circumferential currents are always weak responses and that sensor locations at or near the null in the lowest resonance or half-wavelength responses are troublesome. Based on record lengths and dynamic range of the instrumentation the data are compared only over a frequency range of 5 to 50 MHz.

Even with all these limitations it is evident that this early version of FDTD was an effective tool for predicting exterior aircraft response to EMP.

4.4 INTERIOR ELECTROMAGNETIC SHIELDING

In shielding, electromagnetic energy (at least over selected frequencies) is excluded from a cavity by the cavity walls, which to keep the problem interesting are penetrated by at least one opening or aperture. Because of the energy exclusion we speak of the cavity walls or shell as a shield. An important aspect of cavities with apertures is the shielding provided. Characterizing the shielding provided by a shell and aperture is the objective of the FDTD modeling discussed here. The most important feature we will observe is the highly resonant nature of the interior response of an interior wire.

Let's begin our discussion of shielding with a brief overview of the fundamental considerations. A measure of the shielding effectiveness of a conducting shell forming a cavity pierced by an aperture on the response of an interior wire is the ratio in the frequency domain of the current induced on the wire inside the shell normalized to the current induced on the wire without the shell.[5] Alternately, the square of the frequency domain current integrated over frequency can be used.[6] This is proportional to the energy coupled to the wire. For each wire resonance the integral can be approximated by $A_n^2(\omega_n)\Delta f$, where $A_n(\omega_n)$ is the resonance peak amplitude and Δf the resonance width. Equivalently via Parsival's theorem we could use the time domain current squared integrated over time.

One can distinguish between two shielding regimes, frequencies above aperture cutoff, wherein aperture cutoff is the frequency at which the wavelength equals the aperture circumference, and frequencies below aperture cutoff. Above cutoff the antenna-like resonances have Qs somewhat greater than the corresponding unshielded wire, while below cutoff the Qs are much greater.

We first treat the higher frequency regime at which FDTD works with relative ease and has been validated experimentally. The first definition for characterizing shielding effectiveness is employed here. We then treat the far more difficult low frequency regime, where the behavior of the interior wire is extremely resonant and requires exceedingly long FDTD runs for accurate characterization. The second definition of shielding effectiveness is employed here as the signatures of the resonances (Qs) are so different.

Shielding effects in either regime are found to be far from intuitive and much of the discussion shows how FDTD modeling has modified our understanding of shielding. We will see that very large resources must be expended to accurately characterize shielding effects.

4.4.1 FREQUENCIES ABOVE APERTURE CUTOFF

Let us first consider shielding at frequencies above aperture cutoff. The FDTD modeling presented here on this topic and its experimental validation[7] was performed by one of the authors (Kunz) while at the Lawrence Livermore National Laboratories (LLNL). A generic coupling test object, called the preliminary Livermore universal test object (PLUTO), had its transient interior coupling response measured at the Transient Range Facility. FDTD, in a version substantially duplicated in the listing in Appendix B, was used for the computational modeling.

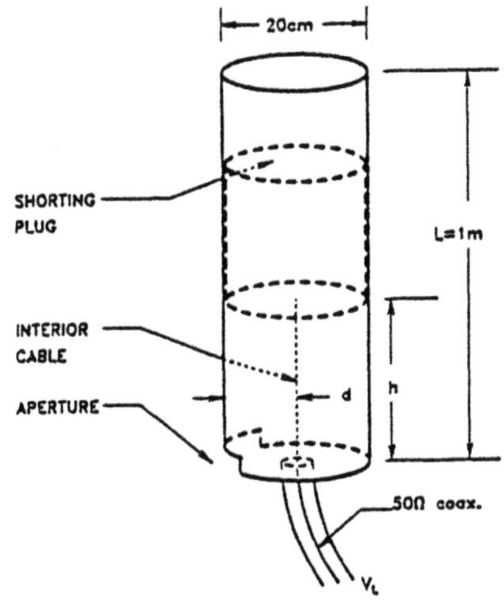

FIGURE 4-12. Preliminary Livermore universal test object (PLUTO).

4.4.1.1 Cavity Shielding Model Description

Before further considering the FDTD calculations let us describe the computational PLUTO model.

PLUTO (Figure 4-12) was a nominally 1-m tall (99.1 cm to be precise) aluminum sheet-metal cylinder of 0.2 m diameter. It was mounted on the floor of the Livermore Transient Range Facility so that its electrical image was formed. Various size apertures on a "face plate" were mounted on the surface of the cylinder next to the floor. A number of aperture sizes (including the ground plane image) was available; the size employed for the computational model was 7.5 × 7.5 cm. This was considered a stressing case because of the relatively large aperture size linking exterior and interior and because of the aperture "fatness". Only fields incident broadside on the aperture with E parallel to the interior wire were treated computationally. Two interior cavity heights were physically employed and computationally modeled, namely 22.5 and 30 cm. These two heights allowed a 22.5-cm wire to span the cavity so that a wire shorted at both ends was present, or to only partially span the cavity so that a wire shorted at one end and open at the other was present. The wires could be situated at any one of five locations (Figure 4-13). All five locations were computationally modeled for the 22.5 cm cavity and the central wire location at Test Point (TP) 2 was computationally modeled for the 30-cm cavity.

The wires were computationally modeled using the relationship

$$E^{scat} = -E^{inc}$$

along the wire run. This is a "thick" full cell wire computational model.

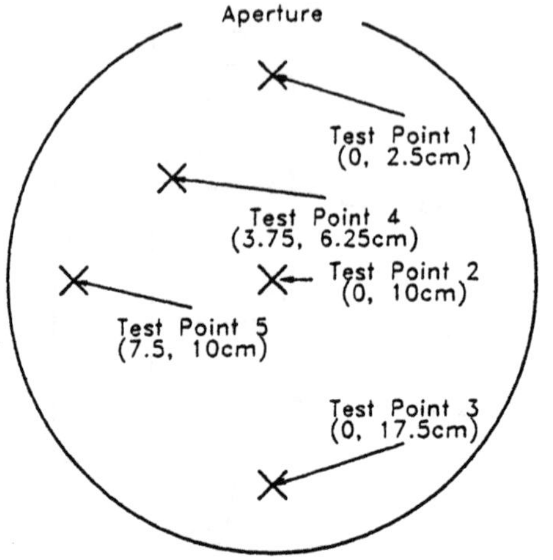

FIGURE 4-13. Test point locations for PLUTO.

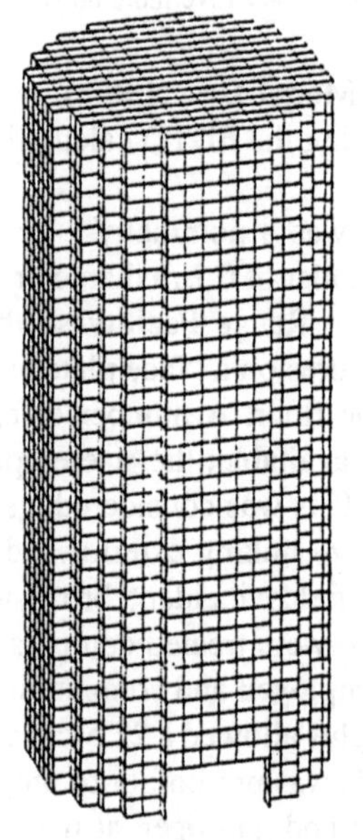

FIGURE 4-14. Finite-difference model of PLUTO.

Methods for "thinning" the wire are discussed in Chapter 10, but were not employed here. The cylindrical shell (Figure 4-14) was computationally modeled out of plates in which the two components of E of a cell face satisfy the perfectly conducting boundary condition given above for the wire where only one component is set. The PLUTO computational model was embedded in a problem space $32 \times 32 \times 88$ cells with cubic cells 1.25 cm to a side.

4.4.1.2 Transient Range Test Facility

The PLUTO experimental object must in turn be illuminated by an incident pulsed electromagnetic field (EMF). This field was provided experimentally by the Transient Range Facility at Lawrence Livermore Laboratory. The Transient Range Facility was undergoing a transition during the time the measurements were taken. The existing monocone was being replaced by an upgraded monocone with a machined "nose piece" at the vertex and with rigidly supported conical side panels. This newer monocone called the EMP engineering research omnidirectional radiator (EMPEROR) allowed operation up to at least 18 GHz, but was not yet available. A long wire was substituted in its place. The advantage was an extremely well-behaved field for short times. The disadvantage was an approximately 12 dB lower field strength. An indoor facility (only later surrounded by an electromagnetic absorber), it allowed measurements with a clear time of ≈ 20 ns. All the data taken under these conditions had to be limited to a 20-ns duration. As seen later, this data truncation obscured some very important response features.

The measured field was approximated in the FDTD calculations as a point-source radiator with a damped sinusoidal temporal behavior

$$E = |E|\,\hat{\phi} \text{ and } |E| = 770 \text{ V/m } (|\vec{r}|/2.4m)$$
$$\cdot \sin\left[2\pi\left(0.625 \times 10^9\right)t\right] e^{-\left(0.625\times10^9\right)t}$$

where $\hat{\phi}$ is the elevation angle. This time-domain computational |E| field model is based on a very simple physical model of the wire as a slender bicone and is approximately twice as large as the experimental time domain field. It is also correspondingly larger in the frequency domain. The transfer function formed between these two sources, the measured and FDTD model, was applied to all the predicted data, allowing a direct comparison of PLUTO measurements and predictions. The upper frequency limit was set by the pulser used in these measurements at approximately 3 GHz. Later measurements with EMPEROR in place used an electromagnetic absorbing shroud and a swept continuous wave (CW) source. They provided frequency domain measurements to 18 GHz in amplitude and phase. These measurements could be inverse transformed to yield time domain data.[8]

In either configuration, thin wire or EMPEROR, it was not feasible to directly measure currents on an interior wire. Commercially available current probes suitable for this type of wire measurement, such as the Tektronix CT 1 respond only up to 1 GHz. To maximize frequency response an equivalent current technique[9] was employed. Antenna modeling concepts were applied to a coaxial cable entering the base of PLUTO with only its center conductor extending beyond the floor. Using time domain reflectometry (TDR) input impedance measurements and the measured coaxial-cable response, the wire

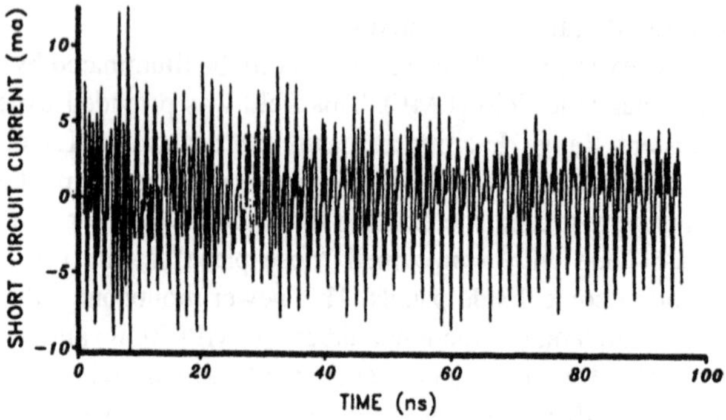

FIGURE 4-15. 100-ns predictions of the short-circuit current at the base of a 22.5-cm wire TP 2 (center).

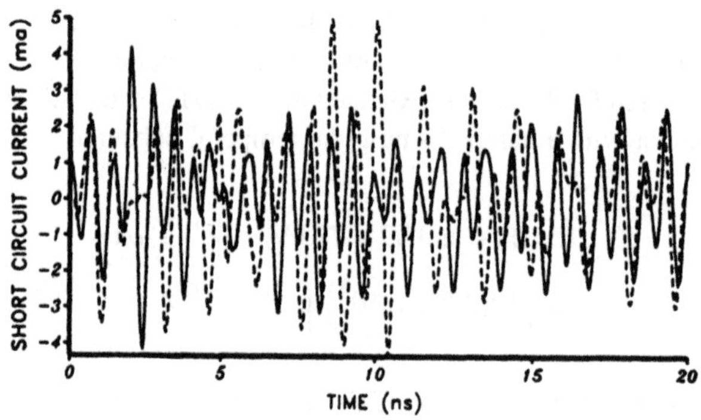

FIGURE 4-16. Overlay of the measured short-circuit current of a 22.5-cm wire at TP 2 (solid) vs. the compensated computed results (dashed).

responses for a wire arbitrarily terminated at its base within PLUTO could be inferred. For the comparisons made here the wire was assumed to be shorted to the floor.

4.4.1.3 Data Evaluation

A 100-ns prediction (Figure 4-15) was made for the 22.5 cm cavity, with the wire at the center of the cylinder at TP 2; the wire was shorted at both ends. The run was 4096 time steps long and took 84 h on a VAX 11/780. The time-domain predicted response was truncated and overlayed with the measured data (Figure 4-16). FFTs (Figure 4-17) of these 20-ns long data records were compared and the agreement was good. We noted at the time that the 100 ns record, with what appears to be ~64 cycles at the lowest resonance, showed little decay. Not surprisingly, then, an overlay of the FFTs of the 100 ns predicted data and 20 ns truncated predicated data (Figure 4-18) show significant differences with the 100 ns record, yielding much sharper peaks with higher amplitudes. What we learned later and what is discussed in the section below on aperture cutoff is that the lowest resonance had a Q in the thousands. This requires a time record more like 10,000 ns long or the better part of

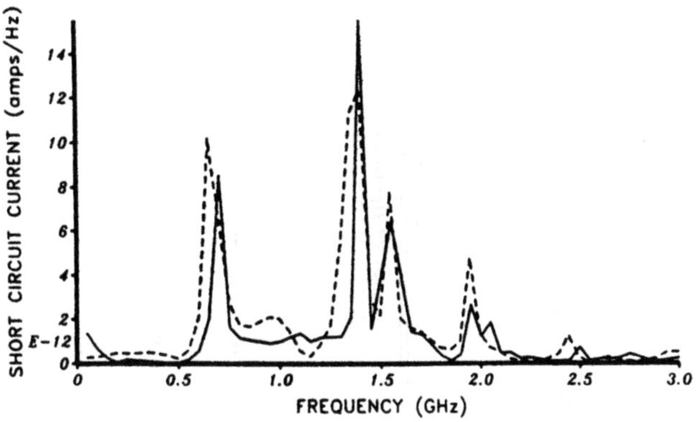

FIGURE 4-17. Fourier transform overlays of the measured short-circuit current of a 22.5-cm wire at TP 2 (solid) vs. the compensated computed results (dashed).

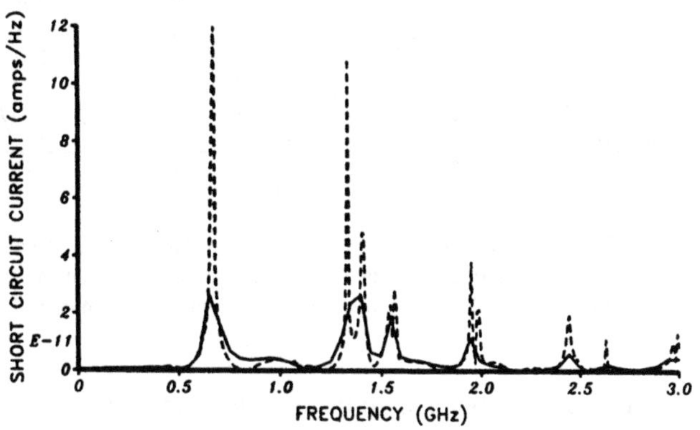

FIGURE 4-18. Illustration of the increase in frequency resolution between a 20-ns record (solid) and a 100-ns record (dashed).

500,000 time steps, with the PLUTO computational model employed here and a run time exceeding 8000 h on a VAX 11/780.

The sharp peaks of the 100 ns prediction can be associated with two types of modes:

1. Wire or antenna-like modes corresponding to the TEM modes of the coaxial geometry with frequency given by

$$f = (2n + 1)\frac{c}{D/Z}$$

where Z is the wire height, c is the speed of light, D is the diameter, and (2n+1) gives odd-order harmonics in keeping with the symmetry.

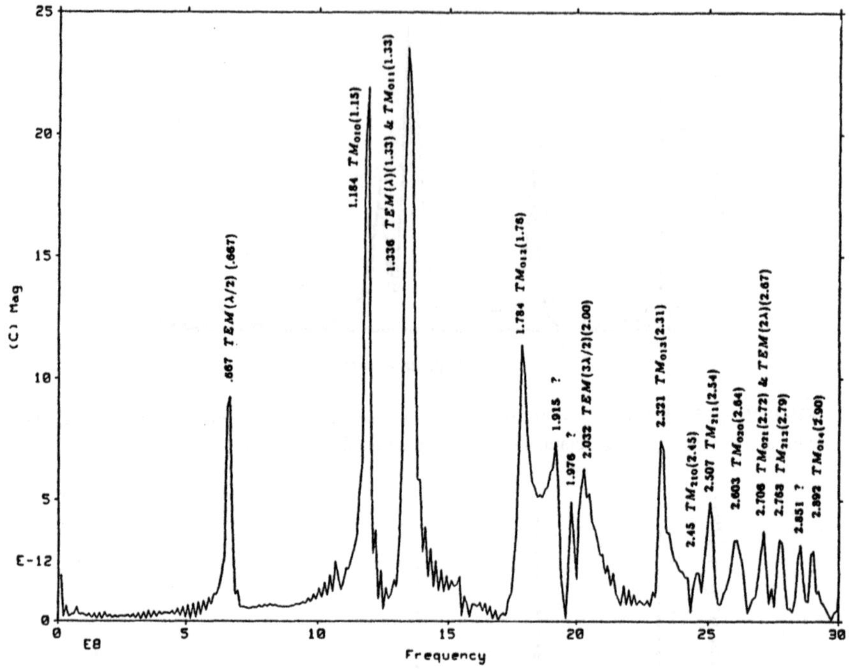

FIGURE 4-19. Wire and cavity modes within the 20.0-cm diameter × 22.5-cm-high cavity inside PLUTO, with the wire shortened at both ends. The first number on the mode represents the finite-difference computational value of the mode, and the parenthetical number is the theoretical value.

2. Cavity modes corresponding to the transverse magnetic (TM) modes of the cavity picked up by the wire acting as a cavity probe and given by[10]

$$f_{mnp} = \frac{c}{2\pi}\left[\frac{x_{mn}{}^2}{R} + \frac{p^2\pi^2}{D^2}\right]$$

where R is the cavity radius and x_{mn} is the nth root of the equation $J_m(x) = 0$.

As a detailed examination of the transform of the 100-ns record shows (Figure 4-19), these modes account for virtually all of the observed structure.

Similar results held for the 30-cm cavity with the 22.5-cm wire open at the end and shorted at the bottom. Here, a 200-ns record was predicted and the spectral peaks (Figure 4-20) were even sharper. A single mode at 2.5 GHz was examined by filtering out the rest of the signal using routine signal processing methods. By examining the rate of decay in this residual signal a Q of approximately 800 was found. We still believe this value to be true, however, had we characterized the peak at 300 MHz in the same way, using

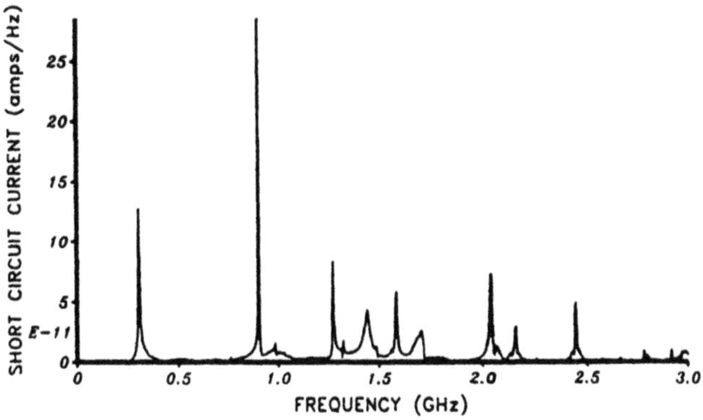

FIGURE 4-20. Fourier transform of a 22.5-cm high wire inside a 30-cm high cavity at TP 2200-ns record.

the 200-ns data record, we would have been very surprised to find an even higher Q at seeming variance with the transform of Figure 4-20. The 200-ns data record provides a $\Delta f = 1/T = 5$ MHz in the transform so that at 300 MHz, a $Q < 100$ is all that can be resolved. The higher Q comes from fitting the data in the time domain, an approach we rely on heavily in the next section.

4.4.1.4 Validation of FDTD Results at Frequencies Above Aperture Cutoff

Time and frequency domain comparisons have already been shown at TP 2 for a 22.5-cm wire shorted at both ends. Similar comparisons for TP 1, 3, 4, and 5 for the same geometry show generally similar good agreement. The code agrees to within 0 to –6 dB in power amplitude and to within ±6 % in matching the frequency of the resonant peaks. We concluded that FDTD is capable of making reasonably accurate interior coupling predictions for modestly complex metallic objects. Even at this early stage of development we noted that frequency domain codes should be used with extreme care if they are to be used at all. We noted that Qs on the order of several hundred imply extremely close frequency spacing requirements for an experiment if the response is to be accurately characterized. Now we know Qs in the thousands are possible with the type of geometry examined here and that even FDTD must be run a long time to fully capture the data.

4.4.1.5 Shielding Effectiveness Characterization Above Aperture Cutoff

When the wire current responses for a shielded and unshielded wire are ratioed in the frequency domain after being smoothed over a modest frequency band using the auto spectral density algorithm,[5] the ratios are near unity above aperture cutoff (Figure 4-21). This had led to the conclusion that the shield provides little or no shielding above aperture cutoff. Below aperture cutoff the data generated in this effort for wires shorted at their base are too short to

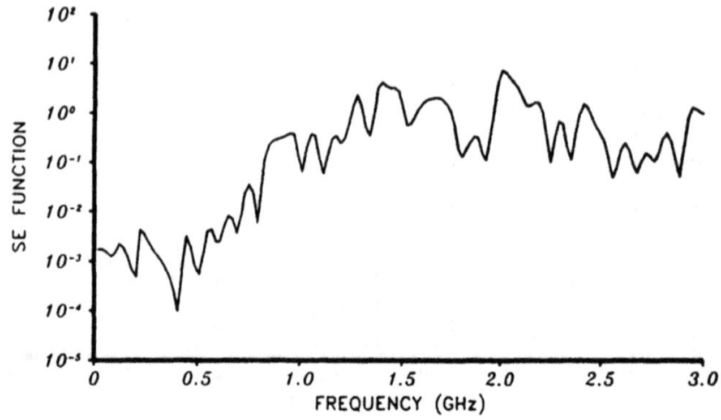

FIGURE 4-21. Shielding effectiveness function of a 22.5-cm wire located at TP 2 in a 30-cm high cavity.

adequately characterize the response below aperture cutoff where the highly resonant shielded response requires very long time records to resolve it.

4.4.2 FREQUENCIES BELOW APERTURE CUTOFF

A perfectly conducting shell about a cavity with a small aperture and an internal wire with varying wire terminations were used in the previous section to model electromagnetic shielding. Experiments show little shielding above aperture cutoff, where the ratio of the wire response with and without the shell determine the shielding. These experiments did not have sufficient resolution to adequately resolve the shielding behavior below aperture cutoff where the wire resonances have extremely narrow bandwidth when the wire is shorted at its base.

FDTD is used here with up to 1 million time steps to accurately characterize wire resonances below aperture cutoff. The results obtained were quite surprising and are briefly summarized here. Only a modest shielding effect is observed below cutoff for perfectly conducting shells with varying size apertures. The amplitudes of the wire resonances rise rapidly as $1/\omega^2$ as frequency decreases. The Q of the wire resonances also increases rapidly with decreasing frequency. These effects are nearly offsetting leaving the energy in each wire resonance only slightly reduced from the unshielded wire response as the aperture size varies. This reduction only increases modestly with smaller aperture size.

4.2.2.1 FDTD Geometry

In place of the PLUTO geometry used in the LLNL measurements and code predictions described in the previous section of this chapter a coarser rectangular cavity modeled by $28 \times 8 \times 8$ FDTD cells (Figure 4-22) was used. With each cell dimension 2.5 cm, this object has roughly the same dimension as the PLUTO model. The cells are twice as large as had been used for the previous FDTD model of PLUTO. This cell size and interior cavity dimension limits the

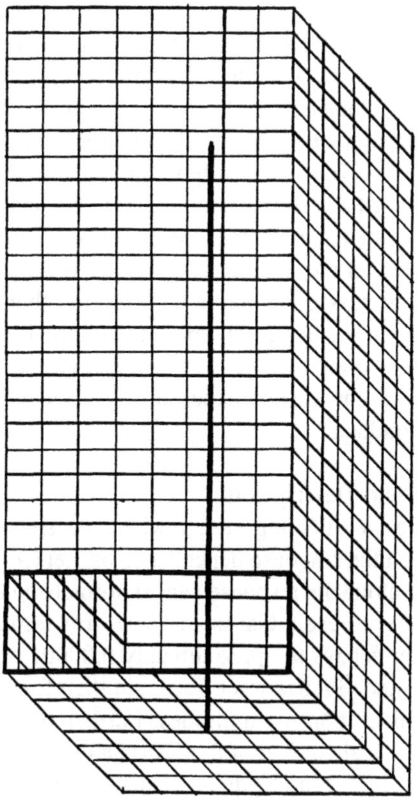

FIGURE 4-22. Coupling model.

rectangular cavity model to only the first four resonances of an internal wire. This is the region in frequency from about the aperture cutoff frequency on down for the aperture sizes we chose, namely 8×4, 6×3, and 4×2 cells. There are enough resonances at and below aperture cutoff to characterize the low frequency amplitude envelope behavior and the resonant width at least approximately. A model with more cells, on the order of the $32 \times 32 \times 88$ cells used to model PLUTO at LLNL or more, would have allowed a more detailed study, but could not be run on the available resources, a VAX 8550. Run times on this machine for a million time steps were approximately 400 h of CPU time.

The wire was centered in the cavity, shorted to the base of the cavity and run nearly to the top of the cavity, stopping four cells short of the top, leaving the wire in an "open" termination condition at the top and "shorted" at the bottom. The antenna-like resonances appeared at approximately 125 , 375 , 625, and 875 MHz, as expected from the 60-cm height of the wire and its terminations. The wire length was roughly two to three times that of PLUTO, depending on PLUTO configuration, accounting for the lower resonant frequencies despite the roughly equivalent cross sections. Aperture sizes (20 × 10 cm, 15 × 7.5 cm, and 10 × 5 cm) for the rectangular cavity were similar in size to the larger apertures used with PLUTO. The rectangular cavity model was configured to give responses similar to what would have been seen with PLUTO had the measurements been much more finely spaced in frequency and the code modeling PLUTO run for a few million time steps.

Our model first used a perfectly conducting metal and then aluminum. We discovered that the inclusion of the conductivity of aluminum had only a negligible effect even for the smallest aperture. To more clearly see the effects of conductivity we increased the ohmic losses of the shield until they equaled the radiation losses for the 4 × 2 cell aperture and the lowest resonance. This occurred when σ was approximately 10^{-6} as large as σ for aluminum. As this was such a large and unrealistic change in σ the remaining discussion only treats the perfect conductor results.

4.4.2.2 Incident Field Description

Because this was a shielding study, when the shielded and unshielded responses are ratioed the response dependence on field amplitude is removed. For either response alone the response can be normalized for the driving field.

Our approach was to use a Gaussian waveform similar to that described in Chapter 3, which decays to approximately 0.00001 of its peak amplitude before being truncated. Truncation effects are therefore on the order of 120 dB down from peak response amplitudes. Further α is selected so that the amplitude of the spectral content of the Gaussian waveform or pulse is down by a nearly equal factor at the Nyquist frequency. Any aliased signal is then insignificant.

4.4.2.3 Response Predictions

Current response at the base of the wire was calculated for the three aperture configurations. The response was characterized in the frequency domain after Fourier transforming the time domain responses as the resonance amplitude $A_n(\omega)$, resonance width Δf_n and Q_n, and resonance energy $U_n(\omega)$ at each of the four resolvable resonances (the highest frequency resonance displayed here corresponds to approximately 16 FDTD cells to a wavelength, a reasonably conservative upper limit in frequency for interior coupling). The response was characterized in the time domain as an extrapolated peak amplitude $A_n(t = 0)$, decay rate α_n, and energy $U_n^{(t)}$. These values were found from filtered (using a rectangular frequency domain window) time domain responses that let only one resonance through at a time.

Our attention was first focused on the 8 × 4 cell aperture which has the fastest decay rate for all the resonances, since as the largest aperture it has the greatest radiation transmissitivity through the aperture. Initially we made runs of 64K, 128K, 256K, 512K, and finally 1 meg time steps. This progression developed as we were forced into longer time records to fully capture the decay of the lowest frequency resonance (Figure 4-23).

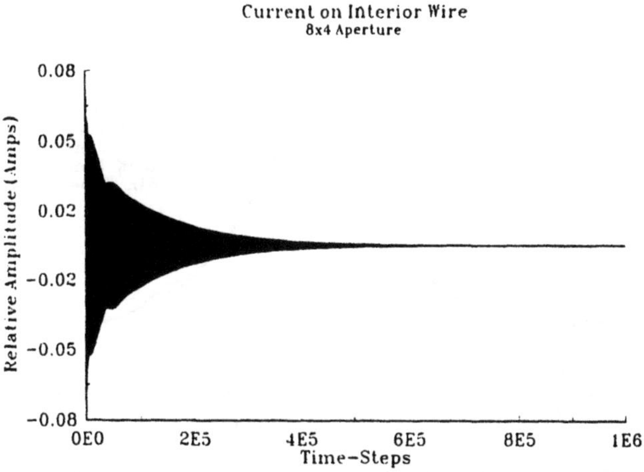

FIGURE 4-23. 1 meg time step FDTD prediction for 8 × 4 cell aperture.

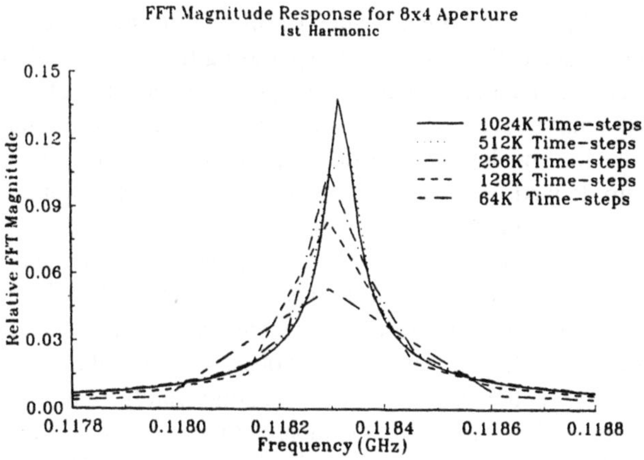

FIGURE 4-24. Effect on transform of longer records for 8 × 4 cell aperture, 1st harmonic.

Transforms of these time domain records yielded a lowest frequency resonance whose amplitude climbed dramatically at first as the record length increased and finally converged to a stable amplitude, but at a much higher level than expected and with a much higher Q, or equivalently a much narrower resonance width (Figure 4-24). This convergence was much faster at the higher frequency resonances that did not require the very long record lengths for accuracy (Figure 4-25).

Nearly identical values of $U_n(\omega)$ and $U_n^{(t)}$ were found using the relations

$$U_n(\omega) = A_n(\omega)^2 \Delta f_n \text{ and } U^{(t)} = \frac{A_n(t)^2}{4\alpha_n}$$

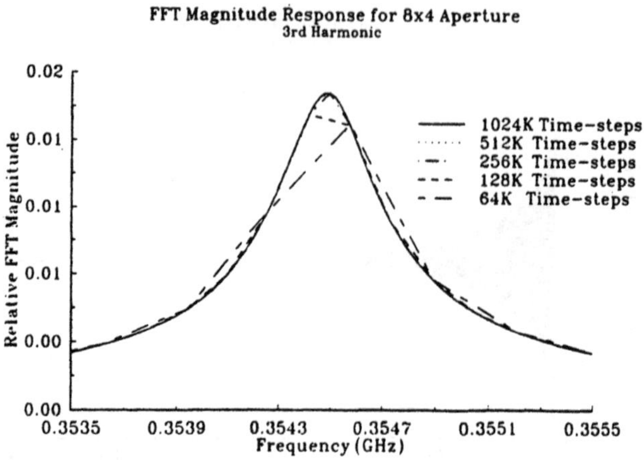

FIGURE 4-25. Effect on transform of longer records for 8 × 4 cell aperture, 3rd harmonic.

When we ran the smaller apertures we found that the lowest frequency resonance had not decayed much after 1 million time steps. Nevertheless we could get energies $U_1^{(t)}$ (from filtered $A_1(f)$ using the above relations) that we were now confident corresponded to the $U(\omega)$ we would have obtained had we extended the runs to even more time steps. Figures 4-23 to 4-25 show for each aperture size the envelopes of the first four resonances from which $A_n(t = 0)$ and α_n were derived. From these data the energy in each resonance for each aperture size was tabulated (Table 4-1).

We cross checked our results where possible in the time and frequency domain to ensure that the surprising trends seen in Table 4-1 are indeed correct, namely that the energy in the resonances increased with decreasing frequency below aperture cutoff. The earlier rolloff observed was indeed an artifact. In fact, the energies in the resonances below aperture cutoff for all three aperture sizes are only modestly reduced as compared to the energy in the corresponding unshielded wire. This observation leads us to state that for a perfectly conducting shell with even a small aperture only modest shielding is afforded from an energetic viewpoint.

What is happening is that the resonances below aperture cutoff become progressively narrower in bandwidth while their amplitudes, previously thought to diminish, grow rapidly. A rough fit to this behavior is an amplitude that behaves as $1/\omega^2$ and a Δf that goes as ω^2 so that energy behaves as $A(\omega)\ ^2\Delta f$ $\alpha\ (1)/\omega^2$, the same as for the unshielded wire.

The Δf proportional to ω^2 can be related to the transmissitivity through an aperture[10] below aperture cutoff which behaves roughly as ω^2. It is the $1/\omega^2$ behavior of $A_n(\omega)$ below aperture cutoff, seen in our 1 meg time step runs (Figure 4-26), that was not expected and that results in roughly equal energies in shielded and unshielded resonances when ohmic losses are not present.

It is important to note that for real systems such as aircraft a much greater conducting surface area is present within the cavity in the form of many wires

TABLE 4-1
Energy Table

	Stand alone antenna	8×4			6×3			4×2		
	$U_n^{(t)} \times 10^6$	$U_n^{(t)} \times 10^6$	$A_n^{(t=0)}$	α_n	$U_n^{(t)} \times 10^6$	$A_n^{(t=0)}$	α_n	$U_n^{(t)} \times 10^6$	$A_n^{(t=0)}$	α_n
1st Harmonic	5,428	1,492	21,400	77,024	1,021	10,000	24,489	687.2	2,753	2,757
3rd Harmonic	277.8	110	15,770	566,890	90.25	7,834	170,000	67.35	2,207	18,082
5th Harmonic	88.8	15.54	14,535	3,397,282	10.96	9,797	2,189,203	24.24	2,982	91,694
7th Harmonic	52.3	5.36	11,845	6,542,383	5.55	10,465	4,935,400	4.09	2,433	361,577
U1 + 3 + 5 + 7	5,846.9		1,622.9			1127.8			782.9	

Note: $U_n^{(t)} \equiv \int_0^\infty I(t^2)\,dt = \int_0^\infty \left[A_n(t=0)\sin(\omega_n t)e^{-\alpha nt}\right]^2 dt \cong \frac{A_n(t=0)^2}{4\alpha_n}$ (amp²/s), with $A_n(t=0)$ in amperes and α_n per second.

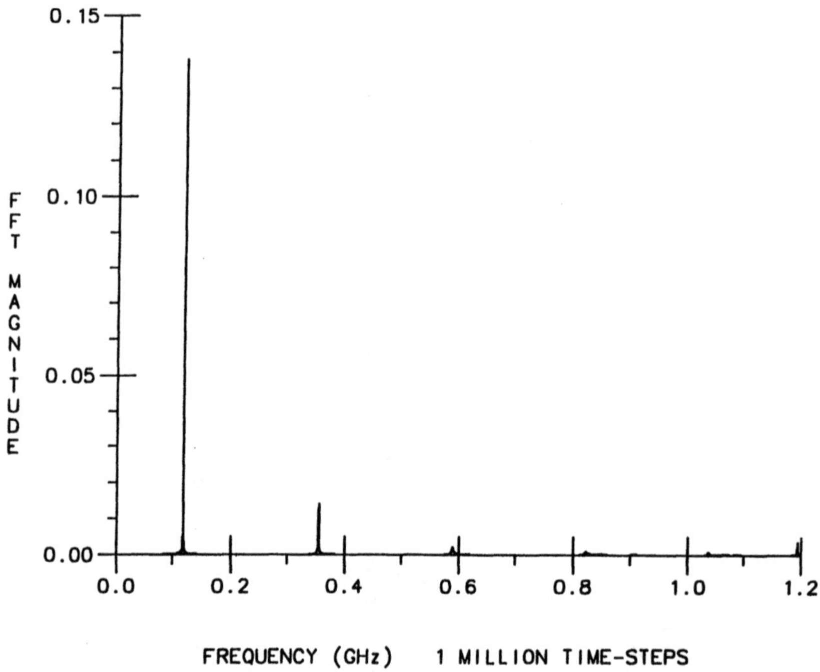

FIGURE 4-26. FFT magnitude response for 8 × 4 aperature.

as well as interior "fill". Clearly more realistic interior geometries and representative conductivities need to be examined using the insight obtained here to more fully characterize shielding effectiveness.

REFERENCES

1. **Lee, K. S. H., Eds.,** *EMP Interaction: Principles, Techniques, and Reference Data,* Hemisphere Publishing, Washington, D.C., 1986.
2. **Baum, C. E. and Baker, L.,** Return-stroke transmission-line model, October 30, 1984. Lightning Phenomenology Notes, LPN-13, Phillips Laboratory, Albuquerque, NM.
3. **Kunz, K. S. and Lee, K. M.,** A three-dimensional finite-difference solution of the external response of an aircraft to a complex transient EM environment. I. The method of its implementation. II. Comparisons of predictions and measurements, *IEEE Trans. EMC,* 20, 328, 1975.
4. **Kunz, K. S.,** Generalized Three-Dimensional Experimental Lightning Code (C3DXL) User's Manual, NASA Contractor Report 166079, NASA, Houston, TX, February 1986.
5. **Kunz, K. S., Hudson, H. G., and Breakall, J. K.,** A shielding-effectiveness characterization for highly resonant structures applicable to system design, *IEEE Trans. EMC,* 28 (1), 18, 1986.

6. **Kunz, K. S., Steich, D. J., and Luebbers, R. J.,** Low frequency shielding effectiveness of a conducting shell with an aperture: response of an internal wire, *IEEE Trans. EMC,* 34 (3), 370, 1982.
7. **Kunz, K. S. and Hudson, H. G.,** Experimental validation of time-domain three-dimensional finite-difference techniques for predicting interior coupling responses, *IEEE Trans. EMC,* 28, (1), 30, 1986.
8. **Kunz, K. S. et al.,** Lawrence Livermore National Laboratory Electromagnetic Measurement Facility, *IEEE Trans. EMC,* 29, (2), 93, 1987.
9. **Breakall, J. K. and Hudson, H. G.,** The equivalent current measurement technique and its comparison with current probe methods, *IEEE Trans. Ant. Prop.,* 34 (1), 119, 1986.
10. **Jackson J. D.,** *Classical Electromagnetics,* 2nd ed., J. Wiley & Sons, New York, 1975.

Chapter 5

WAVEGUIDE APERTURE COUPLING
(Article by P. Alinikula and K. S. Kunz)

5.1 INTRODUCTION

Waveguide coupling through an aperture is a fundamental electromagnetic problem. It has been approximated reasonably well with Bethe small hole theory,[1] and with some more recent modifications of it.[2] However, these analytic solutions do not apply to arbitrary shaped complex waveguide coupling structures. Instead, numerical methods must be used.

FDTD is an appropriate technique for time-domain analysis of passive microwave and millimeter wave structures. The strength of FDTD lies in its capability to model any volumetric structure with what are typically rectangular cells. No requirements for symmetry or smooth surfaces are needed. Better modeling is, of course, achieved by using more cells. Additionally, the FDTD technique, as shown in Chapters 2 and 7, can handle complex materials. Examples are lossy and anisotropic dielectrics as well as ferromagnetic materials. Some modeling of nonlinear materials has also been performed, with results presented in Chapter 11.

Recently, FDTD has been used in modeling microstrip structures,[3,4] and in predicting aperture coupling for a shielded wire.[5] These results suggest that FDTD can be successfully applied to waveguide coupling problems. To show this we will model coupled waveguides using FDTD and demonstrate both propagation and coupling capabilities. The forward and backward couplings are predicted for two geometries: a single circular aperture and a dual circular aperture. The single aperture coupler simulations are performed using different mesh dimensions and different simulation durations to observe the sensitivity of the model and by that means determine a suitable setup for FDTD waveguide simulations. The dual aperture geometry, however, is selected to be the same as in Reference 2 in order to be able to compare the calculated results with measurements.

5.2 APPROACH

In this section the representation of the waveguide geometry with FDTD cells and the method used to launch a waveguide mode in an FDTD waveguide calculation are both discussed.

The problem geometry, shown in Figure 5-1 consists of four sides that are defined to be perfectly conducting, i.e., the tangential electric field components on these planes are forced to be zero. The remaining two sides, the ends of the waveguides, are terminated using Mur's first order approximate absorption condition[6] (discussed in Chapter 4) to absorb the incident waveguide mode fields, thus approximating matched waveguide terminations. This condition produces some reflection for waves that are not normal to the boundary. In a

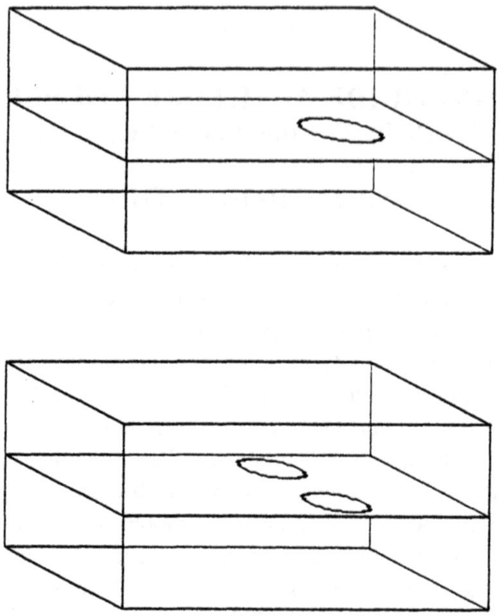

FIGURE 5-1a. Coupled waveguides, single aperture (16-cell diameter hole shown).

waveguide problem, however, the wave is more or less normal and the first order Mur condition is shown to be sufficient at least at frequencies above cutoff.

The metal plate between the two waveguides is defined to be infinitely thin, but effectively it has an approximate thickness of half of the cell. Circular apertures are approximated with square cells. Several different cell diameters, 8, 12, 13, and 16, were used. Naturally, the actual shape is modeled more accurately when more cells are used. The distance of the aperture from the source launcher was selected so that the whole pulse can fit inside the guide before the aperture is encountered. The test locations of the electric field components in the backward and forward propagation directions were selected to be one fourth of the aperture diameter away from the edges of the aperture. Transversely, the test locations are in the middle of the upper waveguide. The coupling coefficient was calculated as the ratio of electric field at the test location inside the upper waveguide and the electric field inside an undisturbed single waveguide.

Two different aperture geometries that were analyzed are shown in Figure 5-1. The first consists of a single centered coupling aperture with circular shape and relatively large size. The diameter of the aperture is 0.375a, where a is the broad side dimension of the waveguide. All the waveguides have a 2:1 ratio between the sides. Three different mesh dimensions were used: $22 \times 16 \times 55$, $32 \times 16 \times 80$, and $43 \times 16 \times 108$ in $\hat{x}$, $\hat{y}$, and $\hat{z}$ and directions, respectively. The corresponding diameters of the apertures were 8, 12, and 16 cells. The second structure has two parallel circular coupling apertures that are slightly

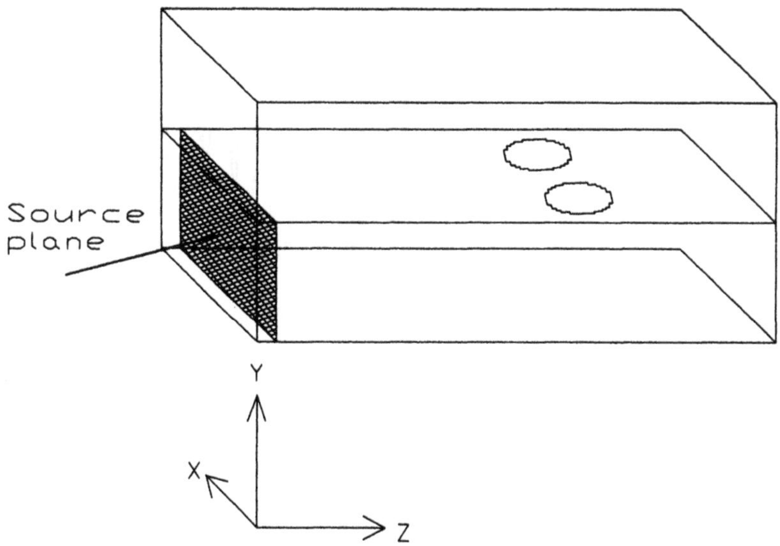

FIGURE 5-1b. Coupled waveguides, double aperture coupling (16-cell diameter holes shown).

smaller in size. Mesh dimensions are $42 \times 26 \times 108$ in $\hat{x}$, $\hat{y}$, and $\hat{z}$ directions. The diameter of the apertures is 13 cells. This geometry was constructed to be identical to that in Reference 2 in order to compare the results.

The incident waveguide mode is introduced into the waveguide using an electric wall behind the source location so that the mode propagates in only one direction. A transverse electric field source is allowed to appear spatially only at a transverse plane at one end of the lower waveguide. The amplitude of the field has a sinusoidal distribution across the plate resulting in a wave that approximates the TE_{10} mode. Temporally the incident field appears in the shape of a truncated Gaussian envelope about a sinusoidally varying signal. The sinusoidal variation is used to minimize the DC power and shifts the spectral content of the pulse upward in frequency. However, a truncated pulse has energy in a band of frequencies around the carrier frequency and therefore some energy is forced inside the waveguide at frequencies below cutoff. This energy is kept to a minimum by using a relatively long Gaussian envelope.

5.3 RESULTS

Single aperture coupler simulations showed that after 8192 time steps the remaining energy in propagating waves was so small that time truncation had a negligible effect for the fast Fourier transformed (FFT) response. The forward and backward coupling coefficients for $32 \times 16 \times 80$-cell geometry with 12-cell hole diameter are shown in figures 5.2a and b.

The sensitivity of the model to mesh dimensions was examined by comparing responses of the single aperture geometry with three different mesh dimen-

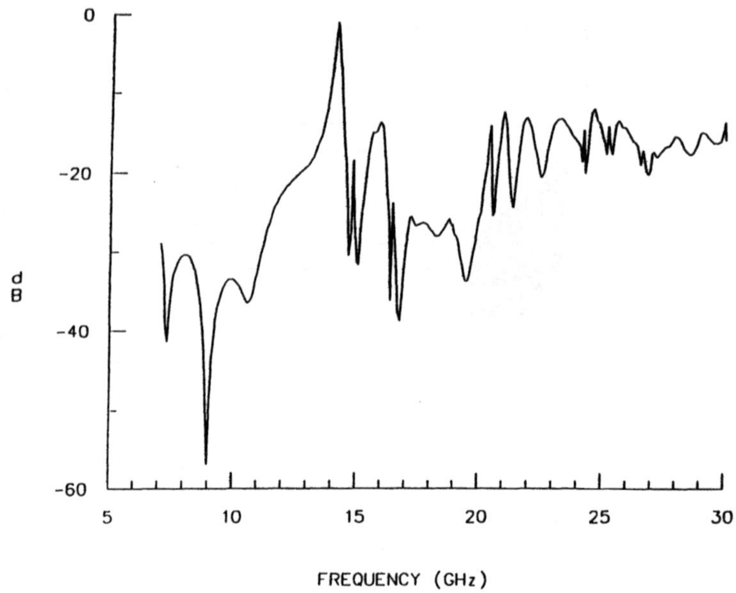

FIGURE 5-2a. Forward coupling 16 × 80 cells, hole 12 cells.

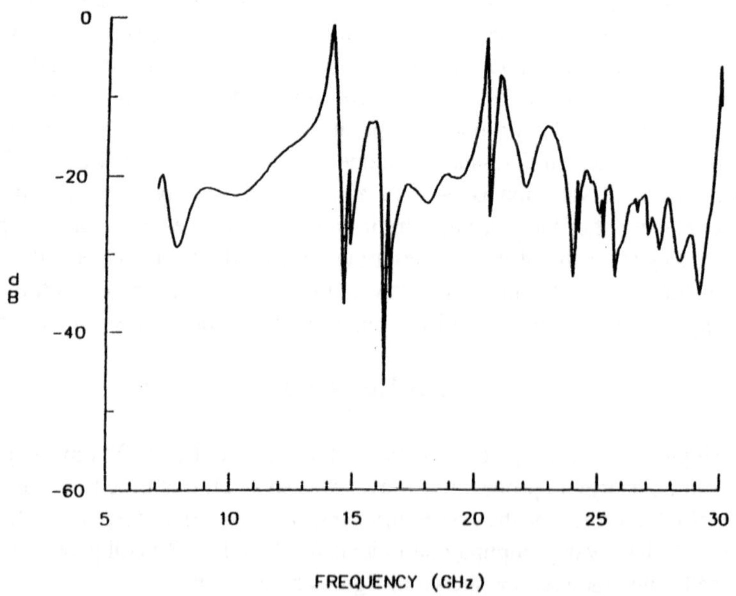

FIGURE 5-2b. Backward coupling, 32 × 16 × 80 cells, hole 10 cells.

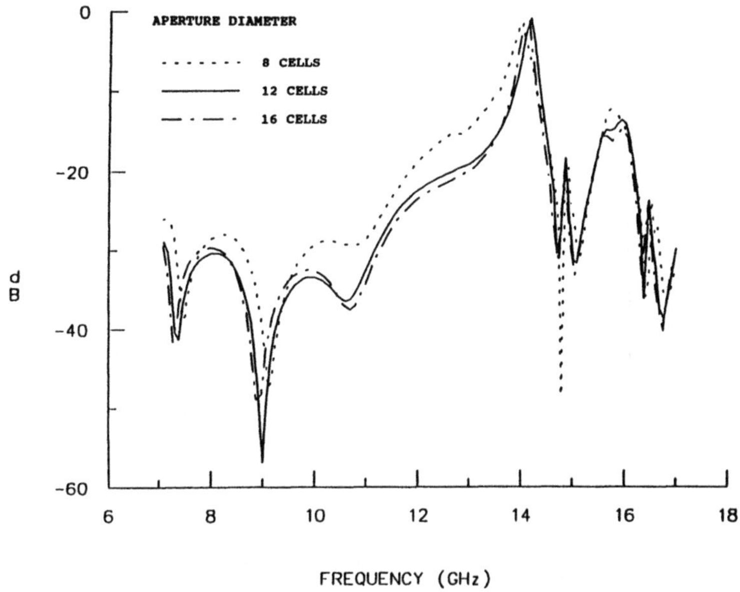

FIGURE 5-3a. Forward coupling, 8-, 12-, 16-cell holes comparison

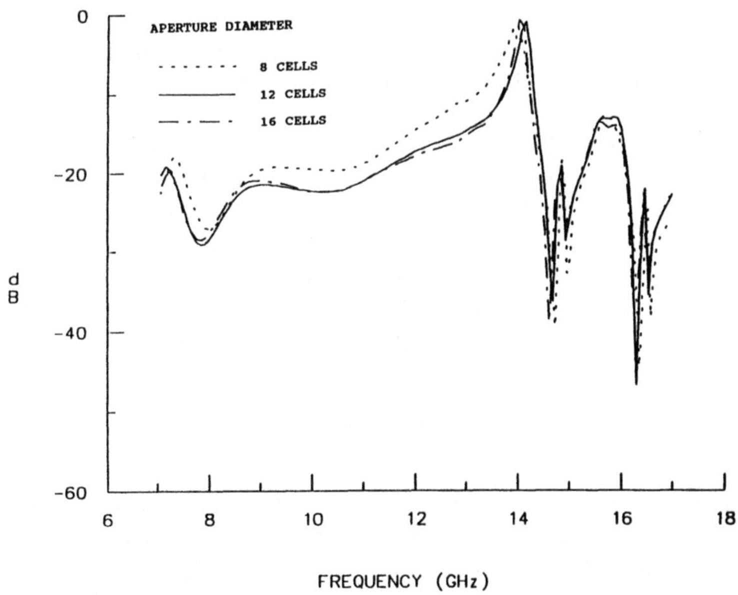

FIGURE 5-3b. Backward coupling, 8-, 12-, 16-cell holes comparison

sions. The forward and backward coupling coefficient comparisons at a narrower frequency band are shown in Figures 5-3a and b. The results show significant difference between the 22 × 16 × 55 and 32 × 16 × 80-cell geometries, whereas the responses for 43 × 16 × 108 and 32 × 16 × 80-cell

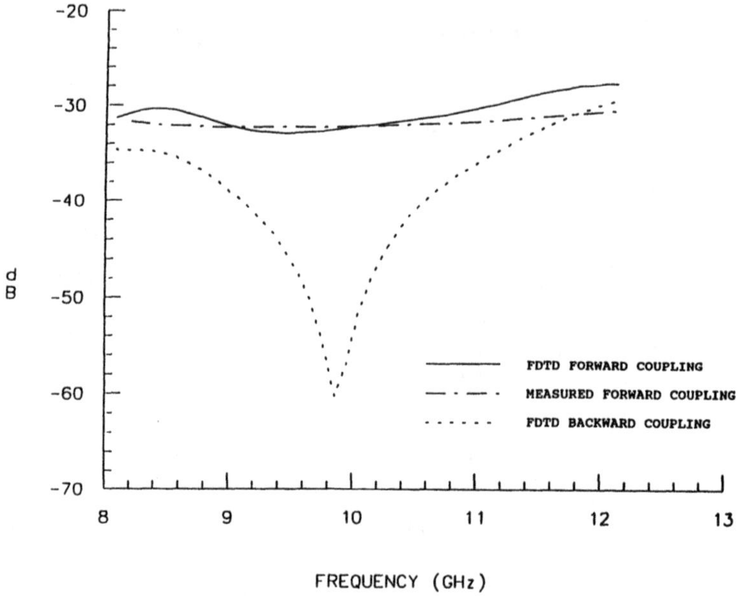

FIGURE 5-4. Two-hole coupler: FDTD results and measurements.

geometries are very close. The results indicate that 32 cells is sufficient for modeling the broad side of the waveguide and 12 cells for the aperture diameter, respectively.

The forward and backward coupling coefficients for a coupler with two apertures is shown in Figure 5-4. The experimental forward coupling results from Reference 2 for the same structure are also shown. The forward coupling coefficient is a very close match to the experiment. Except for the band edges, the simulation results fall inside the measurement accuracy range. Moreover, in Levy,[2] it was observed that the directivity (forward/backward coupling) was larger than 3 dB. Simulations support this with a minimum of approximately 2 dB directivity.

5.4 CONCLUSION

Simple waveguide coupling problems can be analyzed using FDTD. With a well-modeled geometry an accuracy inside the measurement error range can be achieved. Although these simple problems can be approximated reasonably well with analytical expressions, the results show the potential of FDTD to solve more complicated waveguide problems because combinations of dielectric and metal screws and slabs as well as waveguide junctions can be analyzed in the same manner and with the same resources.

ACKNOWLEDGMENT

This chapter and its figures originally appeared as an article by P. Alinikula and K. S. Kunz in *IEEE Microwave and Guided Wave Letters*, Vol. 1, No. 8, August 1981 (reprinted with permission of IEEE).

REFERENCES

1. **Bethe, A.,** Theory of diffraction by small holes, *Phys. Rev.,* 66, 163,1944.
2. **Levy, R.,** Improved single and multi-aperture waveguide coupling theory, including explanation of mutual interactions, *IEEE Trans. Microwave Theory Tech.,* 28, 331, 1980.
3. **Sheen, D. M., Ali, S. M., Abouzahra, M. D., and Kong, J. A.,** Application of the three-dimensional finite-different time-domain method to the analysis of planar microstrip circuits, *IEEE Trans. Microwave Theory Tech.,* 38, 849, 1990.
4. **Moore, J. and Ling H.,** Characterization of a 90° microstrip bend with arbitrary miter via time-domain finite difference method, *IEEE Trans. Microwave Theory Tech.,* 38, 405, 1990.
5. **Kunz, K., Steich, D., and Luebbers, R. J.,** Low frequency shielding effects of a conducting shell with an aperture: response of an internal wire, *IEEE J. Electromag. Compat.,* 34 (3), 370, 1993.
6. **Mur, G.** Absorbing boundary conditions for the finite-difference approximation of the time-domain electromagnetic-field equations, *IEEE Trans. Electromag. Compat.,* 23, 377, 1981.

Chapter 6

LOSSY DIELECTRIC SCATTERING

6.1 INTRODUCTION

In Chapters 4 and 5 we showed results obtained using perfect conductor implementations of scattered field FDTD and compared them to measurements. In this chapter results obtained using scattered field FDTD extended to lossy dielectrics will be shown. Before presenting these results we restate the finite difference equations for a material media developed in Chapter 2. For the results presented in this chapter these equations do not incorporate any frequency dependence in the constitutive parameters, ε, μ, and σ, as discussed in Chapter 8. The materials considered are frequency independent and linear. This formulation can nonetheless treat, at least approximately, most materials commonly encountered in electromagnetic problems.

In order to demonstrate the capabilities of the scattered field formulation applied to lossy dielectrics we treat two lossy dielectric code applications: Mie sphere scattering and electromagnetic (EM) penetration into the human body. The first application demonstrates the validity of the approach. The computationally generated dielectric sphere results[1] overlay the analytic and accepted Mie series response predictions very accurately. The second application[2] indicates the geometric and material level of complexity that can be successfully modeled using this approach, even with limited computer resources.

A review of the lossy dielectric equations is followed by a physical interpretation of the new terms arising from a lossy dielectric material scatterer in place of a perfect conductor. The applications are given next.

6.2 INTERPRETATION OF THE SCATTERED FIELD METHOD

In this section we discuss the physical interpretation of the scattered field approach to FDTD calculations. The equations have already been presented in Chapter 2 and are repeated here for convenience:

$$\frac{\partial H^{scat}}{\partial t} = \frac{-(\mu - \mu_0)}{\mu}\frac{\partial H^{inc}}{\partial t} - \frac{1}{\mu}\left(\nabla \times E^{scat}\right)$$

$$\frac{\partial E^{scat}}{\partial t} = \frac{-\sigma}{\varepsilon}E^{scat} \frac{-\sigma}{\varepsilon}E^{inc} - \frac{(\varepsilon - \varepsilon_0)}{\varepsilon}\frac{\partial E^{inc}}{\partial t} + \frac{1}{\varepsilon}\left(\nabla \times H^{scat}\right)$$

For simplicity we have removed the magnetic conductivity σ^*, as we are interested only in lossy dielectrics in this chapter. In differenced form, as shown in Chapter 2, these become (using E_x and H_y components as examples)

$$\begin{aligned} E_x^s(I,J,K)^n &= E_x^s(I,J,K)^{n-1}\left(\frac{\varepsilon}{\varepsilon+\sigma\Delta t}\right)-\left(\frac{\sigma\Delta t}{\varepsilon+\sigma\Delta t}\right)E_x^i(I,J,K)^n \\ &\quad -\left(\frac{(\varepsilon-\varepsilon_0)\Delta t}{\varepsilon+\sigma\Delta t}\right)\dot{E}_x^i(I,J,K)^n \\ &\quad +\frac{Hz^s(I,J,K)^{n-1/2}-Hz^s(I,J-1,K)^{n-1/2}}{\Delta y}\left(\frac{\Delta t}{\varepsilon+\sigma\Delta t}\right) \\ &\quad +\frac{Hy^s(I,J,K)^{n-\frac{1}{2}}-Hy^s(I,J,K-1)^{n-\frac{1}{2}}}{\Delta z}\left(\frac{\Delta t}{\varepsilon+\sigma\Delta t}\right) \end{aligned} \tag{6.1}$$

$$\begin{aligned} Hy^s(I,J,K)^{n+\frac{1}{2}} &= Hy^s(I,J,K-1)^{n-\frac{1}{2}} \\ &\quad -\left(\frac{\mu-\mu_0}{\mu}\Delta t\right)\dot{H}y^i(I,J,K)^{n+\frac{1}{2}} \\ &\quad +\frac{E_z^s(I,J,K)^n-E_z^s(I-1,J,K)^n}{\Delta x}\left(\frac{\Delta t}{\mu}\right) \\ &\quad +\frac{E_z^s(I,J,K)^n-E_z^s(I,J,K-1)^n}{\Delta z}\left(\frac{\Delta t}{\mu}\right) \end{aligned} \tag{6.2}$$

While these equations may provide little initial insight, they can be interpreted on the basis of incident, scattered, and total field. As defined in Chapter 2, the incident field is the field which exists in the absence of the scatterer, the total field is the field in the presence of the scatterer, and the scattered field is the difference between these two such that the scattered field added to the incident field gives the total field.

Let us first consider a perfect conductor. Within the perfect conductor the scattered electric field is equal to the negative of the incident electric field such that the total electric field is zero. Outside the perfect conductor the scattered electric field excited at the surface of the perfect conductor propagates away from the surface (in free space), and when added to the incident field produces the total field, including all scattering and shadowing effects.

For lossy dielectric, scattered fields are created throughout the scatterer. These scattered fields propagate through the lossy dielectric region so that scattered fields excited within the interior region, not just those on the surface, can affect the total field outside it. For this reason we refer to the difference equations that apply to the lossy dielectric as embodying volumetric boundary conditions, so that terms such as

$$\left(\frac{(\varepsilon-\varepsilon_0)\Delta t}{\varepsilon+\sigma\Delta t}\right)\dot{E}_x^i(I,J,K)^n$$

$$\left(\frac{\mu-\mu_0}{\mu}\Delta t\right)\dot{H}y^i(I,J,K)^{n+\frac{1}{2}}$$

give rise to scattered fields as does the term

$$\left(\frac{\sigma\Delta t}{\varepsilon+\sigma\Delta t}\right)E_x^i(I,J,K)^n$$

The first two terms are symmetric if we reintroduce magnetic conductivity σ^* so that the second term becomes

$$\left(\frac{(\mu-\mu_0)\Delta t}{\mu+\sigma*\Delta t}\right)\dot{H}y^i(I,J,K)^{n+\frac{1}{2}}$$

We choose not to employ σ^* in this chapter, and its inclusion is not necessary unless lossy magnetic materials are being considered.

If either $\varepsilon = \varepsilon_0$ or $\mu = \mu_0$ the terms disappear, an expected result is that the material is no longer dielectric or magnetic. The terms also disappear if the incident field does not vary in time so that $\dot{E}_x^i = \dot{H}_x^i = 0$. If in addition $\sigma = 0$, we have a lossless dielectric and the equations are nearly those of free space except that ε and μ appear in place of ε_0 and μ_0 in the curl portions of the equations for E_x^s and H_y^s used as an example here.

The term in $\dot{E}_x^s$ which depends on σ involves $E_x{}^i$, not $\dot{E}_x^i$, and thus does not disappear as do the terms for the other constitutive parameters for a time invariant field. Stated another way, ε and μ become more important as the time variation becomes more rapid. Taking a Fourier component of $E_x{}^i$ with frequency ω we see that we are simply observing the well-understood behavior of a conductive lossy dielectric (magnetic) material; when $\sigma/\omega\varepsilon(\mu) << 1$, the material is more dielectric (magnetic) than conductive.

A somewhat different insight is obtained by letting either ε or μ go to infinity. Choosing $\varepsilon \to \infty$ the equation for E_x^s becomes

$$E_x^s(I,J,K)^n = E_x^s(I,J,K)^{n-1} - \dot{E}_x^i(I,J,K)^n\Delta t \tag{6.3}$$

or $\dot{E}_x^s = -\dot{E}_x^i$ which implies $\dot{E}_x^{total} = 0$. Therefore, for $\varepsilon \to \infty$ no total field will be found inside the lossy dielectric. For a high, but not infinite ε, the total field only very slowly penetrates the body, moving with the characteristic velocity of the material.

6.3 NEAR ZONE SPHERE SCATTERING

To illustrate the accuracy and capabilities of the scattered field FDTD formulation for lossy dielectrics we determine the near fields scattered by a dielectric sphere and compare them to the exact Mie solution. To accomplish this an FDTD approximation to a dielectric sphere was constructed of cubic cells. The sphere (Figure 6-1) was 16 cells in diameter. Two different values of ε were employed, $\varepsilon = 2\varepsilon_0$ and $\varepsilon = 9\varepsilon_\sigma$. For a sphere of $\varepsilon = (\pi/2)^2\ \varepsilon_0 \cong 2.5\ \varepsilon_0$, electromagnetic signals can propagate around and through the sphere in the same interval. If $\varepsilon < 2.5\ \varepsilon_0$, propagation through the diameter is faster. The values of ε selected yield one case in each region. The cell size was selected so that the sphere diameter d = 2a was unity.

6.3.1 INCIDENT FIELD

The incident field was a uniform plane wave of the form

$$E^i = \hat{x}\,E_0\big(f\big(t-(z+a)\big)/c\big)$$

$$E_0 = 5.92\times 10^4$$

$$f(t) = \left(1-e^{-3.5\times 10^8 t}\right)U(t)$$

$$U(t) = \text{ unit step function}$$

so that f(t) behaves as a step function with a "smoothed" leading edge rising in approximately 10 ns. Note that the waveform first contacts the sphere at $t = 0$, $z = -a$ for the coordinate system and sphere geometry shown in Figure 6-1.

6.3.2 SPECIAL NOTES

The results presented here were obtained with an earlier version of the lossy dielectric FDTD code[1] which was slightly different from the form derived in Chapter 2. It used an exponential differencing scheme (shown below) for two of the field components and included the magnetic conductivity σ^* that is not needed for lossy dielectric materials.

$$\begin{aligned} H_x^s(I,J,K)^{n+1} = {} & H_x^s(I,J,K)^n e^{-\sigma^*\Delta t/\mu} + \left(1-e^{-\sigma^*\Delta t/\mu}\right) \\ & -\Bigg(-H_x^i(I,J,K)^{n+(1/2)} - \frac{(\mu-\mu_o)}{\sigma^*}\dot{H}_x^i(I,J,K)^{n+(1/2)} \\ & - \frac{E_z^s(I,J+1,K)^{n+(1/2)} - E_z^s(I,J,K)^{n+(1/2)}}{\sigma^*\big(Y_0(J+1)-Y_0(J)\big)} \\ & + \frac{E_y^s(I,J,K+1)^{n+(1/2)} - E_y^s(I,J,K)^{\lambda+(1/2)}}{\sigma^*(Z_0(K+1)-Z_0(K))}\Bigg) \end{aligned} \tag{6.4}$$

$$E_x^s(I,J,K)^{n+(1/2)} = E_x^s(I,J,K)^{n+(1/2)} e^{-\sigma\Delta t/\varepsilon} + \left(1 - e^{-\sigma\Delta t/\varepsilon}\right) \cdot \left(-E_x^i(I,J,K)^n - \frac{(\varepsilon - \varepsilon_0)}{\sigma} \dot{E}_x^i(I,J,K)^n + \frac{H_z^s(I,J,K)^n - H_z^s(I,J-1,K)^n}{\sigma(Y(J) - Y(J-1))} - \frac{H_y^s(I,J,K)^n - H_y^s(I,J,K-1)^n}{\sigma(Z(K) - Z(K-1))} \right) \tag{6.5}$$

This scheme reduces to a form nearly the same as derived in Chapter 2 when $\sigma\Delta t/\varepsilon << 1$ so that $e^{-\sigma\Delta t/\varepsilon} \cong 1 - \sigma\Delta t/\varepsilon$. Ostensibly, this approach was taken to handle large values of σ. However, the code becomes unstable in this form for $\sigma\Delta t/\varepsilon > 1$, while the nearly identical formulation of Chapter 2 remains stable for all values of σ, a cautionary tale on the importance of time indices.

Of only minor note is the use of an earlier version of the absorbing outer radiation boundary condition than first order Mur. The code was otherwise functionally identical to the dielectric treatment listed at the end of the chapter.

6.3.3 PREDICTIONS

The analytic results used for comparison[1] were found by inverse Fourier transforming the sinusoidally excited scattered field response $H(\omega)$ given by Stratton[3] and weighted by the Fourier transform of the incident field $F(\omega)$. Thus, the analytic time domain scattered-field response $r(t)$ is obtained from the frequency domain response using

$$r(t) = \int_{-\infty}^{\infty} F(\omega) H(\omega) e^{j\omega t} d\omega \tag{6.6}$$

Figures 6-2a through h compare the FDTD and Rayleigh-Mie analytic solutions in the time domain for $\varepsilon = 9\varepsilon_0$ for the tangential electric and magnetic fields at the four points on the sphere surface in Figure 6-1. Agreement is excellent, especially considering that relatively few cells were used in the computations. Figures 6-3a through h are for $\varepsilon = 2\varepsilon_0$. Except for some high frequency noise, agreement, again, is excellent. This noise[1] can be removed by appropriate filtering or alternatively and somewhat preferably by using a spectrally less "hot" pulse.

6.4 HUMAN BODY ABSORPTION

A extremely challenging problem is prediction of the penetration of electromagnetic fields (EMFs) into a human body. The structure of the body is quite complicated, and the constitutive parameters vary with position. Subsequent to the results discussed here,[2] FDTD has been extensively used for this problem, employing more powerful computers and more detailed modeling of the body,[4,5]

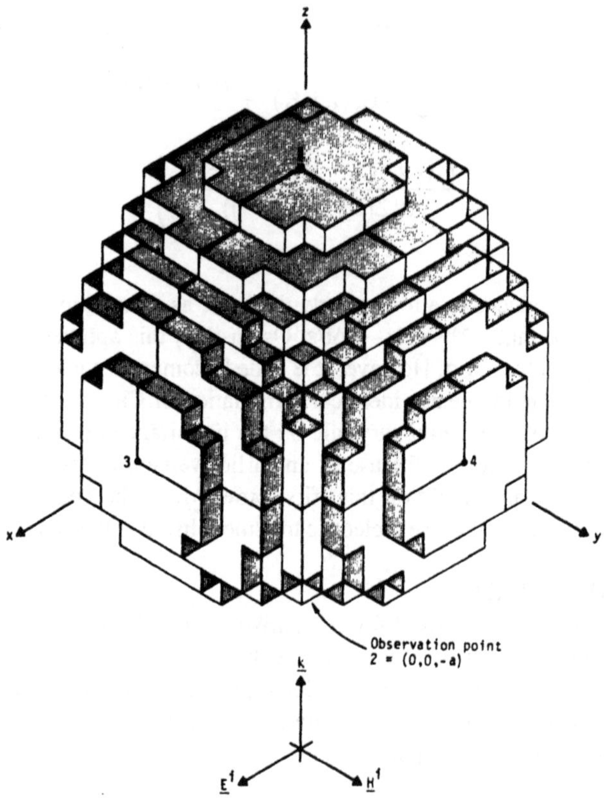

FIGURE 6-1. THREDE model of a dielectric sphere with incident field and observation points indicated.

with a recent effort[6] including the frequency dispersive effects of human tissues using the methods of Chapter 8.

This section illustrates that reasonably accurate results were obtained for this problem using scattered field FDTD with very limited computer resources. In this effort[2] a coarse rendition of a human body was generated using cells 4 × 4 × 6 cm in size in a problem space of 20,736 cells. The body was homogeneous with material properties of average tissue (two thirds muscle). This whole body model yielded responses that were used as the response at the outermost surface of a second more detailed model of a portion of the body. The second model is enclosed in a subvolume of the first model (Figure 6-4a, b). The second model renders the shape of the body more faithfully (Figure 6-5) and can include interior detail (Figure 6-6a through c) not possible with the whole body, along with the appropriate constitutive parameters (Table 6-1).

The approximate responses of the first run are used in lieu of the outer radiation boundary condition in the second run. The second run, at least at low frequencies, incorporates approximately the presence of the whole body because of the stratagem. At higher frequencies the responses in the interior of

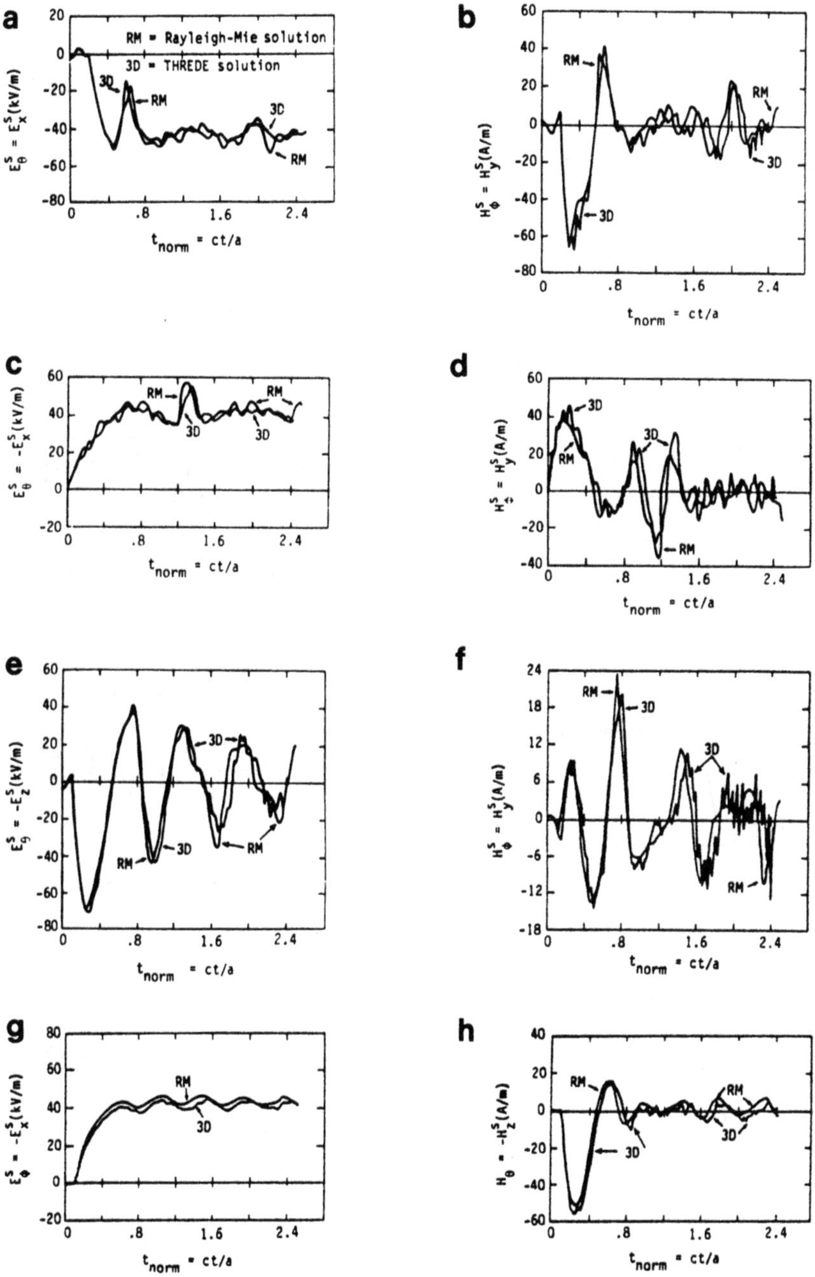

FIGURE 6-2. (a) $E_\theta^s = E_x^s$ at observation point 1 (pole of dark hemisphere) for $\varepsilon = 9\varepsilon_0$. (b) $H_\phi^s = +H_y^s$ at observation 1 for $\varepsilon = 9\varepsilon_0$. (c) $E_\theta^s = E_x^s$ at observation point 2 (pole of illuminated hemisphere) for $\varepsilon = 9\varepsilon_0$. (d) $H_\phi^s = H_y^s$ at observation point 2 for $\varepsilon = 9\varepsilon_0$. (e) $E_\theta^s = E_z^s$ at observation point 3 for $\varepsilon = 9\varepsilon_0$. (f) $H_\phi^s = +H_y^s$ at observation point 3 for $\varepsilon = 9\varepsilon_0$. (g) $E_\phi^s = E_x^s$ at observation point 4 for $\varepsilon = 9\varepsilon_0$. (h) $H_\theta^s = H_z^s$ at observation point 4 for $\varepsilon = 9\varepsilon_0$.

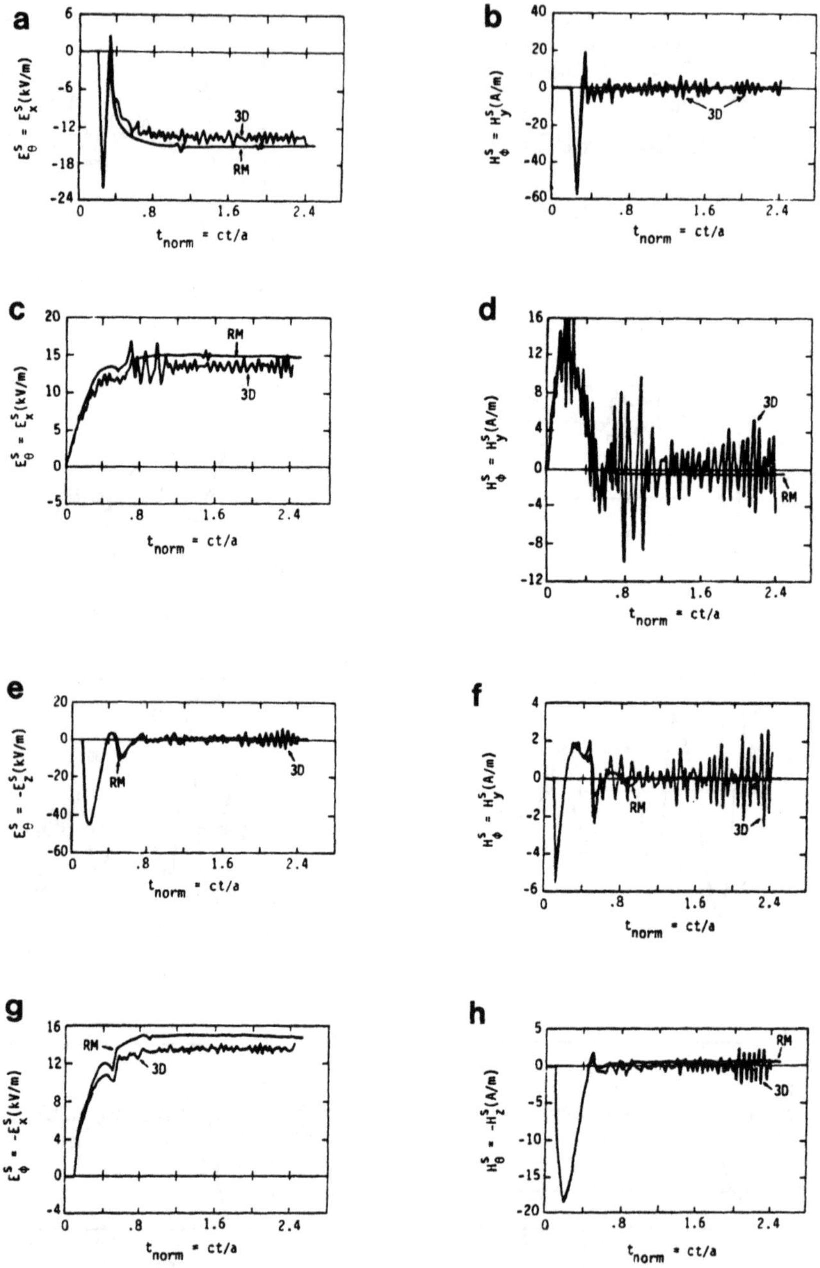

FIGURE 6-3. (a) $E_\theta{}^s = E_x{}^s$ (unfiltered) at observation point 1 (pole of dark hemisphere) for $\varepsilon = 2\varepsilon_0$. (b) $H_\phi{}^s = H_y{}^s$ (unfiltered) at observation point 1 for $\varepsilon = 2\varepsilon_0$. (c) $E_\theta{}^s = -E_x{}^s$ (unfiltered) at observation point 2 (pole illuminated hemisphere) for $\varepsilon = 2\varepsilon_0$. (d) $H_\phi{}^s = +H_y{}^s$ (unfiltered) at observation point for $\varepsilon = 2\varepsilon_0$. (e) $E_\theta{}^s = -E_Z{}^s$ (unfiltered) at observation point 3 for $\varepsilon = 2\varepsilon_0$. (f) $H_\phi{}^s = +H_y{}^s$ (unfiltered) at observation point 3 for $\varepsilon = 2\varepsilon_0$. (g) $E_\phi{}^s = -E_x{}^s$ (unfiltered) at observation point 4 for $\varepsilon = 2\varepsilon_0$. (h) $H_\theta{}^s = -H_Z{}^s$ (unfiltered) at observation point 4 for $\varepsilon = 2\varepsilon_0$.

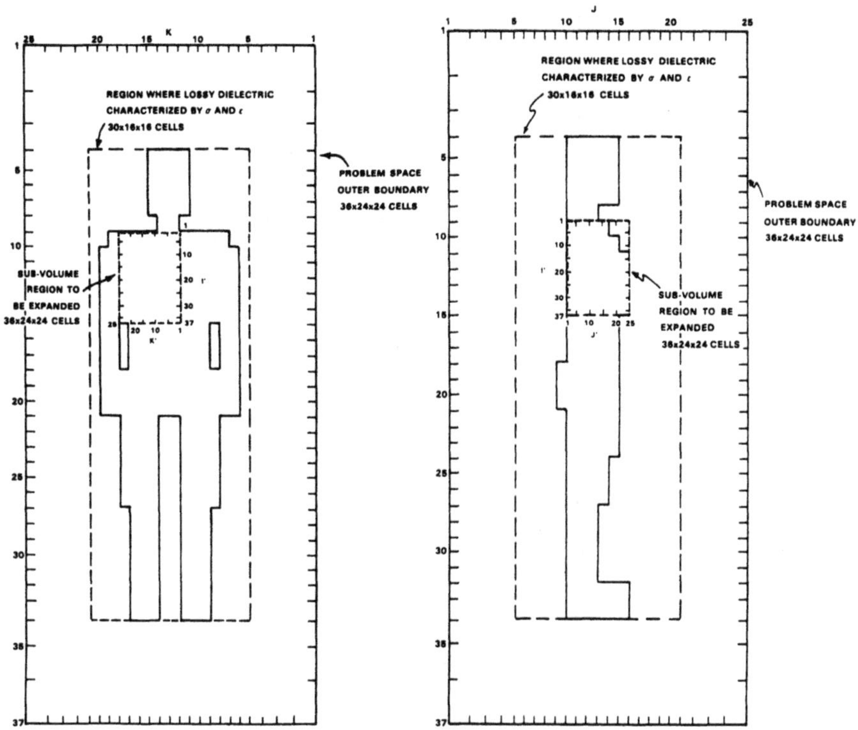

FIGURE 6-4. (a) Human model in a 20,736-cell problem space with the subvolume indicated. The location of the mesh points are given by the indices *(i,j,k)* and *(i′j,′k′)* for the problem space and subvolume, respectively. (a) Front view; (b) side view.

the second run problem space can still be used because the losses in the body materials effectively decouples a small local from its distant surroundings. This technique, called the expansion technique, is discussed in more detail in Chapter 10.

The incident electric field used in this example is a uniform plane wave propagating in the y direction with E polarized in the x direction. This corresponds to the electric-field vector being parallel to the major length of the body and propagating from the front to the back. The time domain behavior of the field is that of a damped sin wave described by

$$E_x^i = E_0 \sin(2\pi\alpha t')U(t')$$

where t' = retarded time = $t - (y-y')/c$, y' = source point, y = observation point, c = speed of light, $U(t')$ = unit step function, E_0 = 1 V/m, and α = oscillation frequency (typically α = 500 MHz).

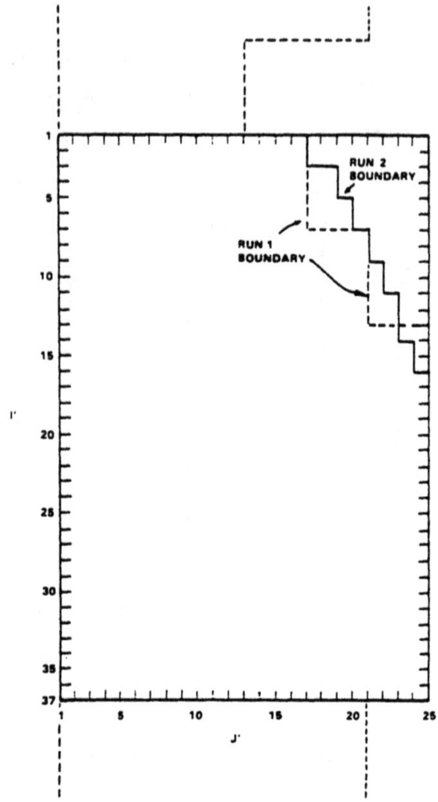

FIGURE 6-5. A side view enlargement of the subvolume showing the refinement of the exterior surface of the chest.

TABLE 6.1
Relative Dielectric Constant and Conductivity of Various Tissues for 350 MHz

	Relative dielectric (e)	Conductivity (s, S/m)
Muscle	38.0	0.95
Lung	35.0	0.73
Fat	4.6	0.06
Bone	8.0	0.05

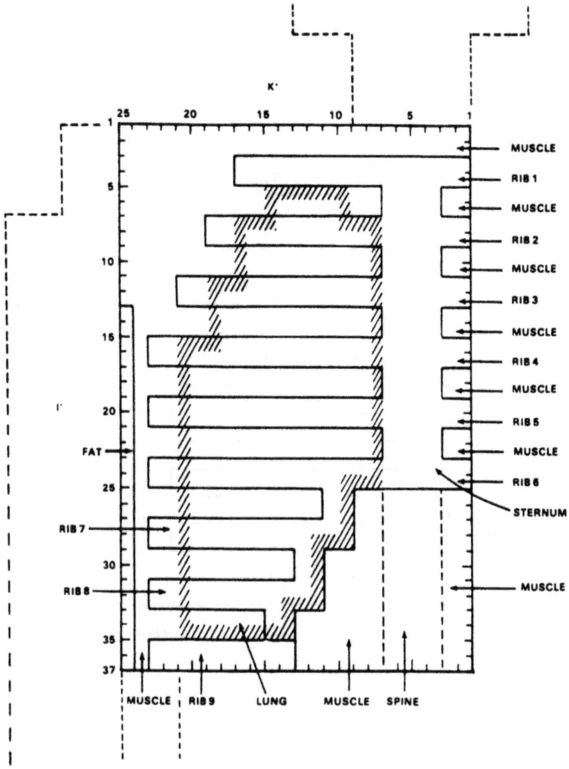

a

FIGURE 6-6. Subvolume inhomogeneous model of the chest: (a) front view; (b) side view; (c) top view.

Predictions were made of the instantaneous specific absorption rates (SAR) according to

$$\text{SAR}(t) = \frac{\sigma}{\rho}\left[\vec{E}^{i}(t) + \vec{E}^{s}(t)\right]^2 \tag{6.7}$$

where ρ is the density in kgm/m^3 of the tissue associated with the cell. Frequency domain information was obtained by applying the FFT to each vector component of $\vec{E}^i$ and $\vec{E}^s$ separately. The normalized SAR(ω) was formed from squaring the magnitude of the resultant total field after normalizing for the incident field spectrum.

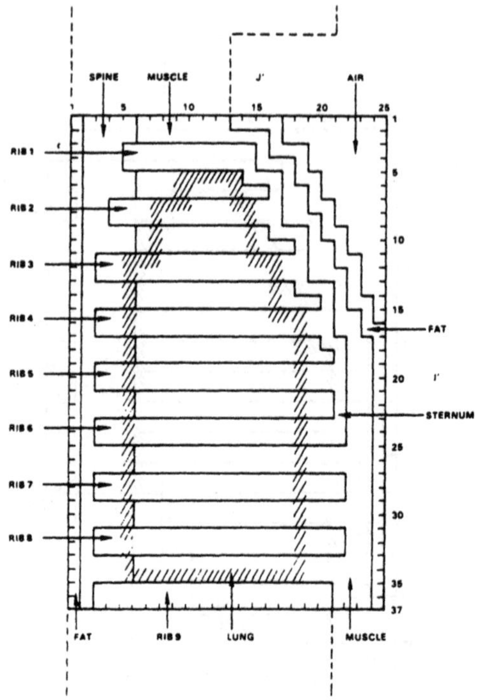

FIGURE 6-6b.

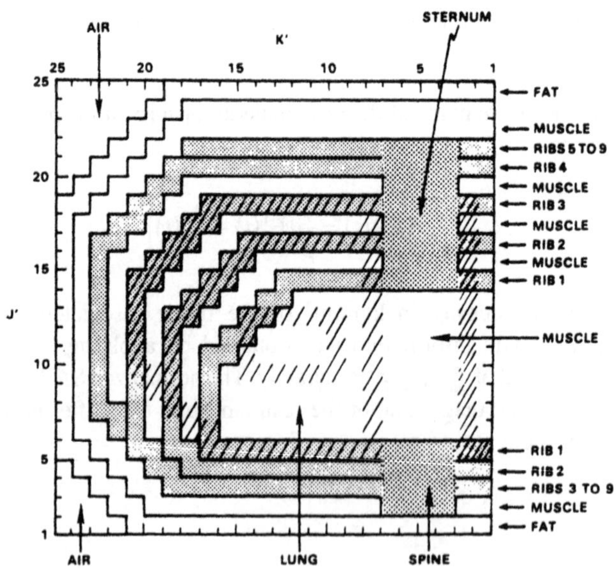

FIGURE 6-6c.

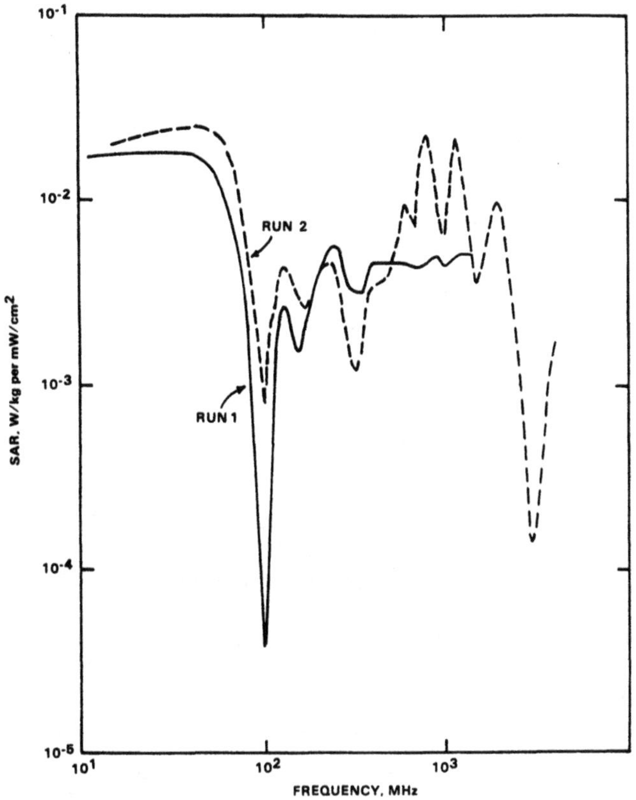

FIGURE 6-7. Frequency-domain SAR in the chest normalized with respect to incident power density. The observation points are (12, 13, 15) for run 1 and (19, 13, 13) for run 2.

Normalized SAR (ω) for the first and second runs were compared (Figure 6-7) when the second run had the same average tissue of the first run ($\varepsilon = 38\varepsilon_0$, $\sigma = 0.95$ S/m). Reasonable agreement between the curves is seen over the frequency range at which they are expected to overlap. The SAR (ω) predicted by the second run for the homogeneous average tissue was compared to measured local values[4] at 350 MHz (Figure 6-8) with very good agreement.

When the inhomogeneous detailed human model is used for the second run the same SAR (ω) prediction for front to back variations (Figure 6-9) differs considerably from the homogeneous model. The reduction near the front surface is likely due to the presence of the sternum in the inhomogeneous model of run 2. When a side to side "scan" is made of the inhomogeneous model of run 2 a considerable SAR increase in the lung is observed compared to the homogeneous run 2 results (Figure 6-10). The enhanced SAR in the lung may be of significant importance, especially for RF-induced hyperthermia in cancer treatment. More detailed results can be obtained with the greatly increased memory and speed of modern computers, and interested readers are

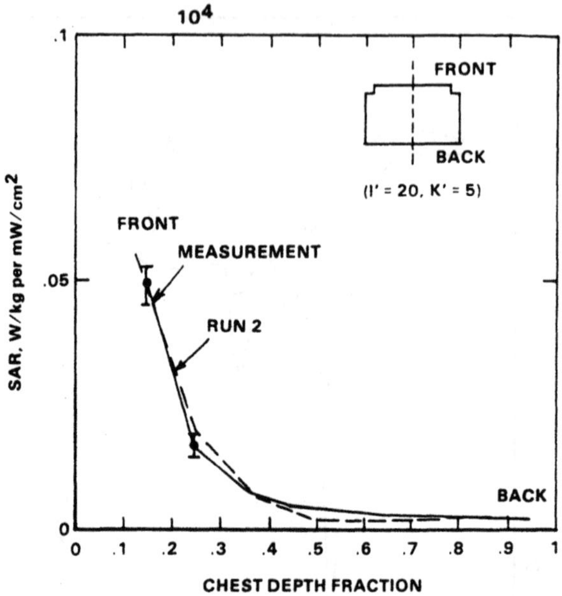

FIGURE 6-8. Comparison of the calculated and measured SAR distribution in the center of the chest (20, j′, 5) as a function of penetration depth. The frequency is 350 MHz and the incident power density is 1 mW/cm^2.

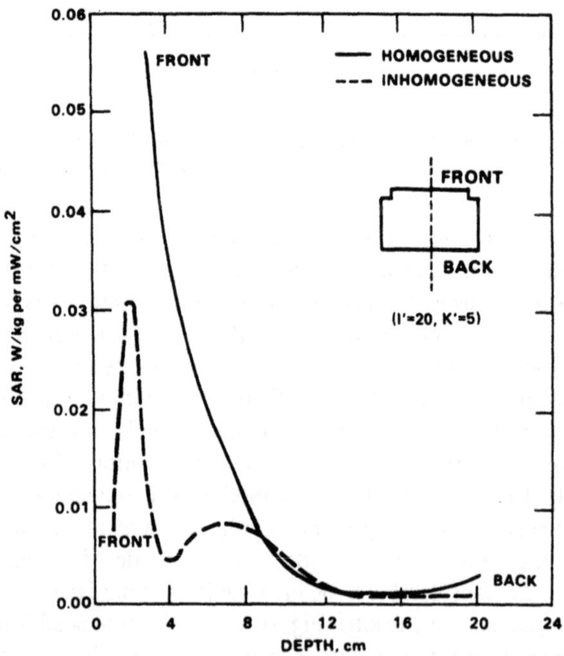

FIGURE 6-9. Comparison of the homogeneous and inhomogeneous SAR distribution in the center of the chest (20, j′, 5) as a function of penetration depth. The frequency is 350 MHz and the incident power density is 1 mW/cm^2.

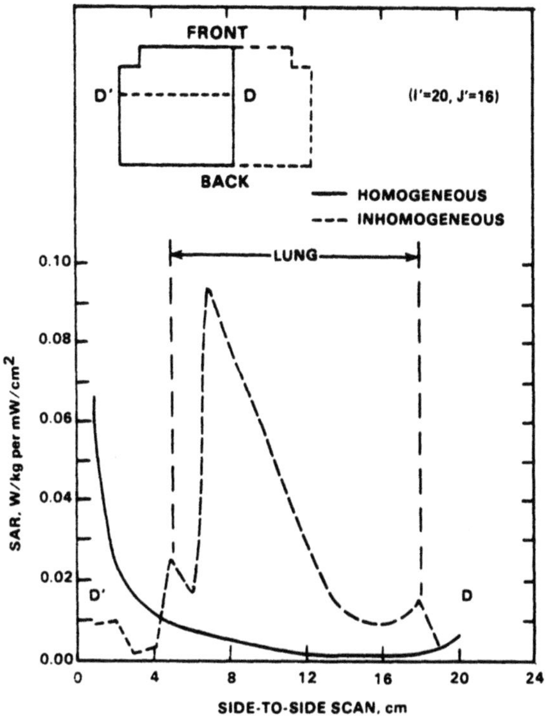

FIGURE 6-10. Comparison of the homogeneous and inhomogeneous SAR distribution for a scan across the chest (20, 16, k′) through the center of the lung. The frequency is 350 MHz and the incident power density is 1 mW/cm^2.

referred to References 5 and 6. However, these early results indicate that even with limited resources scattered field FDTD is capable of providing insight into problems extremely difficult to analyze using other methods.

REFERENCES

1. **Holland, R., Simpson, L., and Kunz, K. S.,** Finite-difference analysis of EMP coupling to lossy dielectric structures, *IEEE Trans. EMC,* 22(2), 203, 1980.
2. **Speigal, R. J., Fatmi, M. B. E., and Kunz, K. S.,** Application of a finite-difference technique to the human radiofrequency dosimetry problem, *J. Microwave Power,* 20 (4), 1985.
3. **Stratton, J. A.,** *Electromagnetic Theory,* McGraw-Hill, New York, 1941, 486–488.
4. **Sullivan, D., Borup, D., and Gandhi, O.,** Use of the finite-difference time-domain method in calculating EM absorption in human tissues, *IEEE Trans. Bioeng.,* 34, 148, 1987.
5. **Sullivan, D., Gandhi, O., and Taflove, A.,** Use of the finite-difference time-domain method in calculating EM absorption in man models, *IEEE Trans. Bioeng.,* 35, 179, 1988.
6. **Sullivan, D.,** A frequency-dependent FDTD method for biological applications, *IEEE Trans. Microwave Theory Tech.,* in press.

PART 3: SPECIAL CAPABILITIES

Chapter 7

FAR ZONE TRANSFORMATION

7.1. INTRODUCTION

All of the applications and examples given in earlier chapters have involved only the electromagnetic fields within or adjacent to the scattering or radiating object such that the fields are contained within the FDTD calculation space. FDTD can also be applied to analyze scattering from radar targets or radiation from antennas, with the desired results being the far zone scattered or radiated fields that lie outside the FDTD space. Near zone to far zone transformations applicable to time harmonic fields are well known. Early FDTD calculations of far zone scattered fields used these frequency domain transformations.[1] Because of this the FDTD far zone results were obtained at only one frequency per FDTD calculation run, and usually obtained with sinusoidal time variation excitation.[2]

The procedure for these single frequency far zone calculations is straightforward. With a sinusoidal time-harmonic source specified, the FDTD calculations are stepped through time until steady-state conditions are reached. The complex time-harmonic electric and magnetic currents flowing on a closed surface surrounding the object are then obtained. This involves very little computer storage, being four complex tangential fields (two electric and two magnetic) or surface currents for each Yee cell face on the closed surface. If these complex fields or currents are written to disk, then in postprocessing the far zone radiated or scattered fields can be calculated in any direction. This a good method to apply when far zone radiation or scattering patterns are desired at only a single frequency.

In order to obtain far zone results at multiple frequencies a hybrid approach is available which uses pulsed excitation for the FDTD calculations, but supplies frequency domain far zone fields. For each frequency of interest a running discrete Fourier transform (DFT) of the time harmonic tangential fields (surface currents) on a closed surface surrounding the FDTD geometry is updated at each time step.[3,4] The running DFTs provide the complex frequency domain currents for any number of frequencies using pulse excitation for the FDTD calculation. This is more efficient than using time harmonic excitation, which requires a separate FDTD calculation for each frequency of interest. It requires no more computer storage (per frequency) for the complex surface currents than the frequency domain far zone transform described above, and like it, provides frequency domain far zone fields at any far zone angle. If far zone results are desired at a few frequencies then the running DFT approach seems to be the optimum choice.

The running DFT method requires more computational effort than the time domain far zone approach described in the following, however, if far zone results are desired at more than a few frequencies, and does not efficiently allow for computation of transient far zone fields. One reason is that the hybrid approach requires complex multiplications at each cell on the transformation surface rather than real multiplications as in the transient approach. A more important reason is that the hybrid approach directly computes the DFT while the fully transient approach allows utilization of the fast Fourier transform (FFT) with a corresponding increase in efficiency.

There are two time domain approaches that roughly correspond to the straight frequency domain and running DFT approaches described above. The most straightforward time domain approach requires saving the time domain tangential fields (surface currents) over a closed surface containing the FDTD geometry for all time steps. With this approach transient far zone fields can be computed in postprocessing for all angles, and frequency domain results computed using FFTs for all frequencies at all angles. This is extremely versatile and provides all possible far zone results available from the FDTD calculation. The difficulty with this method is that except for very small geometries (thin plates or wires, for example), a relatively large amount of computer storage is required to save all the surface current components at all the time steps. If all this information is required, then this method is preferred.

In many situations, however, transient and/or broadband frequency domain results are required at a limited number of angles, for example, transient backscatter for a scattering calculation or transient antenna radiation for a limited number of pattern cuts. For these situations it is economical to directly compute the transient far zone fields at each angle of interest as a running summation.[5] This approach requires storing transient results for each time step for six far zone vector potentials per far zone angle rather than for four tangential field components (surface currents) per cell face on the transformation surface.

To illustrate the storage savings involved consider an FDTD calculation for a $72 \times 72 \times 72$ cell problem space. Because the far zone transform becomes more accurate as the transformation surface is moved farther from the FDTD geometry, let us assume that this surface is four cells in from the outer boundary. In this case it will consist of $64 \times 64 \times 6 = 24{,}576$ FDTD cell faces. Now suppose we desire to calculate far zone antenna patterns in two planes with 360 angles in each. Then for T total time steps the storage required by the running summation approach would be $(6 \times 2 \times 360 \times T) \div [(4 \times 24{,}576 \times T) = 4.4\%$ of that required to save all the transient surface currents.

If frequency domain results are desired they can be efficiently obtained from the far zone transient results with application of the FFT. In this way one FDTD computation using pulse excitation, along with an FFT, produces wideband far zone scattering or radiation results at any number of angles. When utilized in conjunction with FDTD algorithms that include effects of frequency dependent materials (Chapter 8), wideband results can be efficiently obtained even for scattering targets or antennas which contain frequency-dependent materials.

In the remainder of this chapter running summation transient far zone transformations for both 2- and 3-D will be given. While for most electromagnetic analyses the 2-D case is simpler, the derivation of the transient far zone transformation is more straightforward in 3-D. For this reason the 3-D transformation will be given first, followed by that for 2-D. For both transformations only Cartesian coordinates are considered, because these are the most widely used. However, the transformations given in this chapter could be modified for application to other coordinate systems, if desirable.

In addition to the examples included in this chapter, results obtained using the far zone transformation developed here are also given in Chapters 13 and 14 for scattering and antenna applications.

7.2 THREE-DIMENSIONAL TRANSFORMATION

The FDTD approach given in this book is based on a scattered field formulation, which makes obtaining the far zone scattered fields somewhat simpler than if a total field formulation were used. Thus, the scattered field FDTD approach in this book is well suited for obtaining far zone scattered fields.

However, an FDTD code being used for antenna radiation calculations when the antenna is being fed must operate in a total field mode, becasue there is no incident field to excite the problem. This is not a source of any difficulty for scattered field FDTD codes, however, as FDTD codes based on the scattered field formulation can be used easily for computation of antenna radiation by numerically setting the incident field amplitude to zero and adding discrete sources at cell locations corresponding to antenna feed locations (see Chapter 14 for some examples).

In the following discussions in this chapter it is understood that all time domain field quantities are scattered fields as computed by the FDTD methods of this book, and that for antenna (radiation) calculations the incident field is set to zero so that the scattered fields computed by the FDTD code will be identically equal to the total (radiation) fields.

To obtain the 3-D time domain far zone transformation we begin with a frequency domain near zone to far zone transformation as found in Ramo et al.[6] If the scattering or radiating object is surrounded by a closed surface S′, and if $\hat{n}$ is the local surface unit normal, then vector time harmonic equivalent scattered surface currents $J_s(\omega) = \hat{n} \times H(\omega)$ and $M_s(\omega) = -\hat{n} \times E(\omega)$ exist on the surface, where H(ω) and E(ω) are the scattered magnetic and electric fields at the surface.

Reference[6] defines time harmonic vector potentials N (ω) and L (ω) as

$$N(\omega) = \int_{S'} J_s(\omega) \exp(jk\bar{r}' \cdot \hat{r})\, ds' \tag{7.1}$$

$$L(\omega) = \int_{S'} M_s(\omega) \exp(jk\bar{r}' \cdot \hat{r})\, ds' \tag{7.2}$$

with $j=\sqrt{-1}$, k as the wave number, $\hat{r}$ as the unit vector to the far zone field point, $\bar{r}'$ as the vector to the source point of integration, and S′ as the closed surface surrounding the scatterer. After finding N (ω) and L (ω) the time harmonic far zone electric fields are easily obtained from Reference 6.

$$E_\theta = j\exp(-jkR)\left(-\eta N_\theta + L_\phi\right)/(2\lambda R) \tag{7.3}$$

$$E_\phi = j\exp(-jkR)\left(-\eta N_\phi + L_\theta\right)/(2\lambda R) \tag{7.4}$$

where η is the impedance of free space, R is the distance from the origin to the far zone field point, and λ the wavelength at the frequency of interest.

In order to develop the corresponding time domain far zone transformation we need to take the inverse Fourier transforms of the above. However, as written there is frequency dependence in (7.1) to (7.4). To simplify the inverse Fourier transformation process we first define time harmonic vector potentials

$$W(\omega) = j\omega\exp\left(\frac{-j\omega R}{c}\right)N(\omega)/(4\pi Rc) \tag{7.5}$$

$$U(\omega) = j\omega\exp\left(\frac{-j\omega R}{c}\right)L(\omega)/(4\pi Rc) \tag{7.6}$$

Recognizing that $k = 2\pi f/c = \omega/c$ and $\lambda = c/f = 2\pi c/\omega$, where f is the frequency and c the speed of light, the corresponding equations for the far zone electric fields have no frequency dependence and can be written

$$E_\theta = -\eta W_\theta - U_\phi \tag{7.7}$$

$$E_\phi = -\eta W_\phi + U_\theta \tag{7.8}$$

Now considering (7.1) and (7.2), we recognize that the jω multiplier in (7.5) and (7.6) corresponds to a time derivative, and the exponential factors containing jω (or equivalently jk) correspond to time shifts. Thus we can readily inverse transform W(ω) and U(ω) of (7.5) and (7.6) to obtain the time domain vector potentials

$$W(t) = \frac{1}{4\pi Rc}\frac{\partial}{\partial t}\left\{\int_{s'} J_s\left(t + (\bar{r}'\cdot\hat{r})/c - R/c\right)ds'\right\} \tag{7.9}$$

$$U(t) = \frac{1}{4\pi Rc}\frac{\partial}{\partial t}\left\{\int_{s'} M_s\left(t + (\bar{r}'\cdot\hat{r})/c - R/c\right)ds'\right\} \tag{7.10}$$

where $J_s(t) = \hat{n}\times H(t)$ and $M_s(t) = -\hat{n}\times E(t)$ are the time-domain electric and magnetic scattered surface currents on the closed surface S′ surrounding the scatterer. Because there is no frequency dependence in (7.7) and (7.8) they apply in the time domain as well and are used with (7.9) and (7.10) to convert the time domain vector potentials to far zone electric fields.

To apply (7.7) through (7.10) to FDTD calculations we must incorporate the correct temporal and spatial quantization. First, a closed surface must be defined. To simplify the calculations let us define the closed surface S′ as a rectangular box such that all surfaces of the box are Yee cell faces. To make the process of obtaining the correct surface currents as simple as possible, let us further agree to evaluate the tangential electric and magnetic time domain fields at locations corresponding to the center of each Yee cell face which lies on the box surface.

In order to illustrate the process of determining the tangential fields and surface currents let us consider the contribution from a single FDTD cell face which is on the surface of the S′ integration box. The complete far zone scattered field will be obtained by summing similar contributions from all of the tangential field components at the center of each cell face lying on the integration surface at each time step.

The quantization using Yee notation is $t = n\Delta t$, $x = I\Delta x$, $y = J\Delta y$, and $z = K\Delta z$, where n, I, J, and K are integers. Because we are using the Yee unit cell, the field quantities are not specified at any particular surface of the cell. To determine the tangential field values at the centers of cell faces which are on the integration surface S′, field values from different cells will be spatially averaged. This is necessary to obtain sufficient accuracy. Also, the electric and magnetic field values are specified at $^1/_2\Delta t$ time displacements and this must be compensated for as well. The time derivatives in (7.9) and (7.10) will be approximated as finite differences.

When evaluating (7.9) and (7.10) one has a choice of storing the surface currents (or equivalently the tangential fields on the integration surface) vs. time or the far zone W and U vectors. Because for most situations significantly more storage will be required to store the time history of the surface currents than that of the far zone W and U vectors, the surface currents at each time step will be evaluated and their contribution to the future time (delayed by distance) far zone vector potentials W and U will be determined and stored. Some storage could be saved if the theta and phi components of W and U were determined at each time step, as six Cartesian vector components of W and U must be saved as opposed to only two spherical coordinate field components. Because the amount of storage needed to save the six components of W and U is a very small fraction of the storage needed for the FDTD cells themselves, the implementation chosen stores the six Cartesian components of W and U for each time step and evaluates (7.7) and (7.8) after all time steps have been computed.

Corresponding to typical usage, the 1/R amplitude and R/c time delay from the reference point to the far zone field point is suppressed, so that the factor 1/R and the time delay –R/c in (7.9) and (7.10) are not included when these equations are evaluated.

To illustrate the approach we will consider the scattered field component E_x^n tangential to the transformation surface S′ and at the center of a particular I,J,K Yee cell face. The value of E_x^n must be found by appropriate spatial averaging of E_x^n values from cell I, J, K and from adjacent cells. Let us assume that the portion of the integration surface which we are considering has an outward unit normal vector $\hat{n} = \hat{y}$, and that the I,J,K Yee cell is within the integration surface. Then at the center of this cell face $E_x^n = (E_x^n(I,J+1,K) + E_x^n(I,J+1,K+1))/2$ (obtaining the magnetic field components at this point will require averaging four terms). This electric field component will produce a scattered magnetic surface current

$$M_s = -\hat{y} \times E = E_x^n \hat{z} \tag{7.11}$$

so that on this surface E_x^n contributes to U_z only. Our surface area of integration is $\Delta x \cdot \Delta z$, and we approximate E_x^n as constant over this surface. To keep track of the relative time delay for each cell face in the integration, we will locate a spatial reference at the center of cell I_c, J_c, K_c. A cell near the center of the space would provide a convenient reference point. The vector from the reference cell to the center of the Yee cell face we are integrating over is then

$$\bar{r}' = (I - I_c)\Delta x\, \hat{x} + (J + 1/2 - J_c)\Delta y\, \hat{y} + (K - K_c)\Delta z\, \hat{z} \tag{7.12}$$

The value of $^1/_2$ is added because the integration surface of the I,J,K cell face we are considering is one half cell in the y direction away from the cell center. Similar $\pm^1/_2$ cell corrections must be made for all of the cell faces on the S′ integration surface.

Now let us consider how to approximate the time derivative in (7.10) using a centered finite difference. We time shift (7.10) by $(\bar{r}' \cdot \hat{r})/c$, then substitute (7.11) into (7.10). If we then evaluate the time derivative in (7.10) as a centered finite difference we find that E_x^n will contribute to U_z at times $t = (n+^1/_2)\Delta t$ and $t = (n-^1/_2)\Delta t$:

$$U_z\left[(n + 1/2)\Delta t - (\bar{r}' \cdot \hat{r})/c\right] = \Delta x \Delta z\left[E_x^{n+1} - E_x^n\right]/(4\pi c \Delta t) \tag{7.13}$$

$$U_z\left[(n - 1/2)\Delta t - (\bar{r}' \cdot \hat{r})/c\right] = \Delta x \Delta z\left[E_x^n - E_x^{n-1}\right]/(4\pi c \Delta t) \tag{7.14}$$

In order to be accumulated and stored, the far zone fields must be temporally quantized, and the contributions from different cells and time steps must be apportioned into the correct storage locations based on relative time delay. Also, a few more storage locations may be required for the far zone vector potentials than the number of time steps, as fields scattered from parts of the scatter closer to (farther away from) the far zone field point will be advanced (retarded) in time. We must be careful not to have any far zone responses which might occur for times corresponding to negative array subscripts or to subscripts beyond the dimension of the arrays used to store the U and W potential values.

Let us first determine the size of the vector potential arrays needed to store the complete range of possible times. If we define R_f as the distance from the reference cell to the point on the integration surface closest to the far zone field point, and R_b as the corresponding distance farthest from the far zone field point, then for the rectangular parallelepiped integration surface the range of time which must be accomodated for the far zone scattered fields is

$$-\frac{R_f}{c} \le t \le T\Delta t + \frac{R_b}{c} \tag{7.15}$$

where T is the maximum number of time steps in the FDTD calculation. Practically speaking, with the reference cell at the center of the FDTD space, it is simple and safe to take both R_f and R_b as half the maximum diagonal distance of the FDTD computation space. If we now define a shifted far zone time variable

$$t' = t + \frac{R_b}{c} \tag{7.16}$$

then t' will always be positive. After quantizing $t' = m\Delta t$, M storage locations will be required for each U and W vector component, where

$$M = T + \frac{R_b}{c\Delta t} + \frac{R_f}{c\Delta t} \tag{7.17}$$

The array size variable M typically will be only slightly larger than N. Consider for example a situation in which R_b and R_f correspond to 50 FDTD cells each. Then Rb/c and Rf/c will each be approximately 100 time steps (assuming Δt the Courant stability limit) and M will equal T + 200.

One warning is that when computing scattering cross sections or far zone transient radiation, the time offset given in (7.16) must also be applied to any reference signal. For example, in calculating radar cross-section (RCS) the complex FFT of the far zone FDTD field is divided frequency by frequency by the complex FFT of the incident plane wave. For accurate frequency domain phase the time advance of (7.16) must be applied to the time record of the incident wave. Of course, the incident wave must also be sampled at the I_c,J_c,K_c reference location in the FDTD space consistent with (7.12) before (7.16) is applied so that it has the same spatial reference as the far zone fields. If FDTD and frequency domain scattering results are to be compared in phase, then the phase reference of the frequency domain calculations must also be the location of the FDTD spatial reference point. For example, to compare phase between the exact solution for sphere scattering and the FDTD result the FDTD spatial reference at the center of cell I_c,J_c,K_c must be at the center of the sphere. Adjustments using (7.16) also must be made to the time reference of the source within the FDTD space in an antenna radiation calculation if time accurate far zone fields are required.

Returning to the example being considered, we must next find the contributions of E_x^n to the time-quantized U_z. To do this accurately we must consider that the time offset $(\bar{r}' \cdot \hat{r})/c$ will not in general correspond to an integral number of time steps, so that the times when E_x^n contributes to U_z as given in (7.13) and (7.14) will not correspond precisely to quantized times $m\Delta t$. In order to obtain accurate far zone results we must store the contributions to U_z from each integration cell in as time accurate a way as possible. We now consider one approach.

Assume that E_x^n is constant over a time interval Δt. Then on the t' time scale the integer m closest to the center of the time period of duration $2\Delta t$ over which E_x^n contributes to U_z is, from consideration of (7.13) and (7.14),

$$m = \mathrm{GINT}\left[\frac{t_c}{\Delta t} + \frac{1}{2}\right] \tag{7.18}$$

where

$$t_c = n\Delta t - \frac{\bar{r}' \cdot \hat{r}}{c} + \frac{R_f}{c} \tag{7.19}$$

and GINT is the greatest integer function. (For magnetic fields add an additional $^1/_2$ to the argument of (7.18) to compensate for the $^1/_2\Delta t$ time advance.) According to (7.13) and (7.14) E_x^n contributes positively to U_z over the Δt time interval prior to t_c and negatively over the following time interval. This means that U_z time storage locations (m – 1), m, and (m + 1) all correspond to time intervals in which E_x^n is contributing to U_z. In this discussion we consider U_z time storage location m to span the time period from $(m - {}^1/_2)\Delta t$ to $(m + {}^1/_2)\Delta t$. Apportioning the E_x^n contributions by taking the average value of U_z over a particular Δt time span and storing it in the corresponding location we have:

$$U_z[(m+1)\,\Delta t] = -\Delta x \Delta z\, E_x^n \cdot \left[\frac{1}{2} + \left(\frac{t_c}{\Delta t}\right) - m\right] / (4\pi c \Delta t) \tag{7.20a}$$

$$U_z[m\,\Delta t] = \Delta x \Delta z\, E_x^n \cdot \left[2\left(\frac{t_c}{\Delta t} - m\right)\right] / (4\pi c \Delta t) \tag{7.20b}$$

$$U_z[(m+1)\,\Delta t] = -\Delta x \Delta z\, E_x^n \cdot \left[\frac{1}{2} + \frac{t_c}{\Delta t} - m\right] / (4\pi c \Delta t) \tag{7.20c}$$

The above example gives the contribution to the time quantized U_z vector potential from the E_x^n component at the center of the face of cell I,J,K which lies on closed surface S′ and has unit outward normal $\hat{n} = \hat{y}$. The complete vector potentials W and U for all time steps can be obtained by summing the contributions from the other tangential electric and magnetic field components at the center of the same face of cell I,J,K, and then in like manner from all of the cell faces on S′. This must then be repeated for each time step. After all of the time steps are completed, the U and W vector potentials are changed from Cartesian to spherical coordinates, and the time domain far zone electric fields determined using equations (7.7) and (7.8).

Note that determining far zone results at a number of bistatic scattering angles during the same FDTD computation would be quite simple and straightforward, and would not add a significant computational burden.

In order to illustrate the application of the 3-D time domain near zone to far zone transformation, plane wave scattering by a perfectly conducting plate for different incidence angles and polarization will be determined. From these time domain transient scattered fields the frequency domain RCS will then be determined. Examples of far zone scattering results for plates composed of other materials, for other scattering geometries, and of transient far zone radiation from antennas are given in Chapter 14.

The FDTD problem space for these examples is $60 \times 60 \times 49$ cells, with cells being 1-cm cubes. The plate is $29 \times 29 \times 1$ cm, and is located parallel to and 9 cells above the x-y plane, as shown in Figure 7-1. The plate is not centered in the z dimension to reduce unwanted reflections from the Mur second order absorbing boundaries. It is important to avoid situations in which the specular reflection is directed at a corner of the problem space, as this produces the worst performance for the second order Mur absorbing boundary. The time step is 0.0192 ns. The incident plane wave is a Gaussian pulse 1 kV/m in amplitude with an e^{-2} pulse width of 2.46 ns. The integration surface is a parallelepiped four cells in from the boundaries of the problem space.

To illustrate the accuracy of the transformation when there are significant time delays across the integration surfaces, backscatter calculations are shown for the flat plate in Figure 7-1 with a ϕ-polarized plane wave incident from $\theta = 45°$, $\phi = 30°$ (spherical coordinates with θ measured from the positive z axis). The transient far zone cross-polarized scattered electric field is shown in Figure 7-2. The results are Fourier transformed and divided by the Fourier transform of the incident pulse and used to calculate the scattering cross section, with the results shown in Figure 7-3. Results obtained via the method of moments[7] are shown for comparison. Sources of the differences between the two methods include the imperfect FDTD radiation boundary conditions and the finite thickness of the FDTD plate (approximately 1 cell = 1 cm), as opposed to the infinitesimally thin moment method plate.

7.3 TWO-DIMENSIONAL TRANSFORMATION

In the previous section a method for transforming near zone FDTD transient results directly to the far zone in 3-D was presented and demonstrated. In this section the corresponding 2-D transform is determined.[8] The 2-D geometry being considered has no variation in the z direction. Rectangular Yee cells are used to model the space. After deriving the 2-D transformation, some sample results for 2-D far zone scattering by a perfectly conducting infinitely long circular cylinder will be given.

In the 3-D derivation the frequency domain far zone transformation equations were Fourier transformed to the time domain and served as the basis for transforming near zone FDTD fields to the far zone directly in the time domain. In order to simplify the derivation of the 2-D transformation our approach will be to compare the frequency domain far zone vector potential equations for both 2- and 3-D, and by comparing them obtain the factors needed to convert the 3-D time domain far zone transformation to function in 2-D. Because the application we are primarily concerned with is calculating the far zone scattering width, our derivation is directed at obtaining this

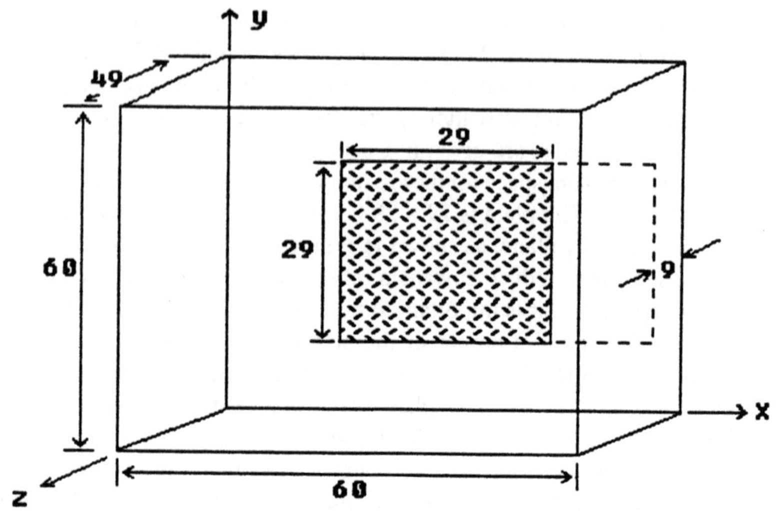

FIGURE 7-1. Geometry of FDTD plate scatterer showing the 29 cell × 29 cell (29 cm × 29 cm) × 1 cell^2 (1 cm^2) conducting plate in the FDTD computation space.

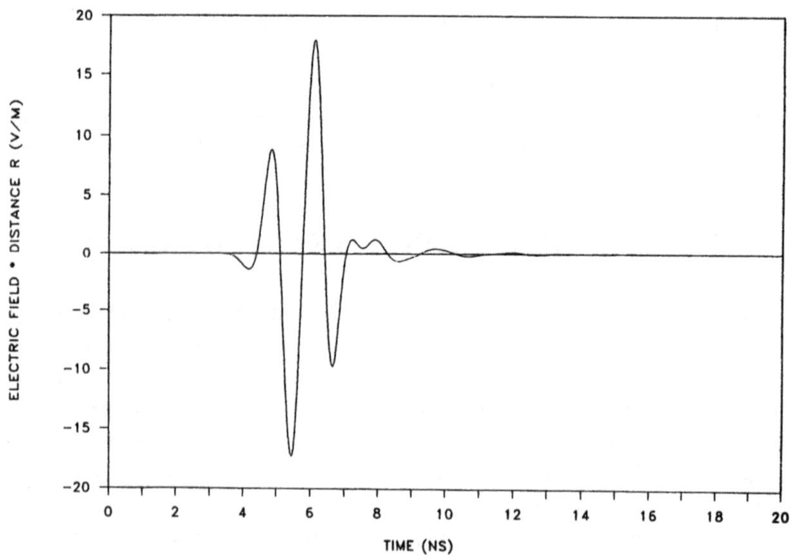

FIGURE 7-2. Far zone cross-polarized backscatter electric field from a 29-cm^2 perfectly conducting plate, 1-cm thick, for a Gaussian pulse ϕ-polarized plane wave incident from $\theta = 45°$, $\phi = 30°$ by FDTD.

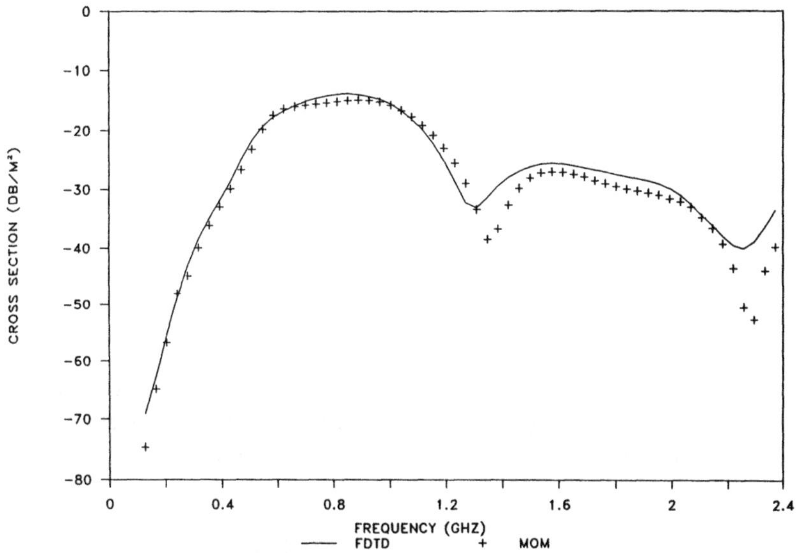

FIGURE 7-3. Far zone cross-polarized backscatter cross section vs. frequency for a perfectly conducting 29 cm^2 plate, 1-cm thick, obtained from the FDTD results of Figure 7-2 using an FFT. For comparison, moment method results for an infinitesimally thin plate are shown.

capability. Of course in these 2-D calculations only three field components (either transverse electric (TE) or TM polarization is assumed) will exist rather than the six components in a 3-D calculation.

We again surround the scatterer with a closed surface S′, and consider that equivalent tangential electric and magnetic time harmonic surface currents may exist on this surface. The 2-D vector potentials which correspond to the 3-D potentials defined in (7.5) and (7.6) are

$$A(\omega) = \sqrt{\frac{j\omega}{8\pi c\rho}} \exp\left(\frac{-j\omega\rho}{c}\right) \int_{s'} J_s(\omega) \exp\left(jk\rho' \cos(\phi - \phi')\right) ds' \quad (7.21)$$

$$F(\omega) = \sqrt{\frac{j\omega}{8\pi c\rho}} \exp\left(\frac{-j\omega\rho}{c}\right) \int_{s'} M_s(\omega) \exp\left(jk\rho' \cos(\phi - \phi')\right) ds' \quad (7.22)$$

where ρ' and ϕ' are the coordinates of the source point of integration, and ρ and ϕ the coordinates of the far zone field point. The corresponding far zone radiated fields are obtained from

$$E_z = -\eta A_z + F_\phi \tag{7.23}$$

$$E_\phi = -\eta A_\phi + F_z \tag{7.24}$$

One can then easily convert to RCS by applying

$$\sigma_{3D} = \lim_{R\to\infty}\left(4\pi R^2 \frac{\left|E^s_{3D}\right|^2}{\left|E^i\right|^2}\right) \tag{7.25}$$

in the 3-D case, where E^s_{3D} is the Fourier transform of either E_θ or E_ϕ of (7.7) or (7.8), and E^i is the Fourier transform of the incident plane wave electric field.

The corresponding 2-D scattering width is defined as

$$\sigma_{2D} = \lim_{\rho\to\infty}\left(2\pi\rho \frac{\left|E^s_{2D}\right|^2}{\left|E^i\right|^2}\right) \tag{7.26}$$

where E^s_{2D} is the Fourier transform of either E_z or E_ϕ of (7.23) or (7.24).

The 3-D approach of Section 7.2 which produces the physically observable far zone fields cannot be conveniently applied in the 2-D case. To understand the reason for this consider that we want to obtain the transient fields scattered by an infinitely long structure. The duration of this transient scattered field can be orders of magnitude longer in time than the duration of the pulsed plane wave which excited it. The mathematical aspect that warns us of this complication is the factor of $\sqrt{j\omega}$ in (7.21) and (7.22).

In order to evaluate the Fourier transform of (7.21) and (7.22) directly in the time domain the $\sqrt{j\omega}$ factor requires a convolution operation. To avoid this our approach will be to modify the 3-D results given previously to provide representative but not physically observable (in the sense of imagining that an infinitely long scatterer could be physically constructed, or at least approximated) 2-D time domain far zone fields. These will be the fields radiated by a unit length of the scatterer, corresponding to the definition of the scattering width. This far zone transient field can then be converted to the actual far zone steady-state frequency domain fields and used for wide band scattering width calculations (assuming pulsed excitation of the time domain FDTD computation). This conversion will involve a simple multiplication in the frequency domain, rather than the more complicated time domain convolution. Should the actual time domain far zone fields be required, they can then be obtained by an additional Fourier transform of these results back to the time domain.

In order to convert our previous 3-D results to 2-D, we compare the two sets of equations. First, comparing (7.7) and (7.8) with (7.23) and (7.24), since the spherical unit θ vector is equal to the negative of the cylindrical unit z vector, (7.7), (7.8), (7.23), and (7.24) correspond exactly, and no adjustment relative to this part of the 2- and 3-D transformations is needed.

Next, comparing (7.5) and (7.6) to (7.21) and (7.22), since the 1/R factor is understood and normalized out of 3-D far zone field calculations while the $1/\sqrt{\rho}$ factor is similarly normalized out of 2-D calculations, no compensation is needed here either.

Finally, consider (7.5) and (7.6) vs. (7.21) and (7.22). The additional dimension of integration in (7.5) and (7.6) is compensated for by computing the fields scattered by a unit width of the scatterer, corresponding to the scattering width per unit length defined in (7.26). This corresponds in (7.5) and (7.6) to having no z variation and integrating the z′ variable over a unit distance. The exponents provide equivalent phase (time) delays and need not be compensated for. Considering the remaining factors, it is easily determined that in the frequency domain, the relationship between far zone electric fields obtained from a 3-D far zone transformation for a unit length of the scatterer with no z variation and the 2-D transient far zone fields is

$$E^s_{2D} = \sqrt{\frac{2\pi c}{j\omega}}\, E^s_{3D} \tag{7.27}$$

With these results the 3-D transient time domain far zone transform given previously can be easily adapted to calculating wideband scattering widths in 2-D as follows:

1. Consider only the field components and corresponding surface currents excited in the 2-D problem. For example, for a TE_z computation only H_z, E_x, E_y, and the corresponding surface currents are included.
2. Calculate the representative transient far zone time domain fields using the 3-D method described earlier, but for a 2-D integration surface which encloses the scatterer. Let Δz, the z coordinate unit cell dimension in the surface integrations ((7.20), for example), equal 1 (meter). Keep in mind, however, that this field is not physically observable. It represents the scattered field radiated from a unit length of the scatterer in the time domain.
3. Fourier transform the result of step (2) and multiply the result by the factor $\sqrt{2\pi c/j\omega}$ in (7.27). This result is the steady-state frequency domain 2-D far zone field, which can then be used in (7.26) to calculate the scattering width as a function of frequency.
4. If the time domain 2-D far zone field is desired, it can be obtained by an additional Fourier transformation of the result obtained in (3) back to the time domain.

In order to demonstrate the capabilities of the above approach to obtaining transient far zone fields in 2-D, scattering widths vs. frequency for a circular perfectly conducting cylinder of radius 0.25 m are calculated. Both TE and TM polarizations are considered. The cells are 0.5 cm squares in a 500 × 500 cell problem space. On a 32-bit 486-based PC running at approximately 1 MFLOPS, each polarization required about 2 h. The 2-D FDTD code was scattered field formulation. The Mur second order absorbing boundary was used.

For both polarizations the incident plane wave was a Gaussian pulse which traveled in the $\hat{x}$ direction, backscatter was calculated, and 2048 time steps were evaluated. In order to clearly show the response, not all time steps are included in the plots. The time domain relative far zone scattered field for the TM polarization is shown in Figure 7-4. The scattering width amplitude and phase obtained from these results by Fourier transforming them and dividing by the Fourier transform of the incident pulse is shown in Figures 7-5 and 7-6 and agrees well with the exact solution. Accurate phase results were obtained by sampling the incident pulse at the center of the cylinder. The corresponding results for TE polarization are shown in Figures 7-7 through 7-9. This is the difficult polarization for approximating a smooth surface with a "staircased" FDTD code, yet the agreement in both magnitude and phase is quite good. A smaller than usual cell size was required to obtain these results, as the 0.5-cm cells correspond at the 3.0 GHz upper frequency limit in the plots to 20 cells per wavelength.

7.4 SUMMARY

An efficient time domain near zone to far zone transformation for FDTD computations has been presented. The approach is to keep a running accumulation of the far zone time domain vector potentials due to the tangential electric and magnetic fields on a closed surface surrounding the scatterer at each time step. At the end of the computation these vector potentials are converted to time domain far zone fields. For scattering calculations, many far zone bistatic directions can be included efficiently during one FDTD computational run. For antenna radiation calculations, transient radiation in many different directions can be efficiently computed during one FDTD calculation, resulting in a transient radiation pattern. However, if radiation patterns at many angles and at only one frequency are required, the approach described in Reference 1 will require less memory and be more efficient. If radiation patterns at many angles and at a limited number of frequencies are desired then the running DFT approach of References 3 and 4 will take less computer time and memory than the time domain methods described in this chapter.

The 3-D far zone transformation given above directly produces the physically observable transient far zone fields. If desired, these fields can be converted to steady-state frequency domain results via FFT. The 2-D time domain far zone transformation gives the fields radiated by a unit length of the scatterer. These fields can be used to determine wide band scattering widths

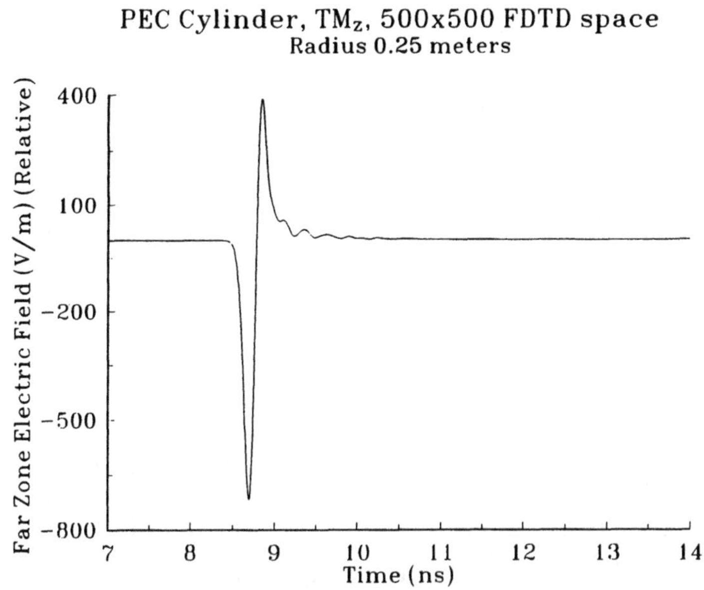

FIGURE 7-4. Far zone relative electric field vs. time for TM_z polarized incident Gaussian pulsed plane wave illuminating a perfectly conducting circular cylinder. FDTD cells are 0.5 cm^2.

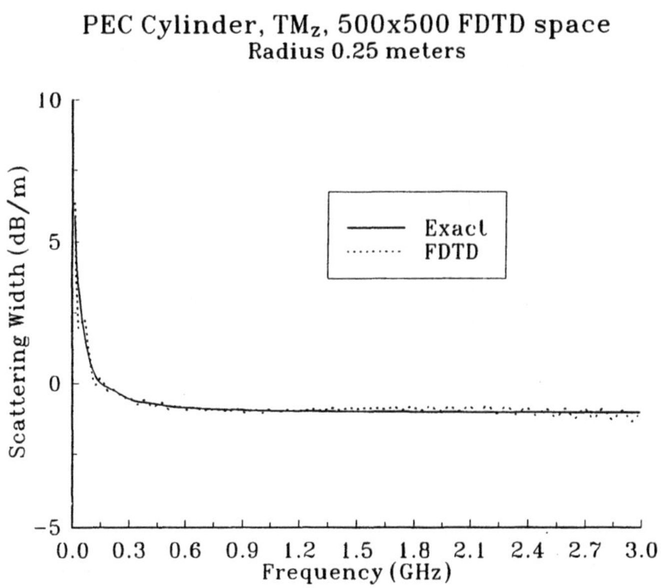

FIGURE 7-5. Scattering width amplitude obtained from far zone time domain results and compared with exact solution. FDTD cells are 0.5 cm^2.

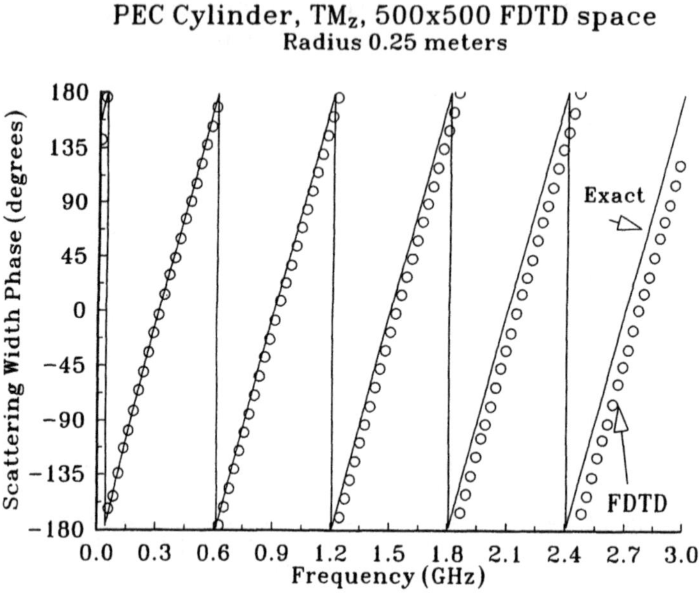

FIGURE 7-6. Scattering width phase obtained from far zone time domain results and compared with exact solution. FDTD cells are 0.5 cm^2.

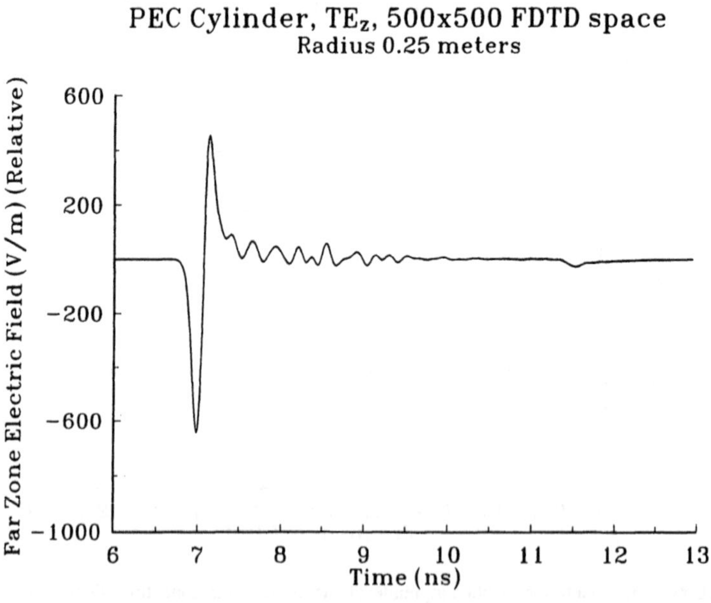

FIGURE 7-7. Far zone relative electric field vs. time for TE_z polarized incident Gaussian pulsed plane wave illuminating a perfectly conducting circular cylinder. FDTD cells are 0.5 cm^2.

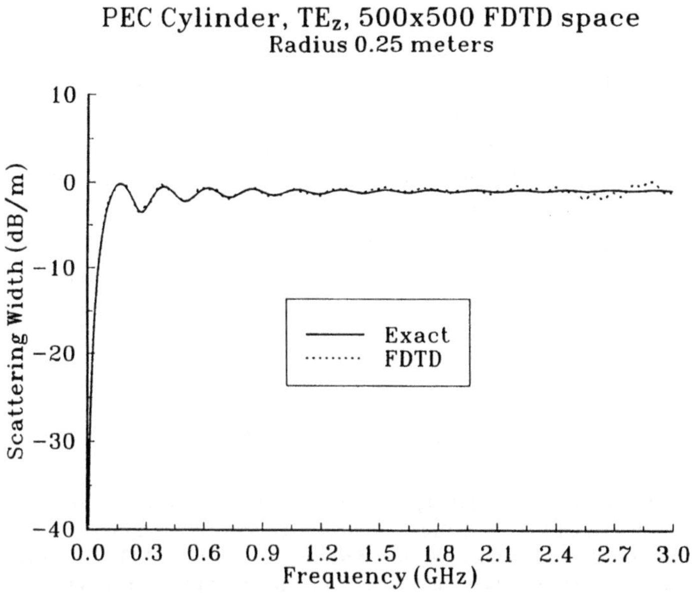

FIGURE 7-8. Scattering width amplitude obtained from far zone time domain results and compared with exact solution. FDTD cells are 0.5 cm².

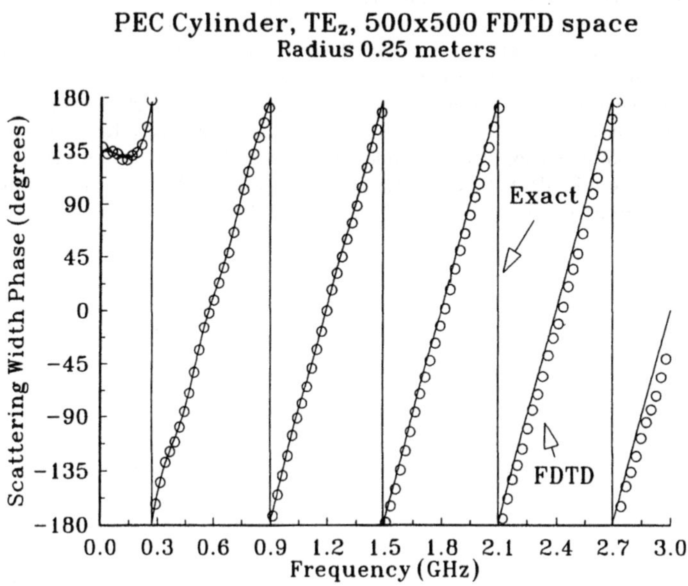

FIGURE 7-9. Scattering width phase obtained from far zone time domain results and compared with exact solution. FDTD cells are 0.5 cm².

after application of a Fourier transformation with the appropriate multiplying factor given in the text. Accuracy was demonstrated for both 2- and 3-D far zone transformations by calculation of wide bandwidth scattering from infinitely long cylinders and plates through pulsed plane wave excitation in the time domain. Using pulse excitation is significantly more efficient than computing many FDTD results using sinusoidally varying excitation if a wide frequency band is of interest. Coupled with the capability to compute FDTD results for frequency dependent materials, wideband results for far zone scattering from targets, including frequency-dependent materials, can be obtained very efficiently.

REFERENCES

1. **Umashankar, K. and Taflove, A.,** A novel method to analyze electromagnetic scattering of complex objects, *IEEE Trans. Electromag. Compat.*, 24, 397, 1982.
2. **Taflove, A., Umashankar, K., and Jurgens, T.,** Validation of FD-TD modeling of the radar cross section of three- dimensional structures spanning up to nine wavelengths, *IEEE Trans. Ant. Prop.*, 33, 662, 1985.
3. **Holland R. and Cho, K.,** Near-to-far field transformation: Numerical results, Quarterly Report, November 15, 1988, Applied Physics, Inc., Albuquerque, NM.
4. **Furse, C., Mathur, S., and Gandhi, O.,** Improvements to the finite-difference time-domain method for calculating the radar cross section of a perfectly conducting target, *IEEE Trans. Microwave Theory and Tech.*, 38(7), 919, 1990.
5. **Luebbers, R., Kunz, K., Schneider, M., and Hunsberger, F.,** A finite-difference time-domain near zone to far zone transformation, *IEEE Trans. Ant. Prop.*, 39(4), 429, 1991.
6. **Ramo, S., Whinnery, J., and Van Duzer, T.,** *Fields and Waves in Communication Electronics*, 2nd ed., John Wiley & Sons, New York, 1984.
7. **Newman, E.,** A user's manual for the electromagnetic surface patch code: ESP Version IV, preliminary version, August 1988, The Ohio State University Research Foundation, ElectroScience Laboratory, Department of Electrical Engineering, Columbus, OH.
8. **Luebbers, R., Ryan, D., and Beggs, J.,** A two-dimensional time domain near zone to far zone transformation, *IEEE Trans. Ant. Prop.*, 40(7), 848, 1992.

Chapter 8

FREQUENCY-DEPENDENT MATERIALS

8.1 INTRODUCTION

We saw in the previous chapter that one of the strengths of FDTD is that results for a wide frequency band can be obtained by applying Fourier transform techniques to transient FDTD results. The level of development of FDTD currently places a limitation on this in that the constitutive parameters must be specified as constants, i.e., μ, ε, and σ must be described by a single number. While this is true for free space, good conductors, and ideal dielectrics, it is only approximately true for most real materials. For some materials over a narrow band of frequencies the approximation is excellent, while for other materials over a wider band of frequencies it is not. For some materials, such as plasmas and ferrites, the permittivity may be zero or negative, so that the FDTD equations we have presented thus far cannot be used at all at certain frequencies as some of the terms become singular.

This is not to say that FDTD with constant constitutive parameters cannot be used for dispersive materials. Even a material with a constant permittivity and conductivity (or permeability and magnetic loss σ^*) will be dispersive. To see this consider the frequency domain ($e^{+j\omega t}$ time convention) equation

$$\nabla \times H = \sigma E + j\omega\, \varepsilon\, \varepsilon_0 E = j\omega\varepsilon_0 \left(\frac{\sigma}{j\omega\varepsilon_0} + \varepsilon \right) E = j\omega\hat{\varepsilon}\, \varepsilon_0 E \qquad (8.1)$$

where $\hat{\varepsilon}$ (the "^" symbol specifically identifying a complex constant) is a complex relative permittivity that includes the combined effect of real σ and ε values, which could be specified in FDTD calculations. Clearly there is a frequency dependence and the media would be dispersive; i.e., in it different frequency components of the electromagnetic field would travel at different speeds.

FDTD as it has been developed thus far can model propagation in dispersive media. However, the frequency dependence of the media must be described by constant real values of conductivity and permittivity as in (8.1). However, many materials have permittivity (or permeability) with different frequency dependence than this, and the remainder of this chapter is concerned with extensions of FDTD that allow accurate wide frequency bandwidth computation of transient electromagnetic fields in these materials. For some materials the frequency-dependent material parameters have been determined analytically, while for others they may be determined from measured data. While our discussion will be concentrated on dielectric materials, magnetic materials can be treated using the same approach.

Making FDTD calculations for frequency-dependent media occupies more computer time and storage than with constant constitutive parameters, and if results at a single frequency or over a narrow frequency band are all that are needed, then constant parameters should be used. However, if the frequency dependence of the material may affect the results over the band of frequencies of interest, then using frequency-dependent FDTD, which requires a recursive convolution, will in most situations be more efficient than making a sequence of separate FDTD calculations and changing the constant constitutive parameter values for each frequency range of interest.

A general discussion of frequency dependent media is outside the scope of this book, and the interested reader is referred to texts concerned with the subject, such as those in References 1 and 2, for a more extensive treatment.

8.2 FIRST ORDER DEBYE DISPERSION

We will begin with the simplest case, that of a material whose complex permittivity is described by an equation with one first order pole. Such materials are often called Debye materials and the equation that describes their permittivity is the Debye equation. For these materials the frequency dependence of the complex relative permittivity is described by

$$\hat{\varepsilon}(\omega) = \varepsilon' - j\varepsilon'' = \varepsilon_\infty + \frac{\varepsilon_s - \varepsilon_\infty}{1 + j\omega t_0} = \varepsilon_\infty + \chi(\omega) \tag{8.2}$$

where ε_s is the static permittivity at zero frequency, ε_∞ is the infinite frequency permittivity, t_0 is the relaxation time, and $\chi(\omega)$ is the frequency domain susceptibility. An extremely common representative material of this type is water, with typical parameters (they change with temperature and pressure) $\varepsilon_s = 81$, $\varepsilon_\infty = 1.8$, and $t_0 = 9.4 \times 10^{-12}$. The corresponding complex relative permittivity as a function of frequency in shown in Figure 8-1. Clearly the permittivity of water over this frequency range is not well approximated by a single constant value.

With this motivation, let us consider how we can extend FDTD to this situation. At least two different approaches can be applied. In one, the relationship between D and E is expressed in the time domain as a differential equation. This equation is then approximated as a difference equation and updated at each time step along with the E and H fields. This approach is considered in a later section of this chapter. For the reasons given in that section the approach we prefer is to express the relationship between D and E in the time domain with a convolution integral

$$D(t) = \varepsilon_\infty \varepsilon_0 E(t) + \varepsilon_0 \int_0^t E(t - \Lambda)\chi(\Lambda)d\Lambda \tag{8.3}$$

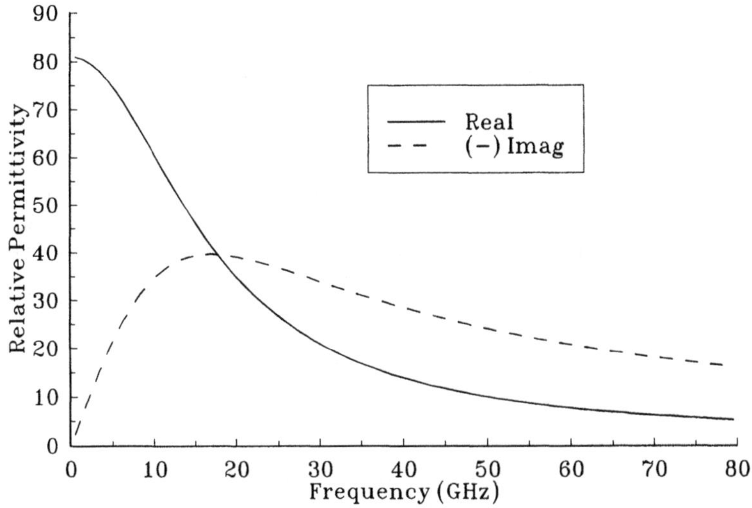

FIGURE 8-1. Complex relative permittivity of water in the frequency domain.

where $\chi(t)$ the time domain susceptibility, is the Fourier transform of $\chi(\omega)$, and zero fields are assumed for negative time. In previous chapters the assumption has been that $\chi(t)$ (and, of course $\chi(\omega)$) is zero, and (8.3) then reduces to the familiar $D = \varepsilon_r\varepsilon_0 E$, where ε_r is equivalent to ε_∞.

Now we consider how we might include the full effects of (8.3) in our FDTD calculations. To simplify this we will temporarily work in total fields, but later in the chapter our results are extended to a scattered field formulation.

We first take the continuous time expression of (8.3) and discretize it, obtaining, with $t = n\Delta t$ as usual,

$$D(t) \approx D(n\Delta t) = D^n = \varepsilon_\infty\varepsilon_0 E^n + \varepsilon_0 \int_0^{n\Delta t} E(n\Delta t - \Lambda)\chi(\Lambda)d\Lambda \tag{8.4}$$

If we make the approximation that all field quantities are constant over each time interval Δt, then the integration becomes in part a summation so that (8.4) is equivalent to

$$D^n = \varepsilon_\infty\varepsilon_0 E^n + \varepsilon_0 \sum_{m=0}^{n-1} E^{n-m} \int_{m\Delta t}^{(m+1)\Delta t} \chi(\Lambda)d\Lambda \tag{8.5}$$

The value of D at the next time step is

$$D^{n+1} = \varepsilon_\infty \varepsilon_0 E^{n+1} + \varepsilon_0 \sum_{m=0}^{n} E^{n+1-m} \int_{m\Delta t}^{(m+1)\Delta t} \chi(\Lambda) d\Lambda \tag{8.6}$$

We now proceed to use (8.5) and (8.6) to eliminate D from the dispersive FDTD equations. To simplify the algebra involved we consider only a 1-D case, but the extension to 3-D is straightforward. For this 1-D case we include only the D_y, E_y, and H_z field components and assume propagation in the ±x directions. Then with x = i Δx we have from the Maxwell curl-H equation

$$\frac{D_y^{n+1}(i) - D_y^n(i)}{\Delta t} = -\frac{H_z^{n+\frac{1}{2}}\left(i+\frac{1}{2}\right) - H_z^{n+\frac{1}{2}}\left(i-\frac{1}{2}\right)}{\Delta x} - \sigma E_y^{n+1}(i) \tag{8.7}$$

The conductivity σ has been included for generality in modeling materials which, unlike (idealized) water, have a conduction current at zero frequency. We need to eliminate the D_y terms from (8.7) in order to be able to solve for E_y^{n+1}. Using (8.5) and (8.6) we find that

$$\begin{aligned} D_y^{n+1}(i) - D_y^n(i) &= \varepsilon_0 \varepsilon_\infty \left[E_y^{n+1}(i) - E_y^n(i)\right] \\ &+ \varepsilon_0 E_y^{n+1} \chi^0 + \varepsilon_0 \sum_{m=0}^{n-1} E_y^{n-m}(i)\left(\chi^{(m+1)} - \chi^m\right) \end{aligned} \tag{8.8}$$

where

$$\chi^m = \int_{m\Delta t}^{(m+1)\Delta t} \chi(\Lambda) d\Lambda \tag{8.9}$$

and where it is assumed that the constitutive parameter values correspond to the material affecting the electric field at spatial location x = i Δx.

To simplify obtaining the update equation for E we regroup (8.8) as

$$\begin{aligned} D_y^{n+1}(i) - D_y^n(i) &= \left(\varepsilon_0 \varepsilon_\infty + \varepsilon_0 \chi^0\right) E_y^{n+1}(i) \\ &- \varepsilon_0 \varepsilon_\infty E_y^n(i) - \varepsilon_0 \sum_{m=0}^{n-1} E_y^{n-m}(i)\left(\chi^m - \chi^{m+1}\right) \end{aligned} \tag{8.10}$$

and also define

$$\Delta\chi^m = \chi^m - \chi^{m+1} \tag{8.11}$$

Now substituting (8.10) and (8.11) into (8.7) and solving for E_y^{n+1} we obtain

$$
\begin{aligned}
E_y^{n+1}(i) = {} & \frac{\varepsilon_\infty}{\frac{\sigma \Delta t}{\varepsilon_\infty} + \varepsilon_\infty + \chi^0} E_y^n(i) \\
& + \frac{1}{\frac{\sigma \Delta t}{\varepsilon_0}\ \varepsilon_\infty + \chi^0} \sum_{m=0}^{n-1} E_y^{n-m}(i) \Delta \chi^m \\
& - \frac{\Delta t}{\frac{\sigma \Delta t}{\varepsilon_0} + \varepsilon_\infty + \chi^0} \left[H_z^{n+\frac{1}{2}}\left(i + \tfrac{1}{2}\right) - H_z^{n+\frac{1}{2}}\left(i - \tfrac{1}{2}\right) \right] / \left(\varepsilon_0 \Delta x\right)
\end{aligned}
\tag{8.12}
$$

The FDTD update equation for the magnetic field is unchanged from that normally used

$$
\begin{aligned}
H_z^{n+\frac{1}{2}}\left(i+\tfrac{1}{2}\right) = {} & H_z^{n-\frac{1}{2}}\left(i+\tfrac{1}{2}\right) \\
& - \frac{\Delta t}{\mu \Delta x}\left[E_y^n(i+1) - E_y^n(i)\right]
\end{aligned}
\tag{8.13}
$$

because we are considering only frequency-dependent dielectric materials. For frequency-dependent magnetic materials a similar derivation would be applied to the curl E Maxwell equation to obtain the corresponding update equation for magnetic fields.

We now proceed with the next step, which is to determine $\chi(t)$ for the Debye material whose frequency domain susceptibility $\chi(\omega)$ was given in (8.2). From a fundamental Fourier transform pair given in Reference 3, this is readily found to be

$$
\chi(t) = \frac{\varepsilon_s - \varepsilon_\infty}{t_o} e^{-\frac{t}{t_0}} U(t)
\tag{8.14}
$$

This susceptibility function is plotted in Figure 8-2, which describes the response of water to an impulsive electric field, impulsive in the sense of a Dirac delta function of time being substituted for E in (8.3). The corresponding electric flux density D instantly rises to its maximum value, then decays exponentially to zero, as described by (8.14).

Another point to be made in connection with (8.14) is that the impulse response is causal, i.e., $\chi(t)$ is zero for negative time so that the material does not respond to the impulse before it occurs. This is a fundamental constraint on the frequency domain susceptibility function $\chi(\omega)$, that it must have a causal Fourier transform. This constraint is equivalent to the more commonly encountered Kramers-Kronig relationship between $\chi(t)$ and $\chi(\omega)$,[3] and cer-

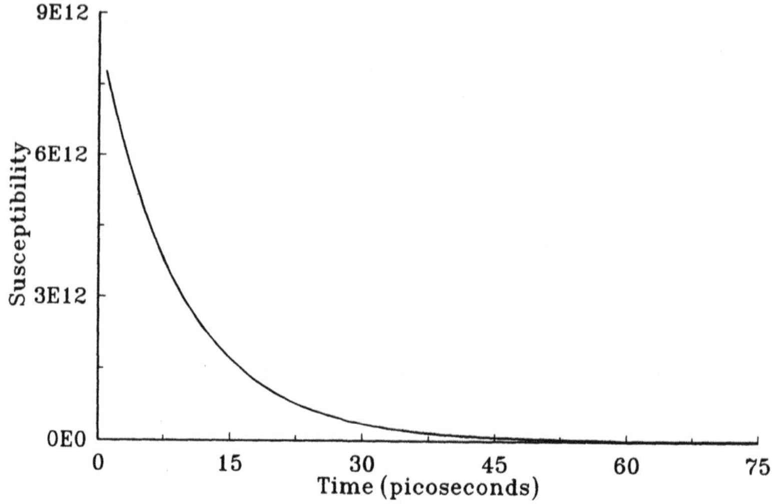

FIGURE 8-2. Time domain susceptibility function for water.

tainly more intuitive. Even Richard Feynman admitted that he did not completely understand the Cauchy theorem.

Having found $\chi(t)$ we can proceed to evaluate the constants needed to update (8.12) at each time step. Applying (8.9) to (8.14) we obtain

$$\chi^m = (\varepsilon_s - \varepsilon_\infty)e^{-\frac{m\Delta t}{t_0}}\left[1 - e^{-\frac{\Delta t}{t_0}}\right] \tag{8.15}$$

and (8.11) produces

$$\Delta\chi^m = (\varepsilon_s - \varepsilon_\infty)e^{-\frac{m\Delta t}{t_0}}\left[1 - e^{-\frac{\Delta t}{t_0}}\right]^2 \tag{8.16}$$

If we now proceed to update (8.12) using (8.15) and (8.16) we notice that in order to evaluate the convolution summation in (8.12) directly we must save all the past values of the electric field, at least far enough back in time to correspond to the time required for χ to decay to a small enough value so that the contributions from later times are small enough to neglect. In addition to the computer storage necessary, the computations required to evaluate the summation must be made at each time step in each FDTD cell. Evaluating the summation in (8.12) directly means that only a relatively small number of FDTD cells can be included in a computation for frequency-dependent materials as the computer time required to update these electric fields will be relatively large. Indeed, it might be faster to use FDTD with a sinusoidal time excitation, and run many computations over the frequency band of interest, changing the constitutive parameters correspondingly with frequency.

However, the exponential time dependence of $\chi(t)$ will allow us to avoid this direct evaluation. We can replace it with a recursive evaluation resulting in a significant savings of both computer memory and calculation time.[4,5] To demonstrate this we will define an accumulation variable Ψ^n in place of the summation as

$$\Psi^n(i) = \sum_{m=0}^{n-1} E_y^{n-m}(i)\Delta\chi^m \tag{8.17}$$

where Ψ^n is a single real variable. To show how this can be updated recursively, we consider the first few time steps. For n = 1 we find that

$$\psi^1 = \sum_{m=0}^{0} E_y^{1-m}(i)\Delta\chi^m = E_y^1\Delta\chi^0 \tag{8.18}$$

and for n = 2

$$\psi^2 = \sum_{m=0}^{1} E_y^{2-m}(i)\,\Delta\chi^m = E_y^2\Delta\chi^0 + E_y^1\Delta\chi^1 \tag{8.19}$$

From (8.16) we easily obtain the relationship

$$\Delta\chi^{\,m+1} = e^{-\frac{\Delta t}{t_0}}\,\Delta\chi^m \tag{8.20}$$

which is due to the special nature of the exponential function that shifting it in time corresponds to multiplying it by a constant proportional to the time shift. Using (8.20) and (8.18), (8.19) can be modified to

$$\psi^2 = E_y^2\Delta\chi^0 + E_y^1 e^{-\frac{\Delta t}{t_0}}\,\Delta\chi^0 = E_y^2\Delta\chi^0 + e^{-\frac{\Delta t}{t_0}}\psi^1 \tag{8.21}$$

from which it follows that

$$\psi^n = E_y^n\Delta\chi^0 + e^{-\frac{\Delta t}{t_0}}\,\psi^{n-1} \tag{8.22}$$

This result allows us to evaluate the summation in (8.12) very efficiently. Instead of storing perhaps thousands of electric field values at each field component location, we need store only the single variable ψ. Instead of multiplying and adding all those potentially thousands of terms in the summation at each time step only the simple update equation of (8.22) need be computed.

To demonstrate FDTD calculations using this method consider the 1-D problem of a pulsed plane wave traveling in vacuum normally incident on a planar vacuum-water interface. For this demonstration a 1-D FDTD space with 1000 cells is used, with the interface located at cell 500. Each cell has a Δx length of 37.5 μm, and the time step is taken at the Courant limit as $\Delta t = \Delta x / c = 0.125$ ps. The excitation is a Gaussian pulse plane wave. The pulse initially has a spatial width of 256 cells between the 0.001 amplitude truncation points. Convenient equations for initial values of the fields at the first time step for exciting this Gaussian pulse so that it propagates only in the positive x direction, where $x = i\,\Delta x$, are

$$E_G(i) = 1000\,\exp\left(\zeta\left((i - i_c)/i_p\right)^2\right) \tag{8.23}$$

$$H_G(i) = 1000\sqrt{\frac{\varepsilon_0}{\mu_0}}\,\exp\left(\zeta\left(\left(i + \tfrac{1}{2} - i_c - \tfrac{0.5\Delta x}{c\Delta t}\right)/i_p\right)^2\right) \tag{8.24}$$

for $i_c - i_p \le i \le i_c + i_p$ and zero otherwise, with 1000 the arbitrarily chosen amplitude of the peak electric field, $\zeta = \ln(0.001)$ and $i_p = 128$. The frequency spectrum of the pulse is shown in Figure 8-3.

A first order Mur[6] absorbing boundary is located at cell 1, while the 500-cell region containing the dispersive medium provided an adequate time delay so that pulses reflecting from the boundary at cell 1000 did not return to the interface before the transients dissipated.

Spatial plots of electric field vs. position for the pulse as it reflects from the interface and propagates in the water are shown in Figures 8-4 to 8-6 after increasing numbers of time steps.

For comparison the same calculation can be done with FDTD, but using constant values of permittivity and conductivity. If we, for example, decide to use the constitutive parameters at 20 GHz, then we find from (8.2) that $\varepsilon' = 34.86$ and $\varepsilon'' = 39.03$. We easily obtain the conductivity σ as $\sigma = \omega\,\varepsilon_0\,\varepsilon'' = 43.43$ S/m. The complex permittivity of a material with these values of ε' and σ is shown in Figure 8-7, which on comparison with Figure 8-1 is clearly not a very good approximation. The FDTD calculation for the pulse propagation example using these constant values of ε' and σ after 600 time steps is shown in Figure 8-8. Comparing with Figure 8-6, note the long tail on the reflected pulse due to the nonzero conductivity at zero frequency, and the relatively low amplitude of the pulse propagating through the water.

The accuracy of the recursive convolution method is indicated by computing the reflection coefficient vs. frequency from the transient fields. The FDTD

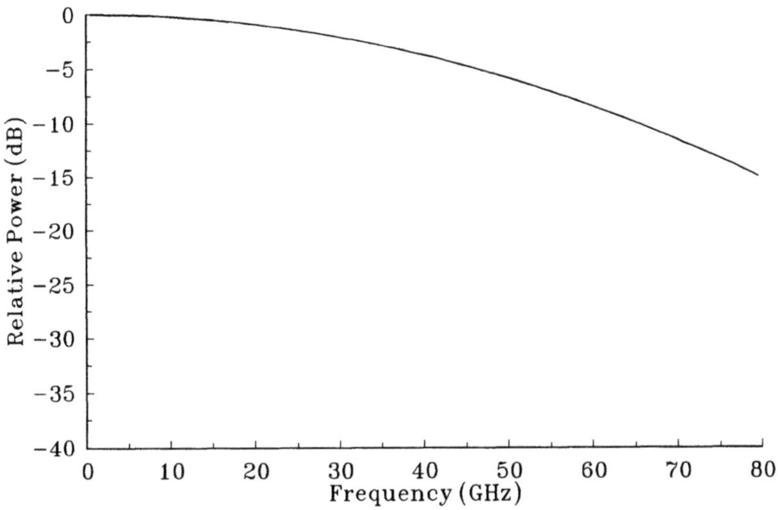

FIGURE 8-3. Frequency spectrum of the incident Gaussian pulse.

computations were continued until the transients at the vacuum-medium interface were dissipated. This required approximately 600 time steps for the recursive convolution method, but 4000 time steps with the usual constant value FDTD method due to the incorrect tail on the reflected pulse. Computation of the reflection coefficient from the transient FDTD results involves taking the Fourier transform of the reflected field vs. time in the vacuum at cell 500 at the interface and dividing by the Fourier transform of the incident pulse as it passes through that cell. The reflected field is obtained by subtracting the incident field (with the material region replaced with vacuum) from the total field (total fields are shown in Figures 8-4 to 8-6) in cell 500 at each time step. The resulting reflection coefficient magnitude is shown in Figure 8-9, along with the exact solution obtained from transmission line methods. The recursive convolution results show excellent agreement with the exact solution, while the constant parameter FDTD results are in error, except at 20 GHz where the constant values are correct.

8.3 FIRST ORDER DRUDE DISPERSION

In this section we will deal with a slightly more complicated material with complex permittivity described by the Drude dispersion relation[1,2]

$$\hat{\varepsilon}(\omega) = 1 + \frac{\omega_p^2}{\omega(j\nu_c - \omega)} \tag{8.25}$$

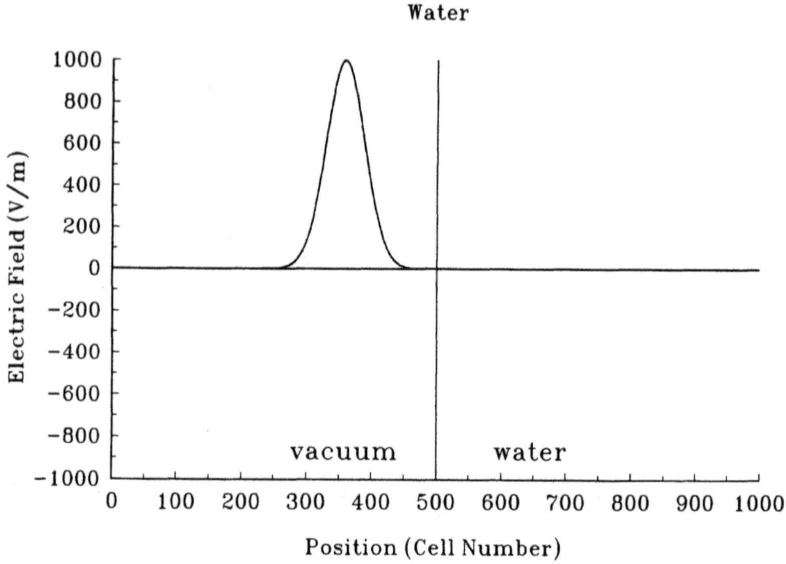

FIGURE 8-4. Electric field vs. position for the Gaussian pulse plane wave incident on a vacuum-water interface after one FDTD time step.

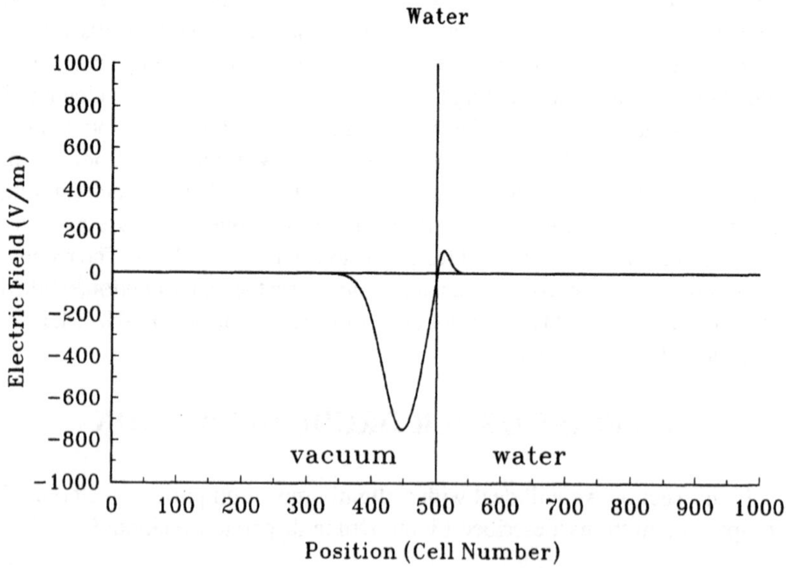

FIGURE 8-5. Electric field vs. position for the Gaussian pulse plane wave incident on a vacuum-water interface after 200 FDTD time steps.

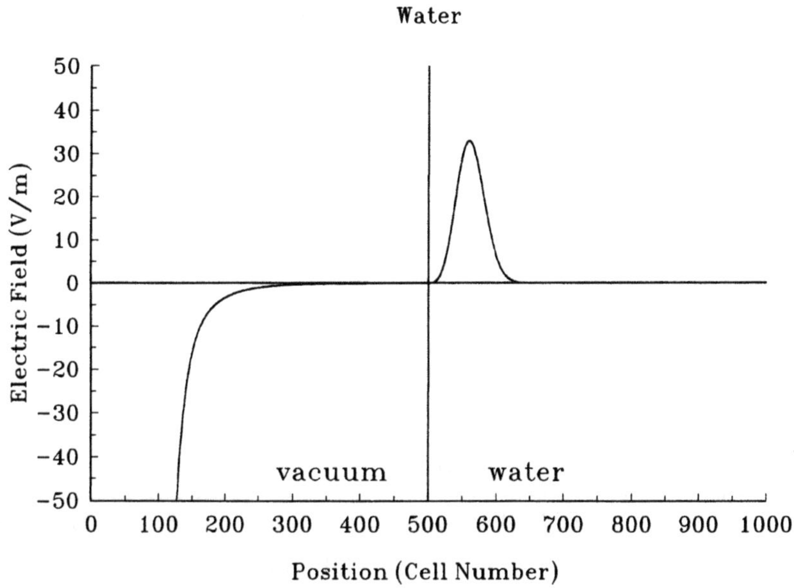

FIGURE 8-6. Electric field vs. position for the Gaussian pulse plane wave incident on a vacuum-water interface after 600 FDTD time steps.

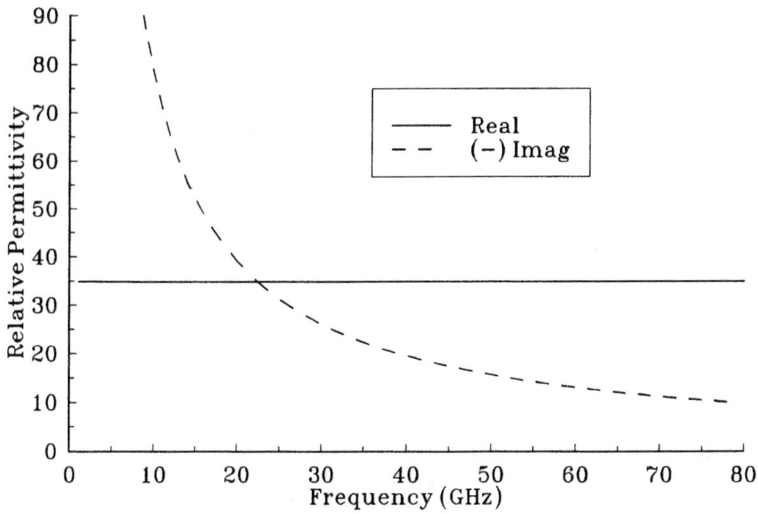

FIGURE 8-7. Complex relative permittivity of constant parameter FDTD "water" in the frequency domain.

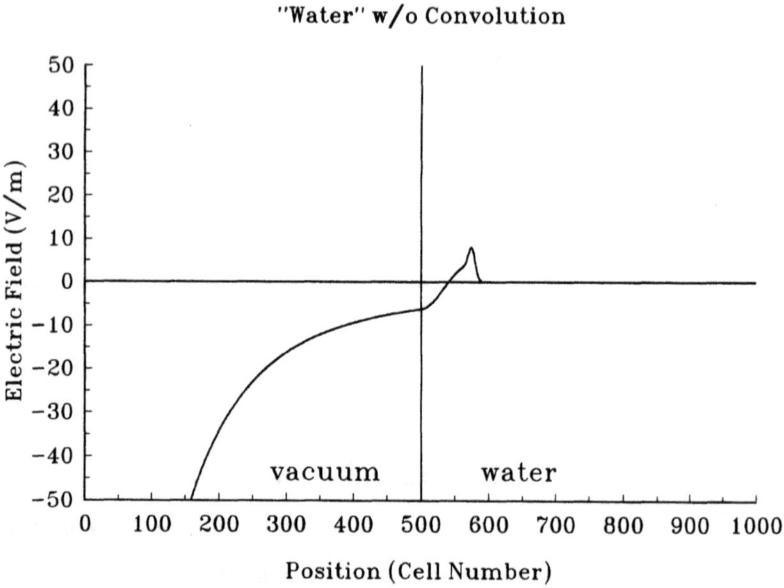

FIGURE 8-8. Electric field vs. position for the Gaussian pulse plane wave incident on a vacuum-"water" interface after 600 time steps computed using nonconvolved FDTD.

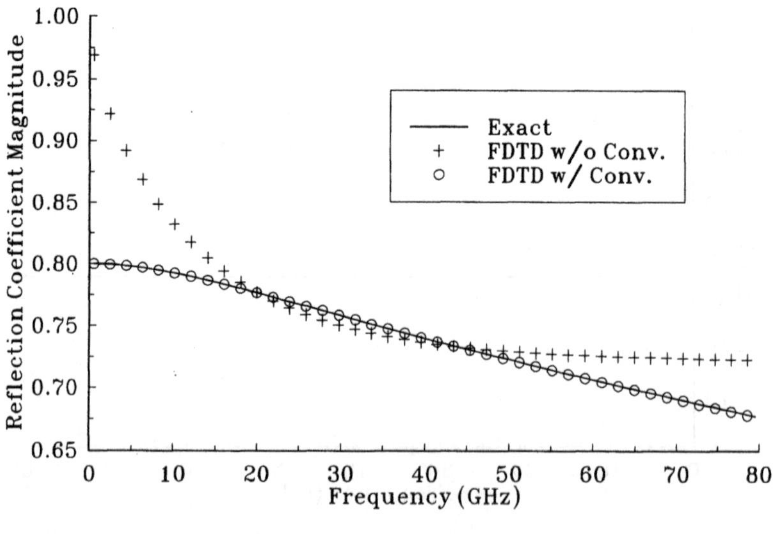

FIGURE 8-9. Reflection coefficient magnitude for the vacuum-water interface computed using convolved and nonconvolved FDTD compared to the exact solution.

where ω_p is the radian plasma frequency and ν_c is the collision frequency. If we do not include a conductivity term, i.e, we assume $\sigma = 0$, then the resulting susceptibility function

$$\hat{\varepsilon}(\omega) = 1 + \frac{\omega_p^2}{\omega\left(j\nu_c - \omega\right)} = \varepsilon_\infty + \chi(\omega) \tag{8.26}$$

has two poles, one at $\omega = 0$ and the other at $\omega = j\nu_c$. The Fourier transform of this susceptibility function is not causal, and the Kramers-Kronig relation for this case must be modified to account for this.[1,2] In Reference 5 we dealt with this situation by using the time domain susceptibility function

$$\chi(t) = \frac{\omega_p^2}{\nu_c}\left[1 - e^{-\nu_c t}\right] U(t) \tag{8.27}$$

which has a Fourier transform equal to $\chi(\omega)$ of (8.26) except at $\omega = 0$, where it differs by $\pi\, \omega_p 2\; \delta(\omega)/\nu_c$, where $\delta(\omega)$ is the Dirac delta function. Rather than use this approach, which involves reevaluating several of the equations in the preceding section, it is simpler to use the conductivity term retained in (8.12) to simplify the susceptibility function. We can expand the complex permittivity function in (8.25) as

$$\hat{\varepsilon}(\omega) = 1 + \frac{\omega_p^2}{\omega\left(j\nu_c - \omega\right)} = 1 - \frac{\left(\omega_p/\nu_c\right)^2}{1 + j\frac{\omega}{\nu_c}} + \frac{\left(\omega_p^2/\nu_c\right)}{j\omega} \tag{8.28}$$

Combining (8.1) and (8.2) we have the relationship

$$\hat{\varepsilon}(\omega) = \varepsilon_\infty + \chi(\omega) + \frac{\sigma}{j\omega\varepsilon_0} \tag{8.29}$$

Comparing (8.28) and (8.29) there will be a correspondence if

$$\varepsilon_\infty = 1 \tag{8.30}$$

$$\chi(\omega) = -\frac{\left(\omega_p/\nu_c\right)^2}{1 + j\frac{\omega}{\nu_c}} \tag{8.31}$$

and

$$\sigma = \varepsilon_0 \frac{\omega_p^2}{\nu_c} \tag{8.32}$$

Having made these assignments, if we compare (8.31) to (8.2) it is evident that we can use the Debye results of the preceding section for Drude materials if we let

$$\left(\varepsilon_s - \varepsilon_\infty\right) = -\left(\frac{\omega_p}{\nu_c}\right)^2 \tag{8.33}$$

and

$$t_0 = \frac{1}{\nu_c} \tag{8.34}$$

and include in our FDTD calculations the conductivity of (8.32). As we should expect, the FDTD update equation obtained in this way corresponds exactly to that obtained with a zero conductivity and the susceptibility function of (8.27), as shown in Reference 5.

Thus, we see that for our purposes a Drude dispersive material is merely a Debye material with a nonzero conductivity. However, the frequency dependence of the resulting complex permittivity is quite different than for a Debye material because the sign of the exponential susceptibility term in (8.31) is negative. To illustrate this, consider the example of an isotropic cold plasma. Our example plasma has a plasma frequency of 28.7 GHz = ($\omega_p/2\pi$) and a collision frequency ν_c of 2.0×10^{10}. The complex permittivity for this plasma, including the effects of the conductivity, is shown in Figure 8-10. The time domain susceptibility, the Fourier transform of (8.31), is shown in Figure 8-11.

From Figure 8-10 we see that the real part of the complex permittivity changes sign near the plasma frequency. Below this frequency, where the real part of the permittivity is negative, the plasma behaves somewhat like a waveguide below cutoff in that electromagnetic waves do not propagate. Above the plasma frequency electromagnetic waves propagate through the plasma, although with some loss, which decreases with increasing frequency as the plasma behaves more like free space, its high frequency limit. While FDTD with constant constitutive parameters can be applied to Debye media at any single frequency, for Drude materials there is some frequency at which the negative real part of the permittivity will cause one of the FDTD multiplying constants to be singular, and at this frequency normal FDTD cannot be used. With the recursive convolution term added, however, FDTD is applicable over the entire frequency range of the Drude permittivity variation, as demonstrated next.

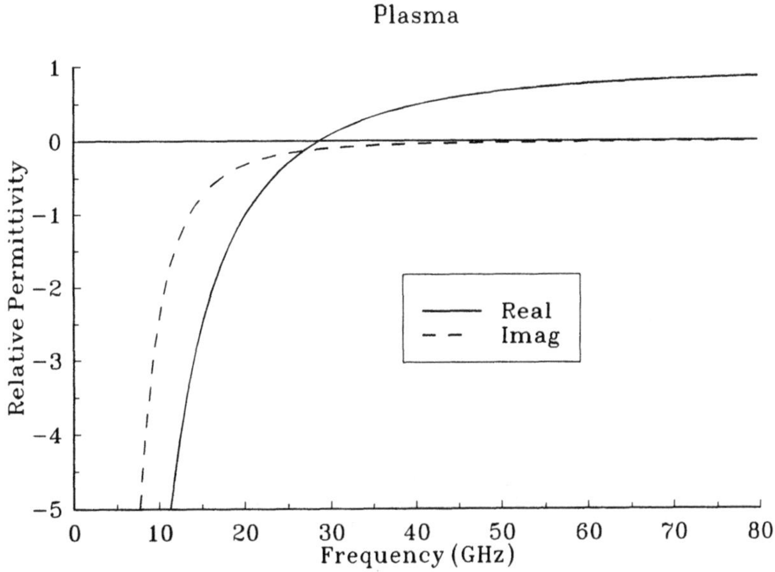

FIGURE 8-10. Complex relative permittivity of plasma in the frequency domain.

This demonstration will be similar to that given for water in the previous section, except that rather than calculate reflection from a half-space, we will calculate both reflection and transmission for a plasma "slab" 1.5-cm thick. The plasma parameters are as given above. The FDTD cells have $\Delta x = 75\ \mu m$. The 1-D problem space consists of 800 cells with the plasma in cells 300 through 500 and free space in the remaining cells. A first order Mur absorbing boundary is used on both ends of the problem space to absorb the reflected and transmitted fields. The FDTD time step is taken at half the Courant limit as $\Delta t = \Delta x/2c = 0.125$ ps, as calculations with Δt at the Courant limit tended to be unstable for this material.

Because for this example there is a nonzero conductivity, an incident Gaussian pulse will excite currents in the plasma slab of extremely low frequency that take exceedingly long times to dissipate, much like the long "tail" on the reflected pulse for the nonconvolved "water" shown in Figure 8-8. This extremely low frequency energy also decreases the accuracy of the resulting FDTD calculations at low frequencies, especially for the transmitted fields as they have very low amplitudes at low frequencies. To avoid this problem we will use a Gaussian derivative pulse instead of a Gaussian, because it has no energy content at zero frequency. Equations for the initial values of the electric and magnetic fields in the computation space at time $t = 0$ for a Gaussian pulse ((8.23), (8.24)) can be modified for the Gaussian derivative pulse using

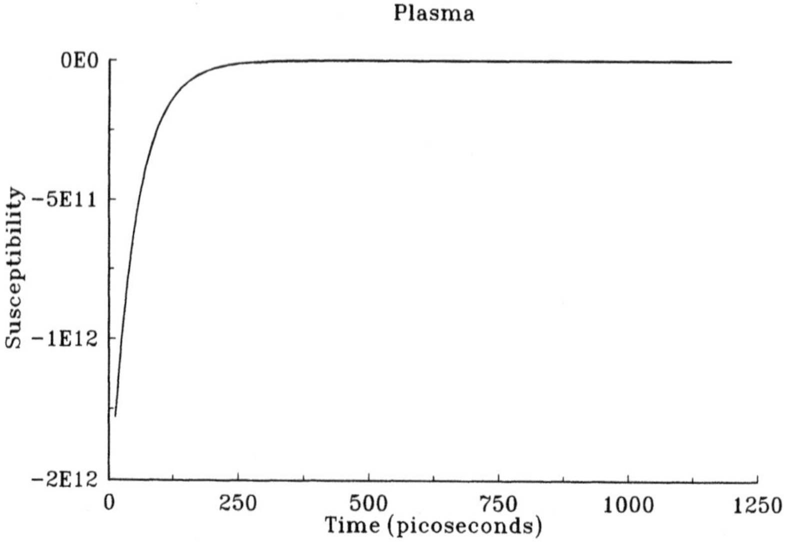

FIGURE 8-11. Time domain susceptibility function for plasma.

$$E_{GD}(i) = E_G(i)\, 2\zeta(i - i_c)/i_p^2 \tag{8.35}$$

$$H_{GD}(i) = H_G(i)\, 2\zeta\left(i + \tfrac{1}{2} - i_c - \tfrac{0.5\Delta x}{c\Delta t}\right)/i_p^2 \tag{8.36}$$

where the GD subscript refers to Gaussian derivative and the G subscript to the Gaussian pulse E and H fields given in (8.23) and (8.24). For this example the Gaussian derivative pulse has a half-width $i_p = 56$ cells. The spectrum of the corresponding Gaussian derivative pulse is shown in Figure 8-12. The incident Gaussian derivative pulse at the first time step is shown in Figure 8-13, and the electric fields in the FDTD computation space after 600 and 1000 time steps are shown in Figures 8-14 and 8-15. The plasma is obviously more highly dispersive than water, with the transmitted and reflected pulses radically different in shape than the incident pulse.

In order to demonstrate the accuracy of recursive convolution FDTD when applied to this highly dispersive media, the reflection and transmission coefficients for the plasma slab will be computed in the frequency domain by taking Fourier transforms of the FDTD time domain results, as was done for water in the previous section, and compared to the exact frequency domain solution. Due to pulse spreading, 9600 time steps were calculated in order to allow the transient response to settle to zero. This corresponds approximately to the full time scale of the susceptibility for the plasma in Figure 8-11. The reflection coefficient magnitude and phase are shown in Figures 8-16 and 8-17. The reflection coefficient results show quite accurate agreement with the exact solution over a dynamic range of over 30 dB, going smoothly through the frequency band where the plasma changes from nonpropagating to propagating and the real part of the permittivity changes sign. The transmission line effects evidenced by the lobing structure above the plasma frequency are also quite accurately computed from the transient FDTD results.

The transmission coefficient results, shown in Figures 8-18 and 8-19, are accurate over a >70-dB dynamic range, with the reversing slope of the phase of the transmission coefficient at low frequencies computed accurately.

8.4 SECOND ORDER DISPERSIVE MATERIALS

Up to this point the complex frequency domain permittivity has been described by either a first order pole or a first order pole plus a conductivity term that is equivalent to a an additional first order pole at $\omega = 0$. While this is adequate to model (at least approximately) a wide range of materials, some materials require one or more second order poles to accurately describe their complex permittivity. In this section we will extend the previous recursive convolution approach to a single second order pole, and in the following section to multiple poles with a demonstration given for a material with two second order poles. This generality should enable our FDTD calculations to accurately model any frequency-dependent material. The simplest second order pole to consider is the Lorentz form and we will begin with this.

The complex permittivity for a frequency dependent material with one second order Lorentz pole can be described as[7]

$$\varepsilon(\omega) = \varepsilon_\infty + \left(\varepsilon_s - \varepsilon_\infty\right)\frac{\omega_p^2}{\omega_p^2 + 2j\omega\delta_p - \omega^2} \tag{8.37}$$

where ω_p is the resonant frequency and δ_p the damping coefficient. We obtain the corresponding time domain susceptibility function $\chi_p(t)$ by considering the Fourier transform pair[3]

$$\chi_p(t) = \gamma_p e^{-\alpha_p t}\sin\left(\beta_p t\right)U(t) \Leftrightarrow \frac{\gamma_p\beta_p}{\left(\alpha_p^2 + \beta_p^2\right) + 2j\omega\alpha_p - \omega^2} \tag{8.38}$$

Comparing these equations we see that (8.38) will provide the correct expression for $\chi_p(t)$ provided that

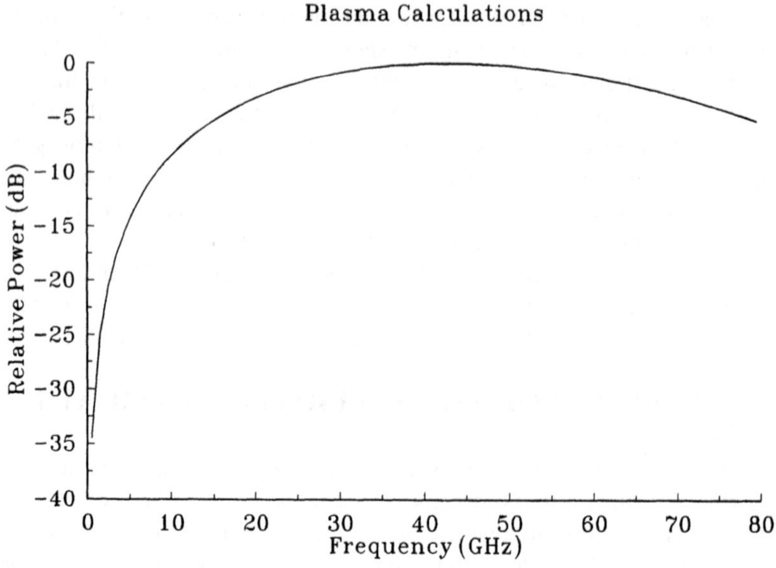

FIGURE 8-12. Frequency spectrum of the incident Gaussian derivative pulse.

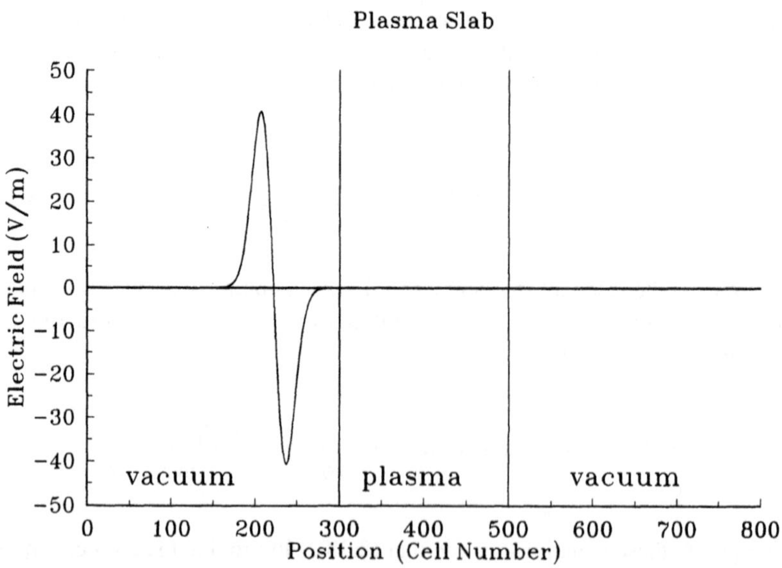

FIGURE 8-13. Electric field vs. position for the Gaussian derivative pulse plane wave incident on the plasma slab after 1 FDTD time step.

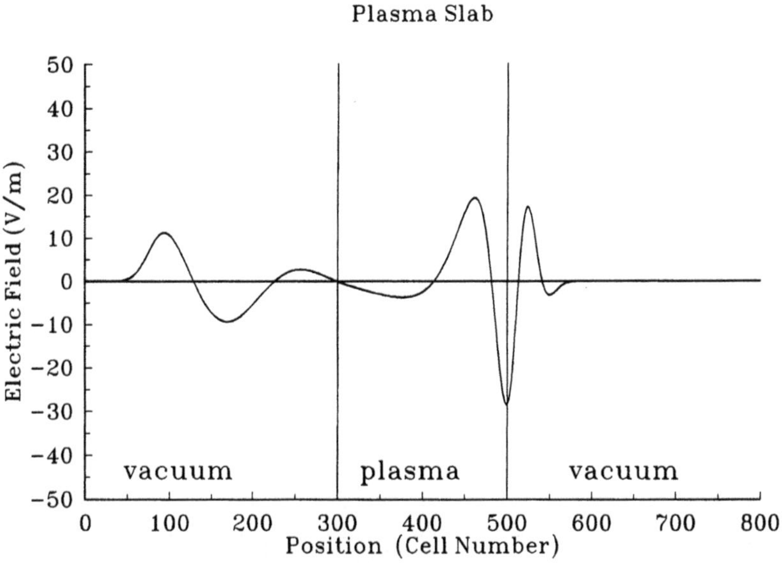

FIGURE 8-14. Electric field vs. position for the Gaussian derivative pulse plane wave incident on the plasma slab after 600 FDTD time steps.

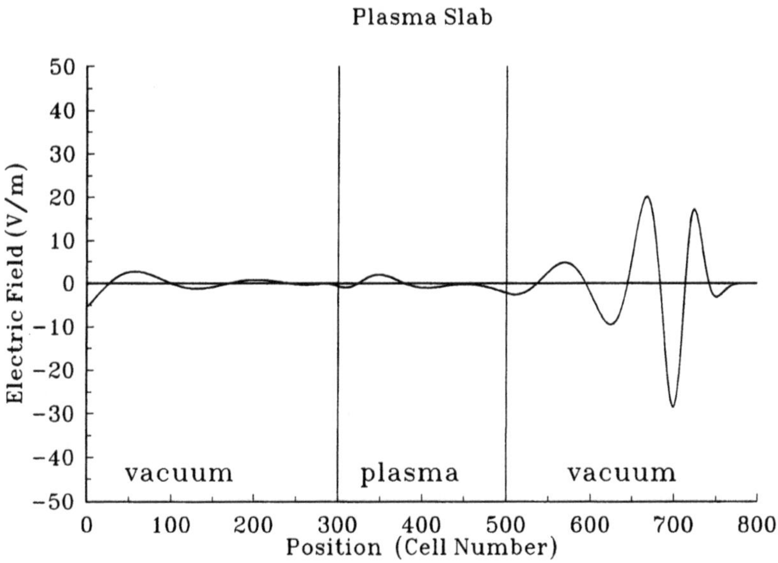

FIGURE 8-15. Electric field vs. position for the Gaussian derivative pulse plane wave incident on the plasma slab after 1000 FDTD time steps.

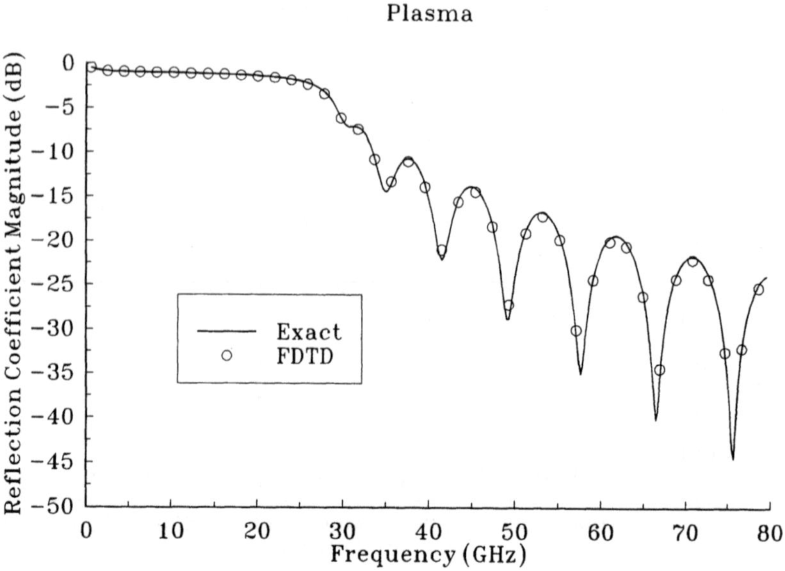

FIGURE 8-16. Magnitude of the reflection coefficient for the plasma slab computed using convolved FDTD and compared with the exact result

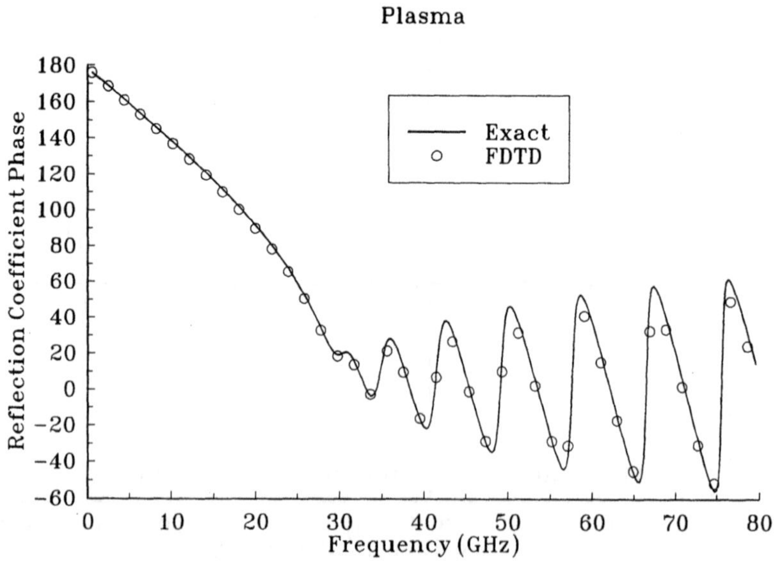

FIGURE 8-17. Phase of the reflection coefficient for the plasma slab computed using convolved FDTD and compared to the exact result.

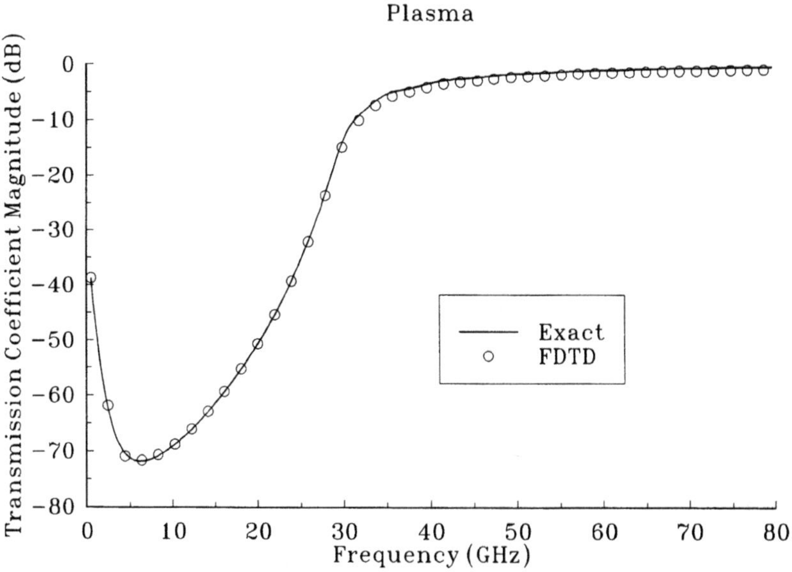

FIGURE 8-18. Magnitude of the transmission coefficient for the plasma slab computed using convolved FDTD and compared to the exact result.

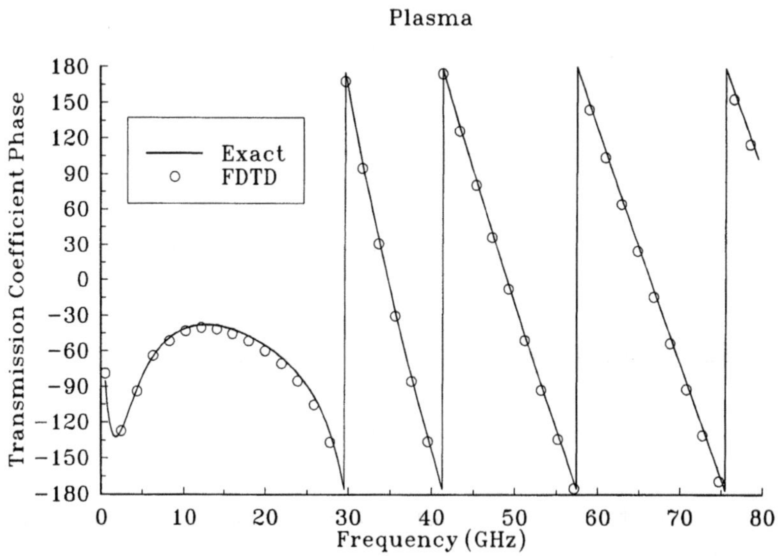

FIGURE 8-19. Phase of the transmission coefficient for the plasma slab computed using convolved FDTD and compared to the exact result.

$$\alpha_p = \delta_p$$

$$\beta_p = \sqrt{\omega_p^2 - \delta_p^2} \tag{8.39}$$

$$\gamma_p = \frac{\omega_p^2(\varepsilon_s - \varepsilon_\infty)}{\beta_p}$$

The time dependence of $\chi_p(t)$ in (8.38) is not in a form for which the corresponding discrete convolution can be updated recursively. However, if we define a complex time-domain susceptibility

$$\hat{\chi}_p(\tau) = -j\gamma_p e^{(-\alpha_p + j\beta_p)\tau} U(\tau) \tag{8.40}$$

so that

$$\chi_p(\tau) = \mathrm{Re}\left[\hat{\chi}_p(\tau)\right] \tag{8.41}$$

where the "^" denotes complex quantities and Re is the real operator, a recursive convolution can be implemented. Applying (8.9) and (8.11) to (8.40) readily produces

$$\hat{\chi}_p^m = \frac{-j\gamma_p}{\alpha_p - j\beta_p} e^{(-\alpha_p + j\beta_p)m\Delta t}\left[1 - e^{(-\alpha_p + j\beta_p)\Delta t}\right] \tag{8.42}$$

and

$$\Delta\hat{\chi}_p^m = \frac{-j\gamma_p}{\alpha_p - j\beta_p} e^{(-\alpha_p + j\beta_p)m\Delta t}\left[1 - e^{(-\alpha_p + j\beta_p)\Delta t}\right]^2 \tag{8.43}$$

with the important feature of (8.43) being the relationship

$$\Delta\hat{\chi}_p^{m+1} = e^{(-\alpha_p + j\beta_p)\Delta t}\Delta\hat{\chi}_p^m \tag{8.44}$$

which allows the recursive evaluation of the convolution summation. If we now define

$$\psi_p^n = \sum_{m=0}^{n-1} E^{n-m} \Delta\chi_p^m = \mathrm{Re}\left[\hat{\psi}_p^n\right] \tag{8.45}$$

then the complex accumulation variable $\hat{\psi}_p^n$ can be updated recursively using the same approach as described previously for the real accumulator variable for first order poles using

$$\hat{\psi}_p^n = E^n \Delta\hat{\chi}_p^0 + e^{(-\alpha_p + j\beta_p)\Delta t} \hat{\psi}_p^{n-1} \tag{8.46}$$

Finally, we obtain the quantities needed to update the electric field in (8.13) by taking the real parts of the complex quantities with

$$\chi^0 = \mathrm{Re}\left[\hat{\chi}_p^0\right] \tag{8.47}$$

and

$$\sum_{m=0}^{n-1} E_y^{n-m} \Delta\chi^m = \mathrm{Re}\left[\hat{\psi}_p^n\right] \tag{8.48}$$

with the summation of (8.48) updated recursively using (8.46).

We now see that a second order pole yields a time domain susceptibility, the convolution of which with the electric fields can be evaluated recursively provided that the recursive accumulation is complex. Other than this complex arithmetic, a susceptibility function with a second order pole is no more difficult to evaluate recursively than with a first order pole.

While the Lorentz second order pole has a Fourier transform given by (8.38), other more complicated second order poles which describe materials such as magnetized plasmas and ferrites have Fourier transforms that also involve the Fourier transform pair

$$\chi_p(t) = \xi_p e^{-\alpha_p t} \cos(\beta_p t)\, U(t) \Leftrightarrow \frac{\xi_p(\alpha_p + j\omega)}{(\alpha_p^2 + \beta_p^2) + 2j\omega\alpha_p - \omega^2} \tag{8.49}$$

in addition to (8.38). In combination these two Fourier transform pairs can be applied to yield a complex time domain susceptibility function for more complicated permeabilities that have the form

$$\begin{aligned}\chi_p(t) &= e^{-\alpha_p t}\left[\xi_p \cos(\beta_p t) + \gamma_p \sin(\beta_p t)\right] U(t) \\ &= \mathrm{Re}\left[\left(\left(\xi_p - j\gamma_p\right) e^{(-\alpha_p + j\beta_p)t}\right)\right] U(t)\end{aligned} \tag{8.50}$$

The only extension from the above is that $-j\gamma_p$ in (8.40) is replaced by $(\xi_p - j\gamma_p)$. Some examples of FDTD calculations for materials with more complicated second order dispersion are included in Chapter 15, which also deals with the anisotropy of these materials.

Rather than give an example result for FDTD calculations for a material with a single second order Lorentz pole we will first extend our results to multiple poles, then give some results for a material with two second order poles.

8.5 MULTIPLE POLES

While most materials have a complex permittivity whose frequency dependence can be described with a single first or second order pole over the frequency band of interest, some materials, artificial dielectrics, and optical materials, for example, have more complicated frequency behavior. For these materials more than one pole may be needed in the permittivity function to describe their behavior. Because we are dealing with linear materials, combining the effects of each pole is a simple matter of adding their separate effects. In this section we extend our previous results to multiple second order Lorentz poles and give an example for a material with two second order poles. Other combinations of poles can be considered in a similar manner.

For this example suppose that P second order Lorentz poles describe the complex permittivity as

$$\varepsilon(\omega) = \varepsilon_\infty + \left(\varepsilon_s - \varepsilon_\infty\right) \sum_{p=1}^{P} \frac{G_p \omega_p^2}{\omega_p^2 + 2j\omega\delta_p - \omega^2} \tag{8.51}$$

with the condition that

$$\sum_{p=1}^{P} G_p = 1$$

where now ω_p is the resonant frequency for the pth pole and δ_p is the damping coefficient for the pth pole. Clearly, we could include other types of first or second order poles in the summation if we wished. The extensions to the results given in the previous section necessary to include multiple poles in the FDTD calculations are quite straightforward. Equations (8.39) through (8.46) still apply, keeping in mind that now the subscript p can take on integer values from

TABLE 8-1
Parameters for the Two Lorentz Poles

$\varepsilon_s = 3.0$	$\varepsilon_\infty = 1.5$
$\omega_1 = 2\pi \cdot 20 \times 10^9$	$\omega_2 = 2\pi \cdot 50 \times 10^9$
$\delta_1 = 0.1\omega_1$	$\delta_2 = 0.1\omega_2$
$G_1 = 0.4$	$G_2 = 0.6$

1 to P so that these equations apply to each pole separately. Equations (8.47) and (8.48) are modified as

$$\chi^0 = \sum_{p=1}^{P} \mathrm{Re}\left[\hat{\chi}_p^o\right] \tag{8.52}$$

and

$$\sum_{m=0}^{n-1} E_y^{n-m} \Delta\chi^m = \sum_{p=1}^{P} \mathrm{Re}\left[\hat{\psi}_p^n\right] \tag{8.53}$$

with each $\hat{\psi}_p^n$ term in the summation over p in (8.53) updated recursively using (8.46).

The demonstration will again involve calculation of the transient fields for a pulsed plane wave normally incident on the boundary between vacuum and the dispersive medium, but for this example the medium is described by two second order Lorentz poles. Referring to (8.51) with $P = 2$, the constants which determine the complex permittivity for the demonstration medium are given in Table 8-1. The conductivity σ is taken as zero. The frequency domain complex relative permittivity for this medium is shown in Figure 8-20. The corresponding time domain susceptibility from (8.38), with the effects of both poles included as $\chi(t) = \chi_1(t)+\chi_2(t)$, is shown in Figure 8-21.

For the 1-D FDTD calculation 1500 spatial cells are used, with the vacuum-medium boundary at cell 500. A first order Mur[6] absorbing boundary is again located at cell 1, while the 1000-cell region containing the dispersive medium provided adequate time delay so that pulses reflecting from the boundary at cell 1500 did not return to the interface before the transients dissipated. The FDTD parameters are the same as for the water example in Section 8.2, with $\Delta x = 37.5$ μm and $\Delta t = 0.125$ ps $= \Delta x/c$, where c is the speed of light in vacuum.

The excitation is a Gaussian pulse plane wave. The pulse is also the same used for the water example, and its spectrum is shown in Figure 8-3.

Spatial plots of electric field vs. position for the pulse as it reflects from the interface and propagates in the dispersive medium are shown in Figures 8-22 to 8-25 after increasing numbers of time steps. The extremely dispersive nature of the medium is clearly indicated.

The accuracy of the FDTD results are again shown by computing the reflection coefficient vs. frequency from the transient fields. The FDTD com-

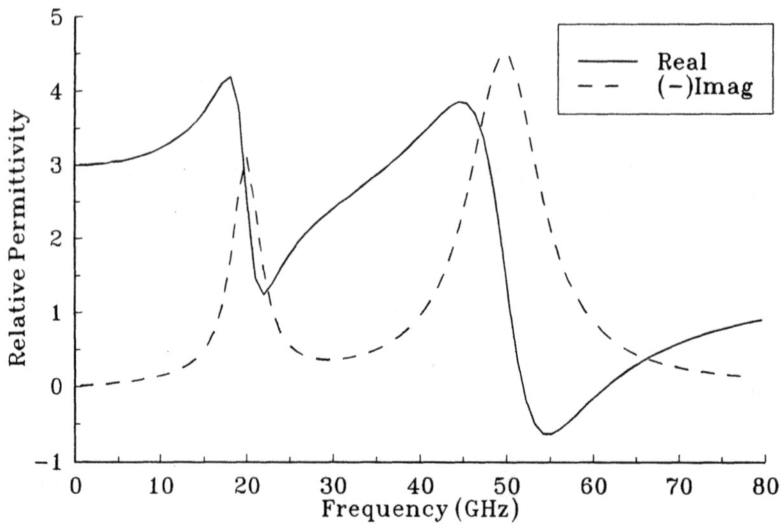

FIGURE 8-20. Complex relative permittivity of two Lorentz pole medium in the frequency domain.

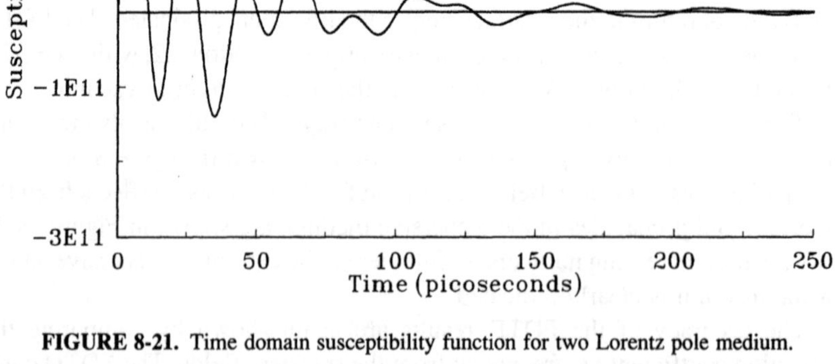

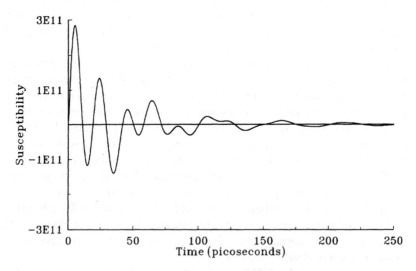

FIGURE 8-21. Time domain susceptibility function for two Lorentz pole medium.

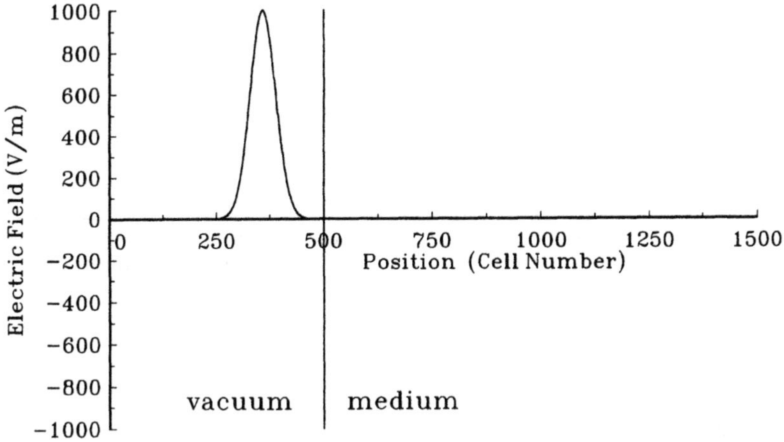

FIGURE 8-22. Electric field vs. position for the Gaussian pulse plane wave incident on the two Lorentz pole medium after one FDTD time step.

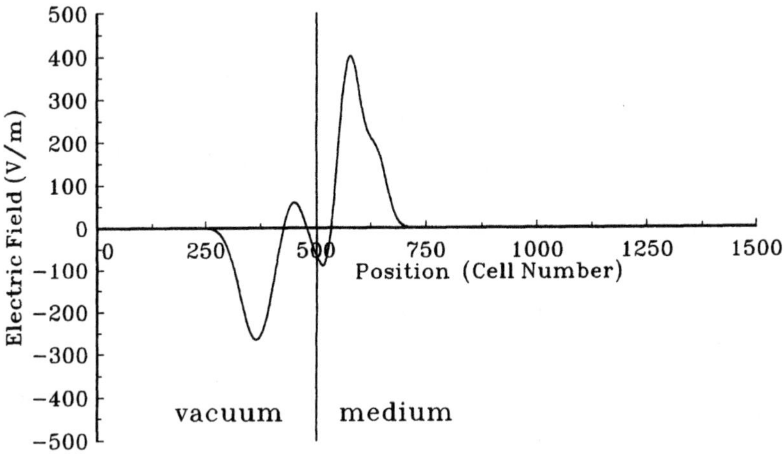

FIGURE 8-23. Electric field vs. position for the Gaussian pulse plane wave incident on the two Lorentz pole medium after 300 FDTD time steps.

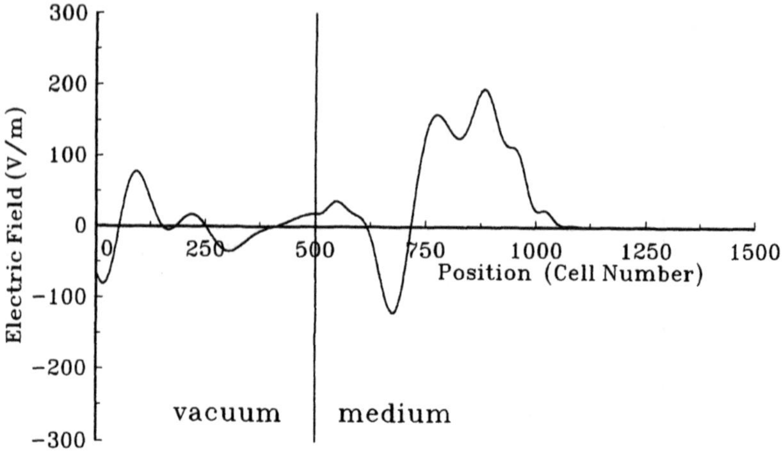

FIGURE 8-24. Electric field vs. position for the Gaussian pulse plane wave incident on the two Lorentz pole medium after 800 FDTD time steps.

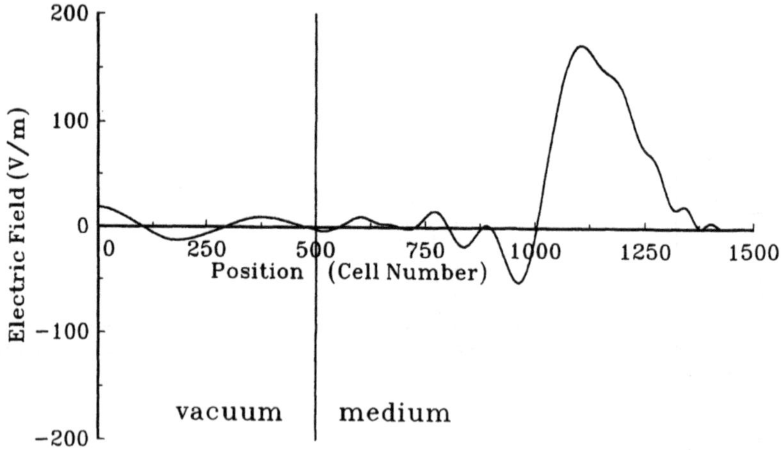

FIGURE 8-25. Electric field vs. position for the Gaussian pulse plane wave incident on the two Lorentz pole medium after 1300 FDTD time steps.

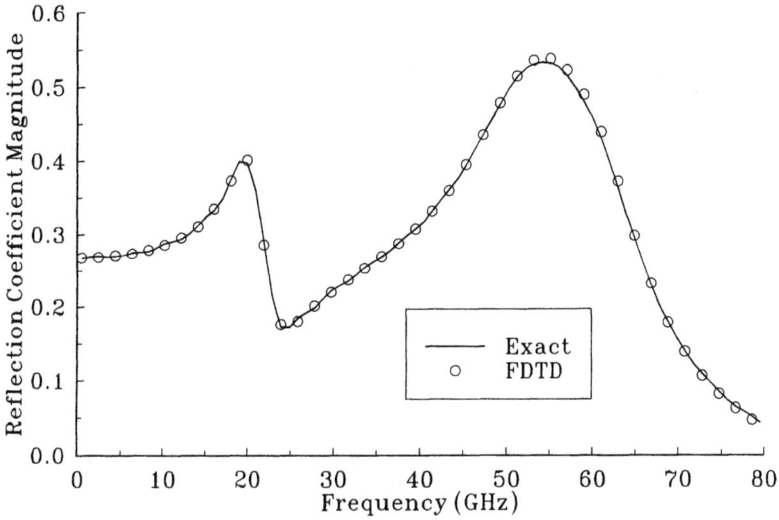

FIGURE 8-26. Magnitude of the reflection coefficient for the two Lorentz pole medium computed using convolved FDTD and compared to the exact result.

putations were continued until the transients at the vacuum-medium interface were dissipated. This required approximately 2000 time steps or 250 ps, which when compared to Figure 8-21 corresponds reasonably well with the duration of the susceptibility response. Computation of the reflection coefficient was accomplished in the same manner as for the water example. The resulting reflection coefficient magnitude and phase are shown in Figures 8-26 and 8-27 and display excellent agreement with the exact solution.

8.6 DIFFERENTIAL METHOD

The previous sections of this chapter presented the recursive convolution approach for efficiently including frequency-dependent materials in FDTD calculations. However, an alternative method exists that expresses the relationship between D and E in these materials with a differential equation rather than a convolution integral.[8,9] This section outlines the differential method and contrasts it with the recursive convolution method.

Let us use for an example a frequency-dependent material with a single second order Lorentz pole. Then, from (8.37) we can express the relationship between D and E in the frequency domain as

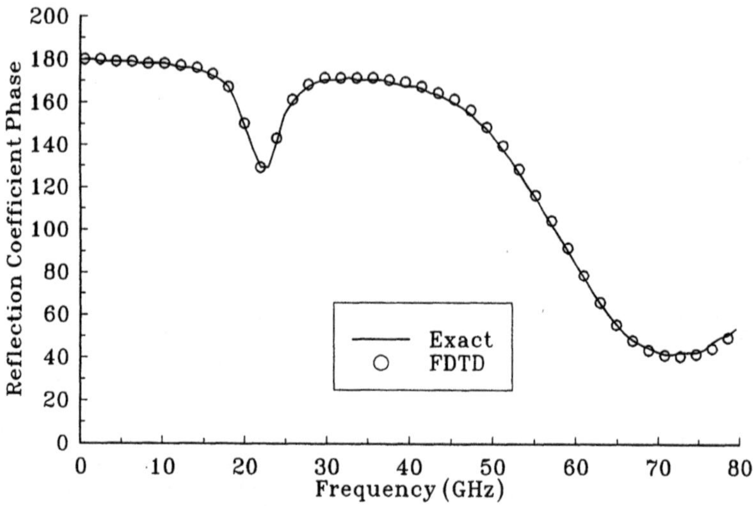

FIGURE 8-27. Phase of the reflection coefficient for the two Lorentz pole medium computed using convolved FDTD and compared to the exact result.

$$\varepsilon(\omega) = \frac{D(\omega)}{E(\omega)} = \varepsilon_\infty + (\varepsilon_s - \varepsilon_\infty)\frac{\omega_p^2}{\omega_p^2 + 2j\omega\delta_p - \omega^2} \tag{8.54}$$

Rather than Fourier transform the susceptibility function for use in a convolution integral we transform this relationship to the time domain as a differential equation, recognizing $j\omega$ as equivalent to a time derivative and $-\omega^2 = (j\omega)^2$ as a second order time derivative. This yields the differential equation[9]

$$\omega_p^2 D + 2\delta_p \frac{\partial D}{\partial t} + \frac{\partial^2 D}{\partial t^2} = \omega_p^2 \varepsilon_s E + 2\delta_p \varepsilon_\infty \frac{\partial E}{\partial t} + \varepsilon_\infty \frac{\partial^2 E}{\partial t^2} \tag{8.55}$$

This differential equation can be approximated as a finite difference equation in E and D[9]

$$\begin{aligned} E^{n+1} = \Big[&\left(\omega_p^2 \Delta t^2 + 2\delta_p \Delta t + 2\right) D^{n+1} - 4D^n \\ &+ \left(\omega_p^2 \Delta t^2 - 2\delta_p \Delta t + 2\right) D^{n-1} + 4\varepsilon_\infty E^n \\ &- \left(\omega_p^2 \Delta t^2 \varepsilon_s - 2\delta_p \Delta t \varepsilon_\infty + 2\varepsilon_\infty\right) E^{n-1} \Big] \\ &/ \left(\omega_p^2 \Delta t^2 \varepsilon_s + 2\delta_p \Delta t \varepsilon_\infty + 2\varepsilon_\infty\right) \end{aligned} \tag{8.56}$$

and is updated at each time step in addition to (8.13) and (8.7) (with zero conductivity to eliminate the dependence of D^{n+1} on E^{n+1} in (8.7)) and solved for D^{n+1}. Thus, at each time step (8.56) would be updated, and this result used to update

$$H_z^{n+\frac{1}{2}}\left(i+\tfrac{1}{2}\right)=H_z^{n-\frac{1}{2}}\left(i+\tfrac{1}{2}\right) -\frac{\Delta t}{\mu\Delta x}\left[E_y^n(i+1)-E_y^n(i)\right] \tag{8.57}$$

with the resulting H field used to find the next time value of D in

$$D_y^{n+1}(i)=D_y^n(i)-\frac{\Delta t}{\Delta x}\left[H_z^{n+\frac{1}{2}}\left(i+\tfrac{1}{2}\right)-H_z^{n+\frac{1}{2}}\left(i-\tfrac{1}{2}\right)\right] \tag{8.58}$$

If we compare the calculations required to update (8.56) to those for (8.46), the recursive convolution update appears to be relatively simple and to require storage of fewer multiplicative constants, even though it requires complex arithmetic. Furthermore, the recursive convolution approach requires less computer storage for previous values of the fields. To apply the differential equation approach, in order to evaluate (8.56) for a material with order-M dispersion, M previous values of D, and M-1, additional previous values of E must be stored[9] relative to nondispersive FDTD calculations. Considering a fourth order dispersion, corresponding to the example given previously with two second order Lorentz poles, the differential method would require storage of seven additional real variables, while the recursive convolution approach required storage of two complex variables. These two complex variables correspond to four real variables in computer storage; so that there is very nearly a 2:1 savings in storage by using the recursive convolution method for this case. As the order of the dispersion increases this ratio approaches 2:1, with the differential method in general requiring 2M-1 real backstore variables while the recursive convolution method requires M/2 complex backstore variables, the equivalent of M real variables.

The explanation for this difference between the two approaches is that the differential equation approach requires finding the flux density D before finding E, while the recursive convolution method does not. Thus, the differential equation method yields additional information (the value of D at each time step), but at the expense of additional calculation and storage. This additional information regarding D may be useful in some situations, for example, in dealing with nonlinear materials, and in such situations the differential equation approach may be preferable. It may also be that due to the greater number of storage locations used (and therefore increased information retained) the differential equation approach may be more accurate if the same number of cells are used. The recursive approach appears simpler to implement and requires less additional computer storage per FDTD cell.

8.7 SCATTERED FIELD FORMULATION

In the preceding sections of this chapter the recursive convolution approach to including frequency-dependent materials in FDTD has been presented. For simplicity these derivations and simple 1-D examples were calculated using total fields. As we have discussed in previous chapters, for many applications a scattered field FDTD calculation is preferred. Even in a scattered field code, one option is to use total fields in cells containing dispersive materials. This works well, but seems to be slightly less accurate in computing scattering from 3-D targets. It, however, has the advantage of simplicity.

This section derives the scattered field form of the recursive convolution method. We follow the approach in Chapter 2, but adapt it to a relationship between D and E that involves a convolution. Thus, expressing this relationship as in (8.3), (2.16) becomes

$$\nabla \times \left(H^{inc} + H^{scat}\right) = \sigma\left(E^{inc} + E^{scat}\right) + \varepsilon_\infty \varepsilon_0 \frac{\partial\left(E^{inc} + E^{scat}\right)}{\partial t} + \varepsilon_0 \frac{\partial}{\partial t}\left[\left(E^{inc} + E^{scat}\right) * \chi(t)\right] \tag{8.59}$$

where "*" denotes convolution. Following the procedure in Chapter 2, we subtract the incident field equation (2.14) from (8.59) obtaining

$$\begin{aligned} \nabla \times H^{scat} = {} & \varepsilon_\infty \varepsilon_0 \frac{\partial E^{scat}}{\partial t} + \sigma E^{scat} \\ & + \varepsilon_0 \frac{\partial}{\partial t}\left[E^{scat} * \chi(t)\right] \\ & + (\varepsilon_\infty - 1)\varepsilon_0 \partial \frac{E^{inc}}{\partial t} + \sigma E^{inc} \\ & + \varepsilon_0 \frac{\partial}{\partial t}\left[E^{scat} * \chi(t)\right] \end{aligned} \tag{8.60}$$

If we apply the same process to this equation as was done in Section 8.2, the result will be an equation equivalent to (8.12), but in updating the scattered component of the electric field instead of the total we obtain

$$
\begin{aligned}
E_y^{\text{scat},\,n+1}(i) &= \frac{\varepsilon_\infty}{\frac{\sigma\Delta t}{\varepsilon_0}+\varepsilon_\infty+\chi^0}E_y^{\text{scat},\,n}(i) \\
&+\frac{1}{\frac{\sigma\Delta t}{\varepsilon_0}+\varepsilon_\infty+\chi^0}\sum_{m=0}^{n-1}E_y^{\text{scat},\,n-m}(i)\Delta\chi^m \\
&-\frac{\Delta t}{\left(\frac{\sigma\Delta t}{\varepsilon_0}+\varepsilon_\infty+\chi^0\right)\varepsilon_0\Delta x}\left[H_z^{n+\frac{1}{2}}\left(i+\tfrac{1}{2}\right)-H_z^{n+\frac{1}{2}}\left(i-\tfrac{1}{2}\right)\right] \\
&-\frac{\sigma\Delta t}{\left(\frac{\sigma\Delta t}{\varepsilon_0}+\varepsilon_\infty+\chi^0\right)\varepsilon_0}E_y^{\text{inc},\,n+1}(i) \\
&-\frac{(\varepsilon_\infty-1)\Delta t}{\left(\frac{\sigma\Delta t}{\varepsilon_0}+\varepsilon_\infty+\chi^0\right)}\frac{\partial E_y^{\text{inc},\,n+1}(i)}{\partial t} \\
&-\frac{\Delta t}{\left(\frac{\sigma\Delta t}{\varepsilon_0}+\varepsilon_\infty+\chi^0\right)}\frac{\partial}{\partial t}\left[E^{\text{inc},\,n+1}(i)*\chi(t)\right]
\end{aligned} \tag{8.61}
$$

The recursive evaluation of the summation of the scattered field with the susceptibility function factor on the right side of (8.61) is performed as discussed previously in this chapter for total fields with no change required. Most of the remaining terms are evaluated in a straightforward manner. However, the last term in (8.61), the derivative of the convolution of the incident field with the susceptibility function, requires some additional effort. We would like to evaluate this in closed form, both to reduce computer time and increase accuracy. While this can be done for a Gaussian pulse, it is simpler to accomplish if we consider a different form of incident pulse, the smooth cosine pulse. We now digress from our implementation of the scattered field form of recursive convolution FDTD to consider this pulse.

The incident field as specified using a smooth cosine pulse is given by

$$
E^{\text{inc}}(\tau)=\frac{10}{32}+\sum_{q=1}^{3}C_q\cos(q\omega_s\tau)\quad 0\le\tau\le T=0 \quad \text{otherwise} \tag{8.62}
$$

where $C_1=-15/32$, $C_2=6/32$, $C_3=-1/32$, and $\omega_s=2\pi/T$, where T is the length of the pulse and τ is the delayed time at the electric field location in question as given in (3.6). Compare this pulse to the Gaussian pulse described in Section 3.3. For consistency in parameter definition let the period T of the smooth cosine pulse be defined in the same way as the length of the truncated Gaussian pulse, i.e., $T = 2\beta\Delta t$. A Gaussian pulse with $\beta = 32$ and α determined as in

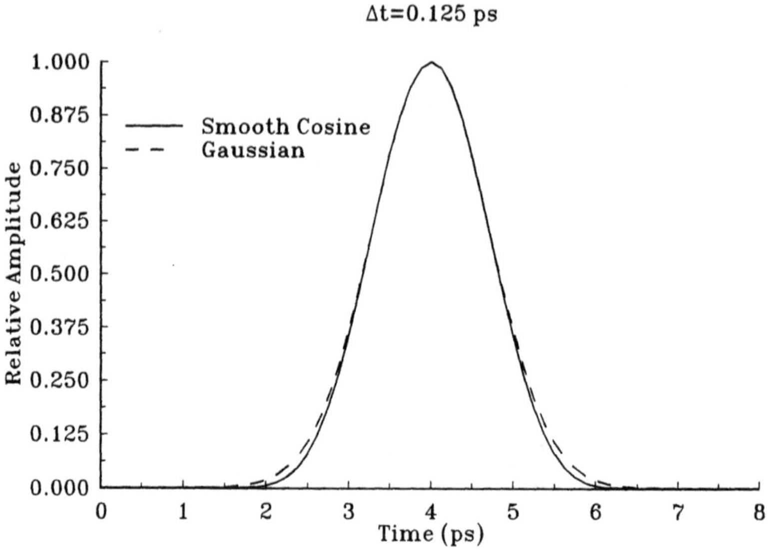

FIGURE 8-28. Smooth cosine pulse used in the evaluation of the scattered field form of the recursive convolution FDTD compared to Gaussian pulse.

Section 3.3 for a time step of $\Delta t = 0.125$ ps is shown in Figure 8-28. The smooth cosine pulse shown in the figure has a period T corresponding to $\beta = 22$. This value of β was determined such that the Fourier transform spectra of the two pulses had the same amplitude at zero frequency. The spectra of the two pulses are shown in Figure 8-29 with the same normalization. It is clear that the smooth cosine pulse does not have as wide a continuous frequency band as the Gaussian for a given pulse duration, and in most situations the Gaussian pulse is preferred. However, the smooth cosine pulse equation is more readily convolved analytically with the susceptibility function and we utilize it here for this purpose.

To further simplify the following derivation let us consider a general susceptibility function

$$\chi(t) = A_p e^{-a_p t} U(t) \tag{8.63}$$

where A_p and a_p may be complex to accommodate second order poles. If this is the case then the real part of the following result is taken. For multiple poles a summation over the p index is understood.

We now proceed to evaluate

$$\frac{\partial}{\partial t}\left[E_y^{\text{inc},\, n+1} * \chi(t)\right] = \frac{\partial}{\partial \tau}\int_0^{\tau} E^{\text{inc}}(\tau - \Lambda)\chi(\Lambda)d\Lambda \tag{8.64}$$

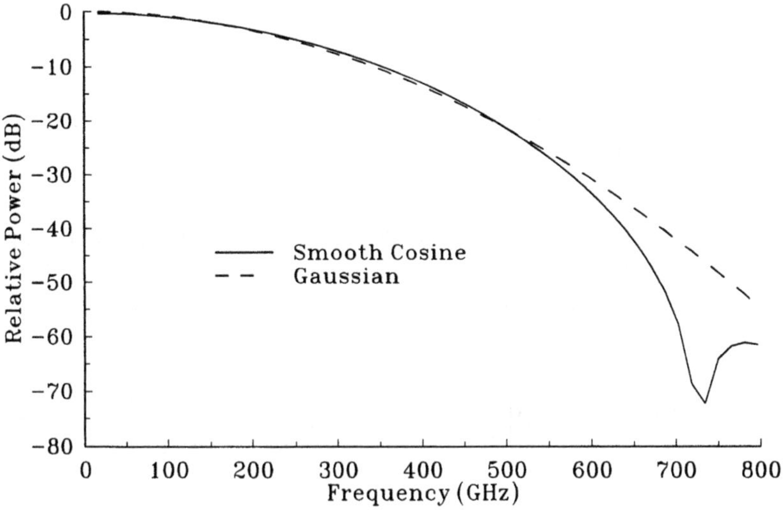

FIGURE 8-29. Spectra of the smooth cosine and Gaussian pulses of Figure 8-28.

Substituting for the incident field from (8.62) and for the susceptibility function from (8.63) we obtain

$$\frac{\partial}{\partial t}\left[E_y^{inc,\,n+1} * \chi(t)\right] = \frac{\partial}{\partial \tau}\left[A_p \int_0^{\tau} \frac{10}{32} e^{-a_p \Lambda} d\Lambda\right] + \frac{\partial}{\partial \tau}\left[\sum_{q=1}^{3} \int_0^{\tau} A_p C_q \cos\left(q\omega_s(\tau - \Lambda)e^{-a_p \Lambda} d\Lambda\right)\right] \tag{8.65}$$

In evaluating (8.65) we must separate our results into three cases depending on the value of τ.

- For $\tau \le 0$:

$$\frac{\partial}{\partial t}\left[E_y^{inc,\,n+1} * \chi(t)\right] = 0 \tag{8.66}$$

- For $0 \le \tau \le T$:

$$\frac{\partial}{\partial t}\left[E_y^{inc,\,n+1} * \chi(t)\right] = A_p \frac{10}{32} e^{-a_p \tau}$$
$$+\sum_{q=1}^{3} \frac{A_p C_q}{a_p^2 + (q\omega_s)^2}\left[(q\omega_s)^2 \cos(q\omega_s \tau) - a_p q\omega_s \sin(q\omega_s \tau) + a_p^2 e^{-a_p \tau}\right] \quad (8.67)$$

Finally, because the lower limits in the integrals in (8.65) change from 0 to $\tau - T$ due to the finite duration of the incident pulse, and because the sine and cosine terms become either 1 or 0 when $\tau = T$, we obtain for $T \le \tau$:

$$\frac{\partial}{\partial t}\left[E_y^{inc,\,n+1} * \chi(t)\right] = A_p e^{-a_p \tau}\left[1 - e^{a_p T}\right]\left[\frac{10}{32} + \sum_{q=1}^{3} \frac{C_q a_p^2}{a_p^2 + (q\omega_s)^2}\right] \quad (8.68)$$

Again, a reminder that the real part of (8.67) and (8.68) is to be taken if the susceptibility function is of the complex form used to describe a second order pole, and a summation over p is understood for multiple poles.

A brief comment on the numerical evaluation of (8.68) should be made. Depending on the duration of the pulse and the corresponding value of the pulse period T, the (real part of the) exponential factor $a_p T$ may become so large that computing results in a numerical overflow. If this is the case then the approximation $e^{-a_p \tau}\left[1 - e^{a_p T}\right] \approx -e^{-a_p(\tau - T)}$ should be used in the evaluation of (8.68). Because $T \le \tau$ this term will never cause a numerical overflow, although it will approach zero rapidly as τ increases. The physical explanation for this is that for large values of the $a_p T$ exponent the material is responding quickly relative to the FDTD computation time scale. A large value of the $a_p T$ exponent (large enough to numerically overflow) is an indication that the dispersive convolution computation may not be required at all (or at least not for that particular pole in materials with multiple poles), because the material can be accurately modeled using constant values of ε and σ. In other words, on the time scale of the FDTD computations the susceptibility function is behaving like a Dirac delta function and the convolution integral reduces to a multiplication of the electric field and a constant amplitude.

To make this point clearer, suppose that for the water example considered in Section 8.2 the time scale of the computation is changed such that both Δx and Δt are multiplied by 1000, with the period T of the incident pulse correspondingly 1000 times longer. Then the $a_p T$ exponent would become 1000 times larger, probably resulting in an overflow in computing $e^{a_p T}$. However, in this calculation the upper frequency limit of the FDTD results has now become 80 MHz rather that 80 GHz, and on this frequency scale the

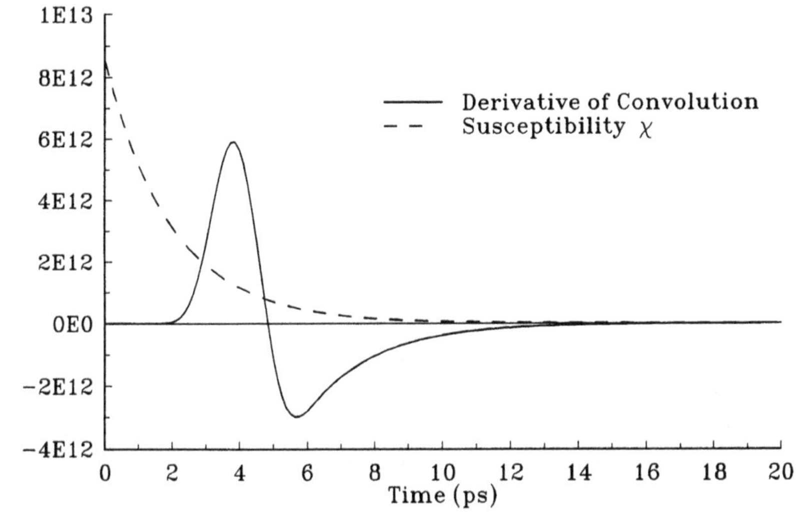

FIGURE 8-30. Derivative of the convolution of the smooth cosine incident pulse of Figure 8-28 with the exponential susceptibility function. Closed form result for use in scattered field form of dispersive convolution FDTD.

dispersive nature of water need not be considered, as the zero frequency values of permittivity and conductivity will yield accurate results without the necessity of evaluating the recursive convolution. The point is that if the factor $e^{a_p T}$ becomes so large as to cause computational difficulties, this indicates that the recursive convolution may not need to be evaluated at all.

Returning to our scattered field implementation, let us illustrate the results of (8.66) to (8.68) by an example calculation. In this example the derivative of the convolution of the smooth cosine incident pulse shown in Figure 8-28 is computed for a susceptibility function with $A_p = 8.5 \times 10^{12}$ and $a_p = 0.50 \times 10^{12}$ as defined in (8.63). The convolution derivative result is plotted in Figure 8-30 along with the susceptibility function, and shows the pulse distortion and memory effect of the convolution process.

As a further illustration of a scattered field computation, the original recursive convolution FDTD calculation for a pulsed plane wave incident on a vacuum-water interface is repeated using the scattered field form. In order to have the smooth cosine pulse approximate in amplitude and duration the Gaussian pulse used for the previous calculation, the peak amplitude is set at 1000 V/m, and the pulse width β parameter is set at 128. After one time step the scattered field is zero because the pulse has not yet begun to interact with the water. After 200 and 600 time steps the scattered electric field vs. position is as shown in Figures 8-31 and 8-32. Comparing to Figures 8-5 and 8-6, we see that in addition to the total field reflected and transmitted pulses there is an

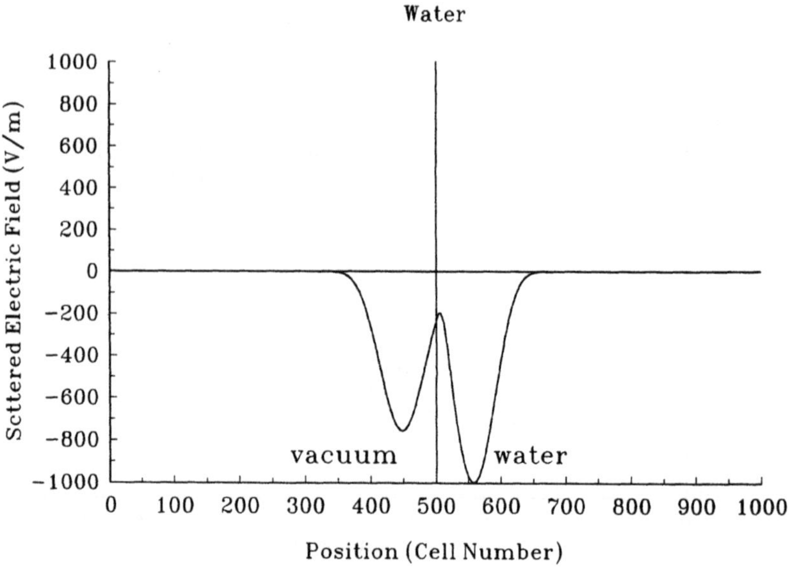

FIGURE 8-31. Scattered electric field vs. position for a smooth cosine pulse plane wave incident on a vacuum-water interface after 200 time steps computed using the scattered field form of recursive convolution FDTD.

additional scattered field transmitted pulse of –1000 V/m amplitude that propagates through the water at the speed of light in vacuum. This pulse exactly (within numerical error) cancels the incident pulse as it should, so that the total field results of the two different approaches will be identical, at least within the numerical precision of the calculations.

The reflection coefficient calculations are simpler than for the total field case, because the incident field at the vacuum-water interface is determined analytically. Also, the reflected field is equal to the scattered field at the interface, so that no subtraction of incident from total field is necessary. The result for the reflection coefficient vs. frequency is identical to that shown in Figure 8-9 and is not repeated.

Adding the additional terms to the FDTD update equation (8.61) clearly increases the computational burden relative to computing in total field. However, (8.67) needs to be evaluated only for the duration of the incident pulse, a relatively small number of time steps relative to the total number in a typical problem. For example, for the pulse used in Figure 8-30 with $\beta = 22$ the period T corresponds to 44 time steps, while a typical FDTD calculation may involve several thousand time steps. Once the incident pulse has passed through any given field point location, (8.68) is evaluated instead, and it only requires calculation of a single exponential function, because the remainder of the equation can be easily reduced to a single numerical constant. The other two additional terms in (8.61) also need to be evaluated only during the duration of the incident pulse. Therefore, the amount of additional computation is not

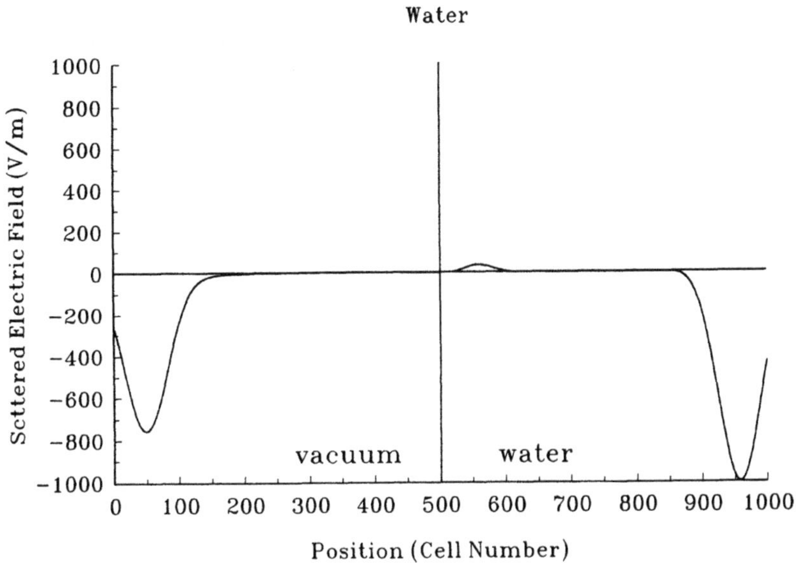

FIGURE 8-32. Scattered electric field vs. position for the smooth cosine pulse plane wave incident on a vacuum-water interface after 600 time steps using the scattered field form of recursive convolution FDTD.

as large as it might appear at first glance, and for many applications, especially scattering problems, the advantages of the scattered field formulation outweigh this slight disadvantage, especially in typical situations where only a small fraction of the FDTD cells are filled with dispersive materials. A 3-D result for scattering from a sphere composed of frequency dependent material obtained using this scattered field implementation is shown at the end of Chapter 13.

REFERENCES

1. **Jackson, J. D.,** *Classical Electrodynamics,* 2nd ed., John Wiley & Sons, New York, 1975.
2. **Landau, L. D. et al.,** *Electrodynamics of Continuous Media,* 2nd ed., Pergamon Press, Elmsford, NY, 1984.
3. **Papoulis, A.,** *The Fourier Integral and it Applications,* McGraw-Hill, New York, 1962.
4. **Luebbers, R., Hunsberger, F., Kunz, K., Standler R., and Schneider, M.,** A frequency-dependent finite-difference time-domain formulation for dispersive materials, *IEEE Trans. Electromag. Compat.,* 32(3), 222, 1990.
5. **Luebbers, R., Hunsberger, F., and Kunz, K.,** A frequency-dependent finite-difference time-domain formulation for transient propagation in plasma, *IEEE Trans. Ant. Prop.,* 39(1), 29, 1991.
6. **Mur, G.,** Absorbing boundary conditions for finite-difference approximations of the time-domain electromagnetic-field equations, *IEEE Trans. Electromag. Compat.,* 23, 1073, 1981.

7. **Balanis, C.,** *Advanced Engineering Electromagnetics,* John Wiley & Sons, New York, 1989.
8. **Kashiwa T. and Fukai, I.,** A treatment of the dispersive characteristics associated with electronic polarization, *Microwave Opt. Technol. Lett.,* 3(6), 203, 1990.
9. **Joseph, R., Hagness, S., Taflove, A.,** Direct time integration of Maxwell's equations in linear dispersive media with absorption for scattering and propagation of femtosecond electromagnetic pulses, *Opt. Lett.,* 16(18), 1412, 1991.

Chapter 9

SURFACE IMPEDANCE

9.1 INTRODUCTION

Surface impedance boundary conditions are employed to eliminate the internal volume of lossy dielectric objects from scattering calculations. In applications of FDTD they can be very important because they also eliminate the need to use small cells, made necessary by shorter wavelengths in conducting media, throughout the solution volume. In this chapter two FDTD implementations of the surface impedance boundary condition are presented. One implementation neglects the frequency dependence of the surface impedance boundary condition, while the other is a dispersive surface impedance boundary condition that is applicable over a very large frequency bandwidth and over a large range of conductivities. Validations are shown in one (1-D) and two dimensions (2-D). Extensions to three dimensions (3-D) should be straightforward. To simplify the derivation we have considered total fields, but the extension to scattered field formulation would involve a straightforward application of the methods of Chapter 2. Because the total fields can be obtained at any time during a scattered field FDTD computation, alternatively for field components involving surface impedance conditions one could use total fields even if the remaining cells in the FDTD space were computing scattered fields.

When FDTD is directly applied to analyze electromagnetic field (EMF) interactions with lossy dielectric objects, both the fields external to the object and those inside the object are calculated. However, for some applications only the exterior fields are of interest. Because the wavelength inside these materials is much smaller than the free space wavelength, accurately computing the internal fields may require a much finer spatial grid within the object than external to it. This greatly increases the computer resources required.

If the material has relatively high loss, the fields do not penetrate very far into the material, and only the fields very close to the surface affect the external response. In this situation the effects of the interior fields on the external fields can be approximated by a surface impedance boundary condition (SIBC). This boundary condition eliminates the need to calculate the fields internal to the object. This reduces the number of cells in the FDTD solution space not only by eliminating cells within the lossy dielectric, but also by allowing larger cells to be used in the exterior region without the need to divide the FDTD space into different regions with different cell sizes and time steps.

Of historical interest, surface impedance boundary conditions were first proposed by Leontovich in the 1940s[1] and were rigorously developed by Senior[2] in a 1960 paper. During the past 30 years, researchers have applied surface impedance concepts in the frequency domain to numerous electromagnetic scattering problems. Time domain surface impedance concepts received little attention until recently. Tesche[3] has investigated surface impedance concepts in an integral equation time domain solution, but presented limited computational time domain results. Based on the approach given by Tesche, Riley and Turner[4] developed an FDTD surface impedance similar to that given here but did not consider recursive evaluation of the required convolutions. This makes their implementation of an FDTD SIBC relatively inefficient. Maloney and Smith[5] have implemented an SIBC in the FDTD method, which does include recursive evaluation of the convolution. However, their implementation has a minor disadvantage relative to that given here because the exponential coefficients for their recursive evaluation method must be determined whenever the conductivity or loss tangent is changed. On the other hand, their implementation does not require that the displacement current is negligible. With the method given here, based on that in Reference 6, the exponential coefficients do not change with changes in permittivity or conductivity. These coefficients are included in this chapter.

In this chapter we use the term "impedance" in relation to the boundary condition, even though when applied in the time domain it is not really an impedance. However, the term "surface impedance" is so widely used that we do apply it in the time domain knowing that confusion is unlikely.

As stated, the motivation for implementing an SIBC in the FDTD method is to reduce the computational resource requirements for modeling highly conducting lossy dielectric objects. In the following discussion we estimate the potential reduction. As FDTD is normally applied to penetrable objects, a small enough cell size must be chosen to resolve the field inside the object at the maximum frequency of interest. Thus, the cell size dimensions are typically

$$\Delta x = \Delta y = \Delta z = \frac{\lambda}{10} = \frac{\lambda_0}{10\sqrt{|\hat{\varepsilon}_r|}} \tag{9.1}$$

where $\hat{\varepsilon}_r$ is the complex relative permittivity of the material and λ and λ_0 are the wavelengths inside the material and in free space, respectively. The complex permittivity for lossy dielectrics is

$$\hat{\varepsilon} = \varepsilon + \frac{\sigma}{j\omega} \tag{9.2}$$

where ω is the radian frequency. The complex relative permittivity is determined using (9.2) as

$$\hat{\varepsilon}_r = \frac{\hat{\varepsilon}}{\varepsilon_0} = \varepsilon_r + \frac{\sigma}{j\omega\varepsilon_0} \tag{9.3}$$

If the material is a good conductor over all frequencies of interest, then the constitutive parameters satisfy the condition

$$\frac{\sigma}{\omega\varepsilon} >> 1 \tag{9.4}$$

Therefore, $\hat{\varepsilon}_r$ can be approximated as

$$\hat{\varepsilon}_r \approx \frac{\sigma}{j\omega\varepsilon_0} \tag{9.5}$$

and we see that under these conditions $\hat{\varepsilon}_r$ is proportional to σ.

If we assume that except for the region containing the lossy dielectric material only free space and perhaps perfect conductor exist, then applying the SIBC would allow us to determine the cell dimensions based on the free space wavelength so that

$$\Delta x_{SIBC} = \Delta y_{SIBC} = \Delta z_{SIBC} = \lambda_0 / 10 \tag{9.6}$$

Thus, by comparing (9.1) and (9.6) we see that the cell dimension would be increased in proportion to $\sqrt{|\hat{\varepsilon}_r|}$ by applying the SIBC. This would allow fewer, larger cells in the FDTD computations, thereby decreasing the memory requirement by a factor of

$$\left(\frac{1}{\sqrt{|\hat{\varepsilon}_r|}}\right)^d$$

where d is the dimension of the FDTD computations (9.1, 9.2, or 9.3).

There would also be considerable savings in computation time. Since the cell dimensions are larger the time step can be made larger. If we assume calculation at the Courant limit then $\Delta t = \frac{\Delta x}{\sqrt{d}c}$ so that the time step size is increased in proportion to $\sqrt{|\hat{\varepsilon}_r|}$. If we further assume that the time length of the response will be the same whether the SIBC is applied, then the total number of time steps will be reduced in proportion to the increase in Δt. Since the total computational time necessary is approximately proportional to the total number of FDTD cells times the total number of time steps (see discussion in Section 3.8), an estimate of the computational savings factor S by which the computation time is reduced is

$$S = \sqrt{d}\left(\frac{1}{\sqrt{|\hat{\varepsilon}_r|}}\right)^{d+1} \tag{9.7}$$

This estimate neglects the increased time which may be necessary to compute the fields in the cells in which the SIBC is applied, but there would be relatively few of these cells, and in compensation all cells interior to the conducting region need not be updated at all.

The assumptions involved in obtaining the estimate in (9.7) may not always be met, but (9.7) gives a rough indication of the computation time savings which may be possible by the application of SIBC.

From this discussion we see that the potential for savings increases with increasing σ. However, as σ becomes larger the response of the material approaches that of a perfect conductor, so that for extremely large values of σ applying the perfect conductor boundary condition may give acceptable results. Thus, the applicability range of SIBC is for conductivities that are large enough so that over the frequency range of interest (9.4) is satisfied, but small enough so that the response of the material differs significantly from that of a perfect conductor.

Having presented the reasons for applying SIBC and an estimate of the potential computational savings, we now proceed to develop the constant and dispersive implementations in FDTD.

9.2 CONSTANT PARAMETER MATERIALS

To implement an SIBC in the FDTD method for materials whose constitutive parameters do not vary with frequency we consider the planar air-lossy dielectric interface as shown in Figure 9-1. The conducting material has permittivity ε, permeability μ, and conductivity σ. We assume that the thickness of the material is large compared to the skin depth. We will also assume that the material is linear and isotropic. Figure 9-1 also shows the 1-D FDTD grid with field locations relative to the material interface which will be used.

The first order (or Leontovich) impedance boundary condition relates tangential total field components and is given in the frequency domain as[1]

$$E_x(\omega) = Z_s(\omega)H_y(\omega) \tag{9.8}$$

where $Z_s(\omega)$ is the surface impedance of the conductor. The frequency domain surface impedance for good conductors is

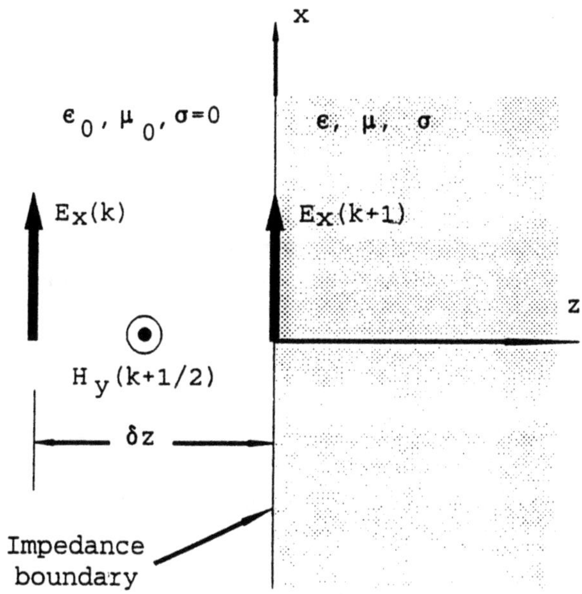

FIGURE 9-1. Geometry for illustrating development of SIBC showing 1-D FDTD grid and planar free space-conductor interface.

$$Z_s(\omega) = (1+j)\sqrt{\frac{\omega\mu}{2\sigma}} = \sqrt{\frac{j\omega\mu}{\sigma}} \tag{9.9}$$

Using (9.9), (9.8) can be rewritten as

$$E_x(\omega) = \left(R_s(\omega) + jX_s(\omega)\right)H_y(\omega) \tag{9.10}$$

where R_s is the surface resistance and X_s is the surface reactance. Consider rewriting (9.10) as

$$E_x(\omega) = \left(R_s(\omega) + j\omega L_s(\omega)\right)H_y(\omega) \tag{9.11}$$

with the resistance and inductance defined by

$$\begin{aligned} R_s(\omega) &= \sqrt{\frac{\omega\mu}{2\sigma}} \\ L_s(\omega) &= \sqrt{\frac{\mu}{2\sigma\omega}} \end{aligned} \tag{9.12}$$

To remove the frequency dependence of the surface resistance and inductance, these quantities are evaluated at a particular frequency and are subsequently treated as constants. Equation (9.11) then becomes

$$E_x(\omega) = \left(R_s + j\omega L_s\right) H_y(\omega) \tag{9.13}$$

This is the required frequency domain constant SIBC. To incorporate this boundary condition into the FDTD algorithm, the time domain equivalent of (9.13) must be obtained. Performing an inverse Fourier transform operation on (9.13) results in

$$E_x(t) = R_s H_y(t) + L_s \frac{\partial}{\partial t}\left[H_y(t)\right] \tag{9.14}$$

This equation defines the time domain constant surface impedance boundary condition.

To implement this constant surface impedance boundary condition in FDTD, space and time are quantized as usual by

$$\begin{aligned} z &\Rightarrow (k\Delta z) \Rightarrow (k) \\ t &\Rightarrow (n\Delta t) \Rightarrow (n) \end{aligned} \tag{9.15}$$

The Faraday-Maxwell law is then used to obtain the H_y component in the free space cell next to the impedance boundary. The impedance boundary condition is formulated on the assumption that the electric and magnetic fields are colocated in space and time. However, on our FDTD grid this is not true. We will neglect the magnetic field spatial offset of one half cell in front of the impedance boundary and show in validation comparisons given later in the chapter that this introduces an error which is proportional to this spatial offset, but that reasonably accurate results nevertheless can be obtained.

Applying the usual finite difference approximation to the Faraday-Maxwell curl equation for the field geometry given in Figure 9-1 yields

$$-(\mu_0 \Delta x \Delta z)\left[\frac{\Delta H_y^n(k+1/2)}{\Delta t}\right] = E_x^n(k+1)\Delta x - E_x^n(k)\Delta x \tag{9.16}$$

Note that $E_x^n(k+1)$ of (9.16) is the electric field component at the impedance boundary. Quantizing space and time in (9.14), neglecting the one half cell spatial offset, and using the result to eliminate $E_x^n(k+1)$ in (9.16) gives

$$\begin{aligned} -(\mu_0 \Delta z)\left[\frac{\Delta H_y^n(k+1/2)}{\Delta t}\right] &= R_s H_y^n(k+1/2) \\ &\quad + L_s\left[\frac{\Delta H_y^n(k+1/2)}{\Delta t}\right] - E_x^n(k) \end{aligned} \tag{9.17}$$

The finite difference time derivatives involving H_y will naturally use the one half time step offsets between E and H fields. However, the $H_y^n(k+1/2)$ term multiplying R_s in (9.17) is time indexed at time step n. In order to apply (9.17) to the usual FDTD fields with time offsets between E and H this term is approximated as

$$H_y^n(k+1/2) \approx \frac{1}{2}\left[H_y^{n+½}(k+1/2) + H_y^{n-½}(k+1/2)\right] \tag{9.18}$$

Using (9.18), and explicitly approximating the finite difference time derivatives on the magnetic fields in (9.17) gives

$$\begin{aligned} &-(\mu_0\Delta z + L_s)\left(H_y^{n+½}(k+1/2) - H_y^{n-½}(k+1/2)\right) \\ &= \frac{R_s\Delta t}{2}\left(H_y^{n+½}(k+1/2) - H_y^{n-½}(k+1/2)\right) - \Delta t E_x^n(k) \end{aligned} \tag{9.19}$$

Solving for $H_y^{n+1/2}(k+1/2)$ in (9.19) yields

$$\begin{aligned} H_y^{n+½}(k+1/2) &= \left[\frac{\mu_0\Delta z + L_s - R_s\Delta t/2}{\mu_0\Delta z + L_s + R_s\Delta t/2}\right] H_y^{n-½}(k+1/2) \\ &- \frac{\Delta t}{\mu_0\Delta z + L_s + R_s\Delta t/2} E_x^n(k) \end{aligned} \tag{9.20}$$

This equation implements the constant surface impedance boundary condition in the FDTD method. It will provide reasonably accurate results for a narrow band of frequencies centered around the frequency for which R_s and L_s are determined. In the next section an FDTD surface impedance implementation valid for a much wider band of frequencies is given.

9.3 FREQUENCY-DEPENDENT MATERIALS

To derive a similar relation to (9.20) for materials whose constitutive parameters vary with frequency which is valid over a wide frequency band, we begin with the same set of underlying assumptions as for the constant surface impedance. The primary exception is that the surface impedance will not be approximated by its value at a particular frequency. The frequency dependence of the surface impedance is inverse Fourier transformed to equivalent time domain form for convolution with the electric field. The SIBC is then implemented in the FDTD method with the required convolution evaluated using the recursive technique presented in Chapter 8.

The first order impedance boundary condition remains unchanged and is given by (9.8). In a similar fashion as Tesche,[3] (9.8) is rewritten as

$$E_x(\omega) = j\omega\left[\frac{Z_s(\omega)}{j\omega}\right]H_y(\omega) \tag{9.21}$$

Defining

$$Z_s'(\omega) = \frac{Z_s(\omega)}{j\omega} \tag{9.22}$$

and substituting (9.9) into (9.22) gives

$$Z_s'(\omega) = \sqrt{\frac{\mu}{j\omega\sigma}} \tag{9.23}$$

Substituting (9.23) into (9.21), a modified SIBC is obtained as

$$E_x(\omega) = Z_s'(\omega)\left[j\omega H_y(\omega)\right] \tag{9.24}$$

The time domain equivalent of (9.24) is obtained via an inverse Fourier transform operation as

$$E_x(t) = Z_s'(t) * \left[\frac{\partial}{\partial t}\left[H_y(t)\right]\right] \tag{9.25}$$

where the "*" denotes convolution,

$$\begin{aligned} E_x(t) &= \mathcal{F}^{-1}\left[E_x(\omega)\right] \\ H_y(t) &= \mathcal{F}^{-1}\left[H_y(\omega)\right] \\ Z_s'(t) &= \mathcal{F}^{-1}\left[Z_s'(\omega)\right] \end{aligned} \tag{9.26}$$

and the $\mathcal{F}^{-1}$ denotes the inverse Fourier transform operation. Note in (9.25) that as $\sigma \rightarrow \infty$, the boundary condition becomes $E_x(t) = 0$ as required for a perfect conductor. To determine $Z_s'(t)$, the Laplace transform variable s=jω is used in (9.23) to obtain

$$Z_s'(s) = \sqrt{\frac{\mu}{\sigma}}\frac{1}{\sqrt{s}} \tag{9.27}$$

Using the Laplace transform pair

$$\frac{1}{\sqrt{\pi t}} = \mathcal{L}^{-1}\left[\frac{1}{\sqrt{s}}\right] \tag{9.28}$$

where L^{-1} the denotes the inverse Laplace transform operation, $Z_s'(t)$ is then determined to be

$$Z_s'(t) = \mathcal{F}^{-1}[Z_s'(\omega)] = \begin{matrix} \sqrt{\dfrac{\mu}{\pi\sigma t}}, & t > 0 \\ 0, & t < 0 \end{matrix} \tag{9.29}$$

This is the required time domain surface impedance function.

In order to implement (9.25) using the results of (9.29) in FDTD, we first consider discrete implementation of the convolution of a time derivative. Explicitly writing (9.25) we have

$$\begin{aligned} E_x(t) &= \int_0^t \sqrt{\frac{\mu}{\pi\sigma\tau}}\left[\frac{\partial}{\partial(t-\tau)}H_y(t-\tau)\right]d\tau \\ &= \sqrt{\frac{\mu}{\pi\sigma}}\int_0^t \sqrt{\frac{1}{\tau}}\left[-\frac{\partial}{\partial\tau}H_y(t-\tau)\right]d\tau \end{aligned} \tag{9.30}$$

The convolution expression in (9.30) must be discretized since H_y is only known at discrete intervals. To aid in this we note that with $\tau = \alpha\Delta t$,

$$\int_{m\Delta t}^{(m+1)\Delta t} \frac{d\tau}{\sqrt{\tau}} = \sqrt{\Delta t}\int_m^{m+1}\frac{d\alpha}{\sqrt{\alpha}}$$

Using this result and assuming H_y is constant over a Δt time interval, (9.30) becomes in finite difference form

$$E_x^n(k+1) = \sqrt{\frac{\mu\Delta t}{\pi\sigma}}\sum_{m=0}^{n-1} Z_0(m)\left[\frac{H_y^{n-m+\frac{1}{2}}(k+1/2) - H_y^{n-m-\frac{1}{2}}(k+1/2)}{\Delta t}\right] \tag{9.31}$$

where

$$Z_0(m) = \int_m^{m+1} \frac{d\alpha}{\sqrt{\alpha}} \tag{9.32}$$

Being able to define the discrete impulse response function Z_0 independently of the size of Δt simplifies the recursive evaluation of the convolution summation as shown in the next section.

We now convert (9.16) to explicit finite differences and substitute (9.31) to eliminate the $E_x^n(k+1)$ electric field component at the surface of the conductor to obtain

$$-\mu_0 \Delta z \left[\frac{H_y^{n+1/2}(k+1/2) - H_y^{n-1/2}(k+1/2)}{\Delta t} \right] =$$

$$\sqrt{\frac{\mu \Delta t}{\pi \sigma}} \sum_{m=0}^{n-1} Z_0(m) \left[\frac{H_y^{n-m+1/2}(k+1/2) - H_y^{n-m-1/2}(k+1/2)}{\Delta t} \right] - E_x^n(k) \tag{9.33}$$

We can now solve for $H_y^{(n+1/2)}$ obtaining

$$\begin{aligned} H_y^{n+1/2}(k+1/2) &- H_y^{n-1/2}(k+1/2) \\ &- \frac{Z_1}{1+Z_1 Z_0(0)} \sum_{m=0}^{n-1} Z_0(m) \left[H_y^{n-m+1/2}(k+1/2) - H_y^{n-m-1/2}(k+1/2) \right] \\ &+ \frac{\Delta t}{(\mu_0 \Delta z)(1+Z_1 Z_0(0))} E_x^n(k) \end{aligned} \tag{9.34}$$

where

$$Z_1 = \frac{1}{\mu_0 \Delta z} \sqrt{\frac{\mu \Delta t}{\pi \sigma}} \tag{9.35}$$

Equation (9.34) is suitable for computer implementation and includes the full convolution with all past magnetic field components. We note that as $\sigma \to \infty$, $Z_1 \to 0$, (9.34) becomes equivalent to the usual FDTD update equation for $H_y(k+1/2)$ with $E_x(k+1) = 0$ at the surface of the perfect conductor.

Direct calculation of this full convolution would be impractical for large 3-D problems, adding considerable computation burden in both storage of past H field values and computation of the summation at each time step. The development of an efficient recursive implementation of (9.34) is the subject of the following section.

9.4 RECURSIVE EVALUATION

In Chapter 8 an approach to evaluate discrete convolutions recursively was presented. The method requires that the time domain susceptibility function be exponentially damped. This recursive evaluation removes the need for the complete time history of field components to be stored and for the complete summation to be evaluated at each time step, resulting in a tremendous computational savings. Because $Z_0(m)$ does not depend on the conductivity and permittivity of the material or on the value of Δt, the exponential coefficients can be used in the application of the SIBC for all materials so long as the constraint of (9.4) for general application of surface impedances is satisfied.

If $Z_0(m)$ can be approximated by a series of exponentials, then the SIBC can be efficiently evaluated using recursion. Figure 9-2 shows $Z_0(m)$ vs. m, and it is of a form such that an accurate approximation by a series of exponentials appears feasible. We thus intend to approximate $Z_0(m)$ as a summation of exponential functions

$$Z_0(m) \approx \sum_{i=1}^{N} a_i e^{m\alpha_i} \tag{9.36}$$

where N is the number of terms in the approximation.

In order to determine the coefficients of the terms in the approximation we will apply Prony's method,[7] as is done in Reference 5. When this is done the resulting coefficients are given in Table 9-1. These coefficients can be used for any values of permeability and conductivity provided that the fundamental constraint of applicability of surface impedance (9.4) is satisfied.

Figure 9-2 also shows the Prony approximation to $Z_0(m)$ with N = 10 using the coefficients in Table 9-1. Other numbers of terms were considered, but N = 10 provides an accurate approximation with a reasonable number of terms. Thus, using (9.37) with N = 10 in equation (9.34) gives

$$\begin{aligned} H_y^{n+½}(k+1/2) &= H_y^{n-½}(k+1/2) \\ &- \frac{Z_1}{1+Z_1 Z_0(0)} \sum_{i=1}^{10} \sum_{m=1}^{n-1} a_i e^{m\alpha_i} \left[H_y^{n-m+½}(k+1/2) - H_y^{n-m-½}(k+1/2) \right] \\ &+ \frac{\Delta t}{(\mu_0 \Delta z)(1+Z_1 Z_0(0))} E_x^n(k) \end{aligned} \tag{9.37}$$

where (9.37) is now a form suitable for recursive evaluation.

Following the same procedure as in Chapter 8, we define the accumulator function Ψ for the ith exponential term as

$$\Psi_i^n = \sum_{m=1}^{n-1} a_i e^{m\alpha_i} \left[H_y^{n-m+½} - H_y^{n-m-½} \right] \tag{9.38}$$

Now consider the first several terms. Because the summation starts at m = 1 and goes only to n – 1 we find that $\psi_i^0 = \psi_i^1 = 0.$ Now for n = 2 we find

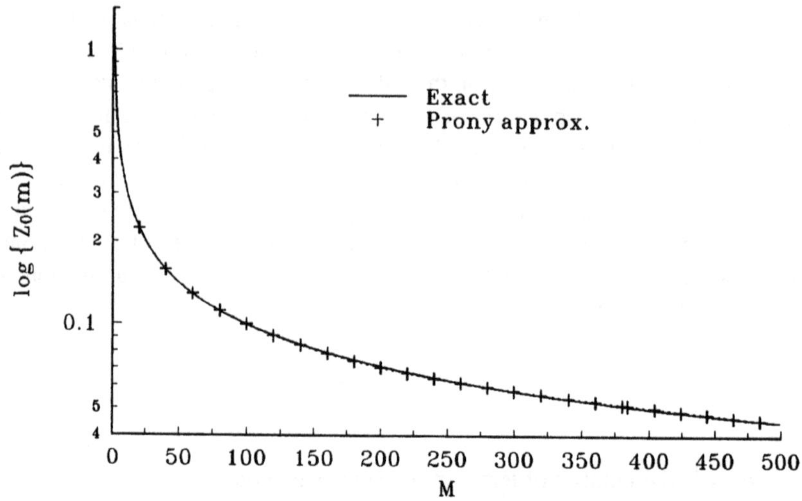

FIGURE 9-2. FDTD dispersive SIBC discrete time domain surface impedance function $Z_0(m)$ vs. m and 10-term Prony approximation.

TABLE 9-1
Prony Coefficients for Ten-Term Exponential Series Approximation to the Discrete Impulse Response Function Z_0 (m)

i	a_i	α_i
1	0.79098180D–01	–0.11484427D–02
2	0.11543423D+00	–0.13818329D–01
3	0.13435380D+00	–0.54037596D–01
4	0.21870422D+00	–0.14216494D+00
5	0.98229667D–01	–0.30128437D+00
6	0.51360484D+00	–0.56142185D+00
7	–0.20962898D+00	–0.97117126D+00
8	0.11974447D+01	–0.16338433D+01
9	0.11225491D–01	–0.28951329D+01
10	–0.74425255D+00	–0.50410969D+01

$$\psi_i^2 = a_i e^{\alpha_i}\left[H_y^{3/2} - H_y^{1/2}\right] \tag{9.39a}$$

and for n = 3

$$\begin{aligned}\psi_i^3 &= a_i e^{\alpha_i}\left[H_y^{5/2} - H_y^{3/2}\right] + a_i e^{2\alpha_i}\left[H_y^{3/2} - H_y^{1/2}\right] \\ &= a_i e^{\alpha_i}\left[H_y^{5/2} - H_y^{3/2}\right] + e^{\alpha_i}\psi_i^2\end{aligned} \tag{9.39b}$$

so that in general

$$\psi_i^n = a_i e^{\alpha_i}\left[H_i^{n-1/2} - H_y^{n-3/2}\right] + e^{\alpha_i}\psi_i^{n-1} \tag{9.40}$$

Substituting (9.38) into (9.37) we obtain

$$\begin{aligned} H_y^{n+1/2}(k+1/2) &= H_y^{n-1/2}(k+1/2) \\ &- \frac{Z_1}{1+Z_1 Z_0(0)} \sum_{i=1}^{10} \psi_i^n(k+1/2) \\ &+ \frac{\Delta t}{(\mu_0 \Delta z)(1+Z_1 Z_0(0))} E_x^n(k) \end{aligned} \tag{9.41}$$

where (9.40) is used to evaluate the Ψ_i accumulators at each time step before (9.41) is evaluated. Note that only one additional past value of magnetic field is required for this. For consistency in the numerical evaluation of the Z_0 (m) function at m = 0 it is approximated as

$$Z_0(0) = \sum_{i=1}^{10} a_i \tag{9.42}$$

9.5 DEMONSTRATIONS

Two demonstrations of the application and accuracy of the constant and dispersive FDTD SIBCs will be given. The first is a 1-D demonstration of calculating plane wave reflection for normal incidence on a conducting half space. The second involves calculating the scattering width from a 2-D conducting cylinder.

In order to make the 1-D calculations, (9.20), (9.40), and (9.41) were implemented in a 1-D total field FDTD code for the geometry shown in Figure 9-1. The problem space size is 301 cells, the impedance boundary is located at cell 300, and the electric field is sampled at cell 299. The maximum frequency of interest for each problem was 10 GHz. The incident electric field is a Gaussian pulse with maximum amplitude of 1000 V/m and has a total temporal width of 256 time steps. The incident pulse contains significant energy to 12 GHz. Two computations were made, with $\sigma = 2.0$ S/m and $\sigma = 20.0$ S/m. The loss tangents at 10 GHz are 3.599 and 35.99, respectively. The permittivity and permeability for the lossy dielectric were those of free space. The cell size and time step were 750 μm (40 cells per free space wavelength at 10 GHz) and 2.5 ps, respectively. The surface impedance for the FDTD constant SIBC was evaluated at 5 GHz, so that at this frequency we would expect the two methods to agree. For each FDTD computation, a reflection coefficient vs. frequency was obtained by first dividing the Fourier transform of the reflected field by the transform of the incident field at cell 299. The incident field was obtained by running the FDTD code with free space only and recording the electric field at cell 299. The reflected field is then obtained by subtracting the time domain incident field from time domain total field.

The results are compared with the analytic surface impedance reflection coefficient computed from

$$|R| = \frac{|Z_s(\omega) - \eta_0|}{|Z_s(\omega) + \eta_0|} \tag{9.43}$$

where $Z_s(\omega)$ is given by (9.8) and η_0 is the free space wave impedance. The phase of the FDTD reflection coefficient was corrected to account for the round trip phase shift of one cell since the FDTD reflection coefficient is computed from electric fields recorded one cell in front of the impedance boundary.

The high conductivity surface impedance of (9.8) is an approximation to the general surface impedance for lossy dielectrics given by

$$Z_s(\omega) = \sqrt{\frac{j\omega\mu}{\sigma + j\omega\varepsilon}} \tag{9.44}$$

with no restriction on the conductivity in (9.44). However, for a surface impedance to be applied in general it must be much greater than the impedance of free space (so that the wave vector inside the material is approximately normal to the surface independent of the direction of the wave vector outside the material), and further, the material must be lossy enough so that all energy entering it remains inside. Thus, the restriction in (9.4) is usually satisfied in practical applications of surface impedance. The advantage of using (9.8) over (9.44) for the FDTD SIBC implementation is that the resulting time domain impulse response is independent of the conductivity. The exponential approximation needs to be performed only once the conductivity is not changed each time; the Prony coefficients of Table 9-1 can be used in all applications of this SIBC.

Figures 9-3 and 9-4 show the FDTD constant and recursive SIBC reflection coefficient magnitude and phase results vs. the analytic SIBC results for σ = 2.0 S/m. Notice the agreement between the dispersive and exact solution is excellent, and the maximum error is about 0.01 at 10 GHz in Figure 9-3. The constant SIBC agrees at 5 GHz as expected.

Figures 9-5 and 9-6 show the FDTD constant and recursive SIBC reflection coefficient magnitude and phase results vs. the analytic SIBC results for σ = 20.0 S/m. The dispersive SIBC shows excellent agreement with the exact solution.

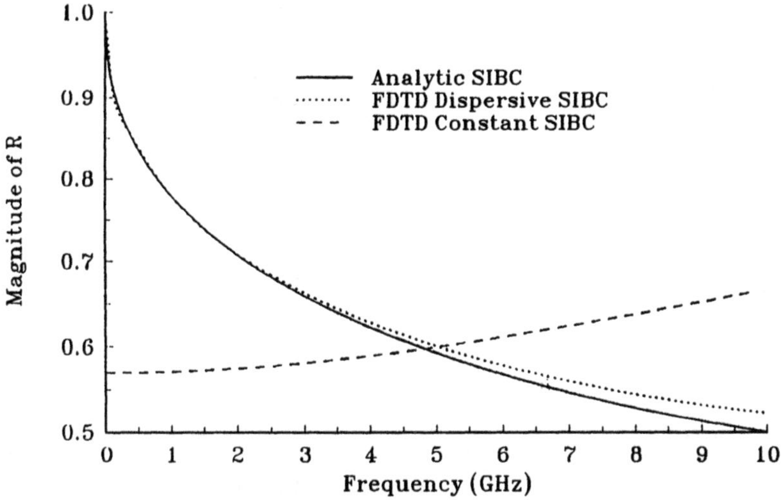

FIGURE 9-3. Reflection coefficient magnitude vs. frequency for normal incidence plane wave calculated for σ = 2.0 S/m using FDTD constant and dispersive SIBC compared to exact solution.

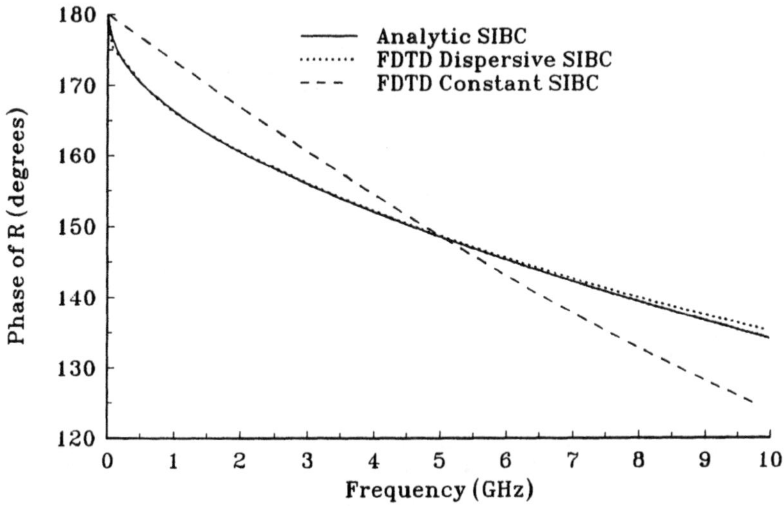

FIGURE 9-4. Reflection coefficient phase vs. frequency for normal incidence plane wave calculated for σ = 2.0 S/m using FDTD constant and dispersive SIBC compared to exact solution.

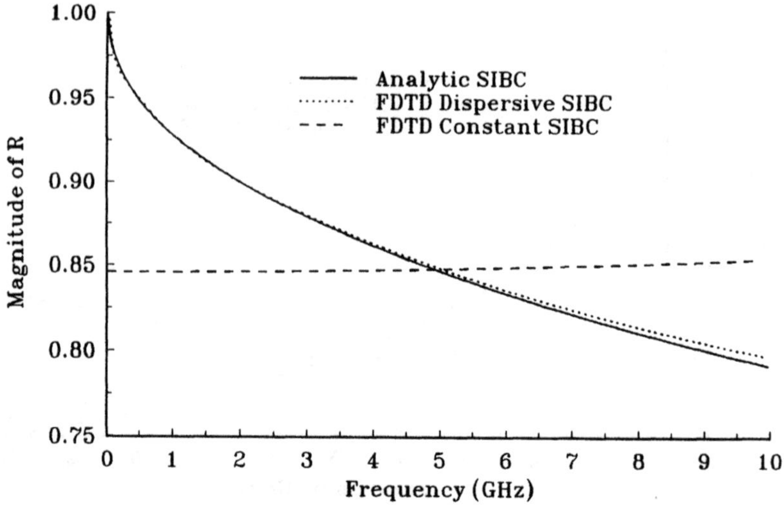

FIGURE 9-5. Reflection coefficient magnitude vs. frequency for normal incidence plane wave calculated for $\sigma = 20.0$ S/m using FDTD constant and dispersive SIBC compared to exact solution.

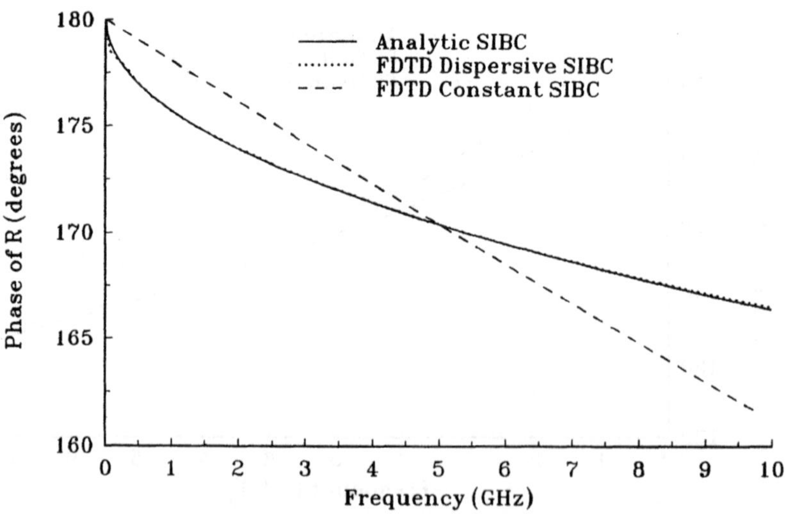

FIGURE 9-6. Reflection coefficient phase vs. frequency for normal incidence plane wave calculated for $\sigma = 20.0$ S/m using FDTD constant and dispersive SIBC compared to exact solution.

Note that as the conductivity increases the accuracy of the SIBC improves. For these conductivities of 2 and 20 S/m, the demonstration cell size corresponds to 26 and 9 cells per wavelength inside the conducting material at 10 GHz, respectively. Thus, the cell size is not reduced from that which would be required if the entire space were gridded with cells small enough to resolve the fields inside the conductor. These conductivities were chosen to illustrate the accuracy which can be obtained at the lower limit of the high conductivity approximation necessary for application of the SIBC. For higher conductivities with a fixed cell size the accuracy improves as the cell size becomes a larger fraction of the wavelength inside the conductor.

Because cell size is a factor in applying the SIBC, let us consider this further. The FDTD SIBC implementation is an approximation, therefore, some amount of divergence between the SIBC curves and the analytic solution is to be expected. To observe this, the same 1-D test problems as above (using the dispersive SIBC only) were reevaluated with larger cell sizes equal to twice and four times the original cell size, and with the time step size proportionately reduced. This is equivalent to having 20 and 10 cells per λ_0 in the free space region, respectively, and a corresponding doubling of the time step size with each cell size change. Figures 9-7 and 9-8 show the FDTD dispersive SIBC reflection coefficient magnitude and phase results vs. the exact solution for $\sigma = 2.0$ S/m using the original cell size and the larger cell sizes. Notice that for each doubling in cell size and time step, the error between the SIBC result and the exact solution approximately doubled. In other tests of the SIBC the cell size was kept constant while the time step size Δt was reduced. In these tests the error did not change appreciably with reduction in Δt, indicating that the error in the SIBC implementation is proportional to Δz over the range of cell sizes examined here. This is probably due to the assumption that E and H fields at the surface are colocated, while in fact they are spatially separated by one half cell. The constant SIBC exhibited similar behavior at the 5 GHz tie point for larger cell sizes, confirming that this error is due to the spatial offset between E and H at the surface rather than to errors in the dispersive SIBC recursive convolution implementation.

While the 1-D demonstration illustrates the improvement in wide bandwidth accuracy than can be obtained from the dispersive SIBC relative to the constant implementation, it is limited to normal incidence. In the next demonstration oblique incidence is considered in a fully 2-D calculation of the scattering width vs. frequency for an infinitely long square cylinder for two incidence angles, $\phi = 0.0°$ and $\phi = 30.0°$. These calculations were made using a 2-D TM scattered field FDTD code as described in Chapter 2. In order to apply the SIBC the total field was determined at each time step at each SIBC field location, the SIBC applied, and the result reconverted to scattered field. Because only a relatively small number of cells are actually at the surface of the cylinder this approach was not really that inefficient. The cylinder was 0.099 m^2 and had parameters $\varepsilon = \varepsilon_0$, $\mu = \mu_0$, and $\sigma = 20.0$ S/m.

To illustrate the advantages of the SIBC, the cylinder was modeled in two ways. The first was a normal FDTD computation with a fine grid size of 14

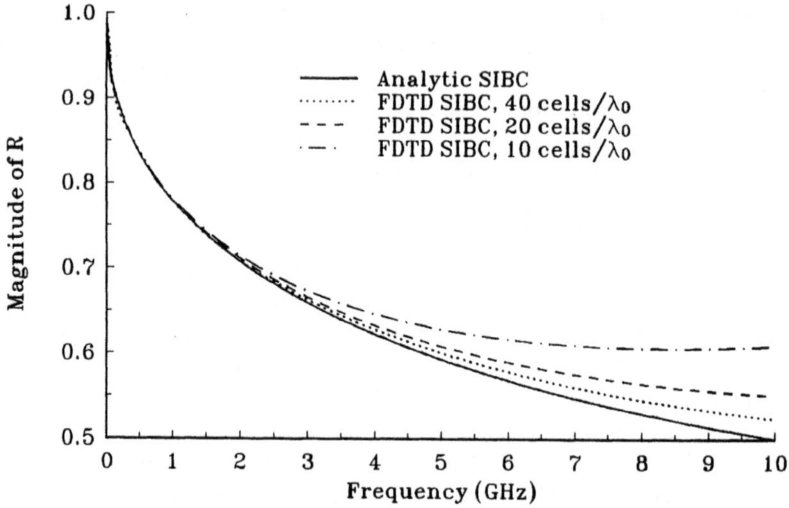

FIGURE 9-7. Reflection coefficient magnitude vs. frequency for normal incidence plane wave calculated for $\sigma = 2.0$ S/m using FDTD dispersive SIBC with original and larger cell size compared to exact solution.

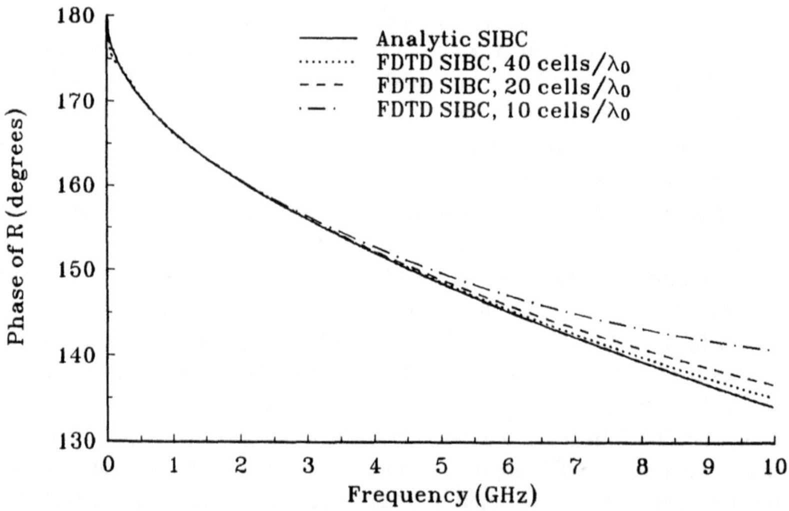

FIGURE 9-8. Reflection coefficient phase vs. frequency for normal incidence plane wave calculated for $\sigma = 2.0$ S/m using FDTD dispersive SIBC with original and larger cell size compared to exact solution.

cells per λ (at 10 GHz) inside the conducting cylinder, and the second was a SIBC computation with a grid size of 10 cells per λ in free space (at 10 GHz). Figure 9-9 shows the 2-D field components and the cylinder dimensions (in cells) for the FDTD and SIBC computations. For the full FDTD computation, the cylinder was modeled using 198 cells in the x and y directions, the cell size was 500 μm, and the time step was 1.67 ps. For the SIBC computation, the cylinder was modeled using 32 cells in the x and y directions, the cell size was 0.003 m, and the time step was 10 ps. The SIBC cell size corresponds to only about two cells per wavelength in the conducting material at 10 GHz. For both computations, there is a 100-cell border between the cylinder and the secnd order Mur absorbing boundary. Thus, the total number of cells in the SIBC calculation is only 34% of that for the full FDTD calculation. The total number of time steps for each calculation was 1024, and an incident Gaussian pulse with total pulse width of 64 time steps was chosen. The near zone scattered fields were transformed to far zone fields by the 2-D near zone to far zone transformation described in Chapter 7.

Figures 9-10 and 9-11 compare the scattering width magnitude vs. frequency results of the full FDTD computation and the SIBC computation for $\phi = 0°$ and $\phi = 30°$. The agreement between the two methods is quite reasonable over the entire frequency bandwidth, especially considering the much larger cells and time step size used in the SIBC calculations. The computation time using the SIBC was approximately 30% of the full FDTD calculation, which is very nearly equal to the reduction in the total number of cells.

To summarize, in this chapter we have presented both constant and dispersive surface impedance boundary conditions valid for good conductors. The corresponding time domain impedance boundary conditions have been derived and their validity demonstrated by 1-D computation of the reflection coefficient at an air-conductor interface over a wide frequency bandwidth. While the constant SIBC is simpler, the accuracy advantage of the dispersive SIBC implementation for wide bandwidth calculations was demonstrated. The applicability of the SIBC to 2-D scattering problems was demonstrated by scattering width computations for an infinite square cylinder. For both the 1- and 2-D cases, the dispersive FDTD results were shown to be in good agreement with exact results over the entire bandwidth. Considerable computational savings were illustrated in the wide bandwidth 2-D example. This is due in part to a recursive updating scheme which permits efficient application of a dispersive surface impedance boundary condition to practical scattering problems.

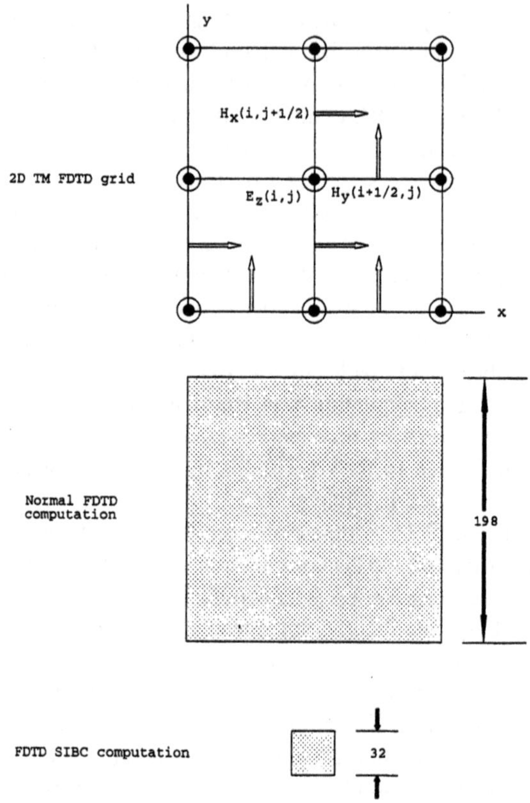

FIGURE 9-9. Two-dimensional geometry for scattering width computations from an infinitely long square cylinder with $\sigma = 20.0$ S/m using normal FDTD and FDTD dispersive SIBC.

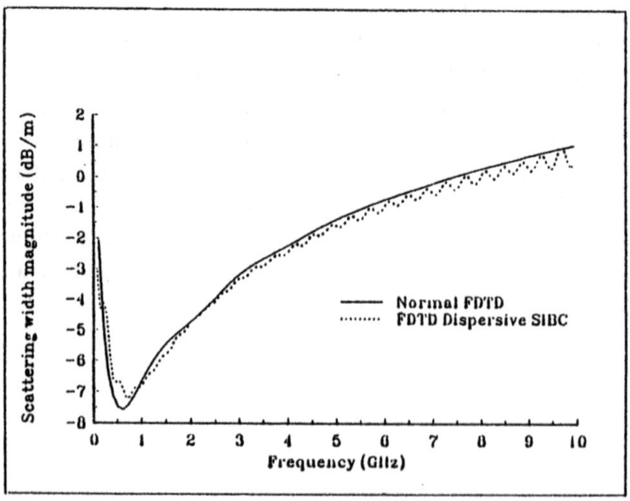

FIGURE 9-10. Scattering width magnitude vs. frequency at incidence angle $\phi = 0.0°$ from an infinitely long square cylinder 0.099 m on a side with $\sigma = 20.0$ S/m using normal FDTD and FDTD dispersive SIBC.

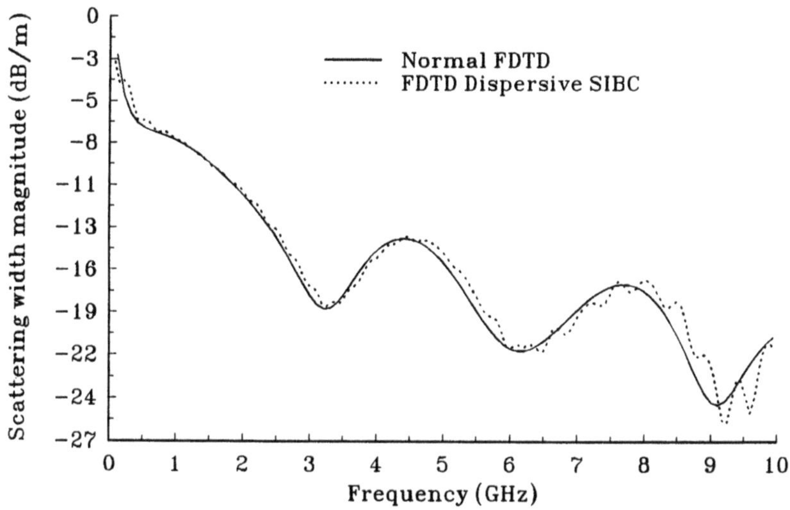

FIGURE 9-11. Scattering width magnitude vs. frequency at incidence angle $\phi = 30.0°$ from an infinitely long square cylinder 0.099 m on a side with $\sigma = 20.0$ S/m using normal FDTD and FDTD dispersive SIBC.

REFERENCES

1. **Leontovich, M. A.,** On the approximate boundary conditions for electromagnetic fields on the surface of well conducting bodies, in *Investigations of Propagation of Radio Waves,* B. A. Vedensky, Ed., Academy of Sciences USSR, Moscow, 1948.
2. **Senior, T. B. A.,** Impedance boundary conditions for imperfectly conducting surfaces, *Appl. Sci. Res. B,* 8, 418, 1960.
3. **Tesche, F. M.,** On the inclusion of loss in time-domain solutions of electromagnetic interaction problems, *IEEE Trans. Electromagn. Compat.,* 32, 1, 1990.
4. **Riley, D. R. and Turner, C. D.,** The inclusion of wall loss in finite-difference time-domain thin-slot algorithms, *IEEE Trans. Electromag. Compat.,* 33(4), 304, 1991.
5. **Maloney, J. G. and Smith, G. S.,** The use of surface impedance concepts in the finite-difference time-domain method, *IEEE Trans. Ant. Prop.,* 40(1), 38, 1992.
6. **Beggs, J. H., Luebbers, R. J., Yee, K. S., and Kunz, K. S.,** Finite difference time domain implementation of surface impedance boundary conditions, *IEEE Trans. Ant. Prop.,* 40(1), 49,1992.
7. **Hildebrand, F. B.,** *Introduction to Numerical Analysis,* McGraw-Hill, New York, 1956, 378.

Chapter 10

SUBCELLULAR EXTENSIONS

10.1 INTRODUCTION

The computational cost of the FDTD technique scales directly with the number of cells. Often the object being analyzed has important structural features, thin wires, or narrow slots (for example) that are very small in at least one dimension as compared to the main body of the scattering or coupling object. Large cells that allow the main body to be accurately rendered are inadequate when it comes to the "small" structure features. Reducing the cell size throughout the FDTD computational space is one method for dealing with this situation, but it is computationally expensive, and a method that may not even be practical if computer resources are inadequate.

Two ways around this problem are discussed here. The first approach uses large FDTD cells throughout the computation space, but approximates the small geometry elements by modifying the equations for the large cells that contain them. For example, a surface impedance like that considered in the previous chapter, may be used to include material layers thinner than the FDTD cells. Another variation involves special equations for calculating the fields in the vicinity of wires thinner than the FDTD cell size. Effects of lumped circuit elements which are contained within one FDTD cell may also be included by modifying the field equations for that cell. Development of this approach often involves application of Maxwell's equations in integral rather than differential form, but the finite difference equations can be obtained easily from the integral form of Maxwell's equations, and they are more easily modified to include geometry variations within an FDTD cell.

The second approach is a replacement of the region about the structured feature of interest with a finer grid, what has been referred to as the expansion technique. This approach is effective in that a finer grid is only employed where needed. The effort lies in combining in a physically reasonable way the two geometries: the first or coarse model with the second or finer and spatially smaller model. The discussion in Chapter 6 of human body modeling utilizes this approach. In this chapter the expansion technique is presented from the standpoint of modeling interior coupling through a relatively small aperture into an aircraft.

10.2 INTEGRATION CONTOURS

In modeling structures that are small as compared to the FDTD cell size (thin wires, for example), a useful approach is to apply the integral form of Maxwell's equations in the derivation of the FDTD equations rather than the differential form. The integrations are carried out so as to include the sub-cell

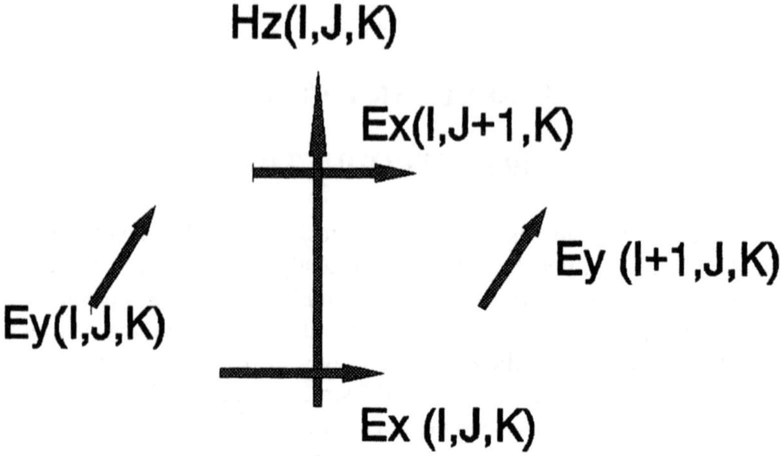

FIGURE 10-1. Yee cell field locations relative to Hz(I,J,K).

geometry and field effects, and the corresponding FDTD update equation is then obtained. In this section the fundamental approach is given, and in the next section it is applied to a thin wire. The concept of applying integration contours in FDTD was developed by Taflove and Umashankar in References 1 and 2.

To illustrate the approach consider an Hz component of magnetic field and the encircling electric field components that are shown in Figure 10-1. The usual Yee notation is used, and the field locations are separated by Δx, Δy, and Δz as usual. We now apply the Maxwell integral equation

$$\oint_C \overline{E} \cdot \overline{dl} = -\mu \frac{\partial}{\partial t} \int\int_S \overline{H} \cdot \overline{ds} \tag{10.1}$$

to the fields in Figure 10-1, the integral contour being a square with sides passing through the four electric field components. With the assumption that the fields are uniform along each side of the contour, we obtain the result

$$\begin{aligned} &\left[E_x(I,J,K) - E_x(I,J,+1,K)\right]\Delta x + \left[E_y(I,+1,J,K) - E_y(I,J,K)\right]\Delta y \\ &= \mu \frac{\partial}{\partial t} H_z(I,J,K)\Delta x \Delta y \end{aligned} \tag{10.2}$$

If we now divide through by $\Delta x \Delta y$ and rearrange terms slightly, the result is the usual FDTD update equation

$$\begin{aligned} &-\mu \frac{\partial H_z(I,J,K)}{\partial t} \\ &= \frac{E_x(I,J,K) - E_x(I,J+1,K)}{\Delta y} + \frac{E_y(I+1,J,K) - E_y(I,J,K)}{\Delta x} \end{aligned} \tag{10.3}$$

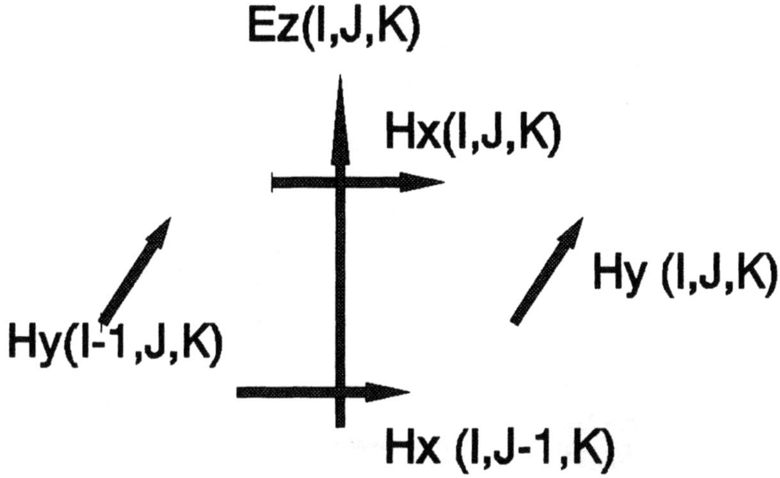

FIGURE 10-2. Yee cell field locations relative to Ez (I, J, K).

where we recognize the two terms on the right-hand side as the finite difference form of the curl operation, which is where our derivation began in Chapter 2. If we apply the other Maxwell integral equation

$$\oint_C \overline{H} \cdot \overline{dl} = -\varepsilon \frac{\partial}{\partial t} \iint \overline{E} \cdot \overline{ds} + \iint \sigma \overline{E} \cdot \overline{ds} \tag{10.4}$$

to the fields in Figure 10-2 in a similar fashion we obtain

$$\begin{aligned}\left[H_x(I,J-1,K) - H_x(I,J,K)\right]\Delta x + \left[H_y(I,J,K) - H_y(I-1,J,K)\right]\Delta y \\ = \varepsilon \frac{\partial}{\partial t} E_z(I,J,K)\Delta x \Delta y + \sigma E_z\,(I,J,K)\Delta x \Delta y\end{aligned} \tag{10.5}$$

which, as expected, reduces to the usual FDTD equation

$$\begin{aligned}\varepsilon \frac{\partial E_z(I,J,K)}{\partial t} + \sigma E_z(I,J,K) \\ = \frac{H_x(I,J-1,K) - H_x(I,J,K)}{\Delta y} + \frac{H_y(I,J,K) - H_y(I-1,J,K)}{\Delta x}\end{aligned} \tag{10.6}$$

The point of this exercise is to show that we can obtain the FDTD difference equations by using either the differential or the integral forms of the Maxwell equations applied to the field positions in the Yee cell.

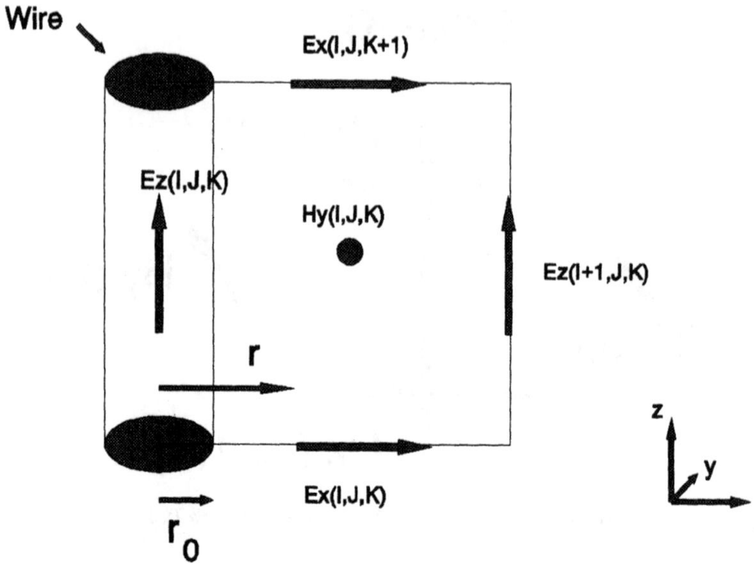

FIGURE 10-3. Field locations and geometry for thin wire.

In the next section of this chapter this is exploited to modify the FDTD update equations so as to include a subcellular thin wire.

10.3 THIN WIRES

In antenna and coupling applications a common geometry to be modeled is a thin wire. Often these wires are much smaller in radius than other geometry features, and it is desirable to avoid sizing the FDTD cells small enough to accurately model the thin wire. On the other hand, approximating the wire as being the same size as a much larger FDTD cell may yield poor results, as both antenna impedance and coupling are sensitive to the wire radius.

In this section we develop a simple approach to include, at least approximately, the effects of a wire with radius smaller than the FDTD cell dimensions on the FDTD update equations for the adjacent magnetic fields. The approach here follows that described by Umashankar and Taflove.[1] An alternative approach has been described by Holland and Simpson.[3]

The geometry is shown in Figure 10-3. A conducting circular wire of radius r_0 is positioned to align with and center on the Ez(I,J,K) field component. The wire is assumed to have a radius smaller than 0.5 Δx, and since Δx must be considerably smaller than the wavelength for FDTD to be applied, therefore, the wire radius also must be much smaller than a wavelength. This justifies the assumption that the total normal electric and circumferential magnetic fields in the vicinity of the wire have a $1/r$ dependence, where r is the radial distance from the center of the wire. In the following total fields are assumed, but the approach can be extended to scattered fields if desired.

With the above assumptions, we approximate the spatial dependence of the fields in the vicinity of the wire as

$$H_y(r,J,K) \approx H_y(I,J,K) \cdot \frac{\Delta x}{2r} \tag{10.7}$$

within the contour,

$$E_x(r,J,K) \approx E_x(I,J,K) \cdot \frac{\Delta x}{2r} \tag{10.8}$$

along the upper and lower integration contours, with Ez(I,J,K) = 0 all along the wire axis, and Ez(I+1,J,K) assumed to be uniform along the right contour.

If we now apply the Maxwell Faraday's law equation (10.1) to the contour passing through the four electric field locations we obtain

$$\begin{aligned} &0 + \int_{r_0}^{\Delta x} E_x(I,J,K+1)\frac{\Delta x}{2}\frac{dr}{r} \\ &-E_z(I+1,J,K)\Delta z - \int_{r_0}^{\Delta x} E_x(I,J,K)\frac{\Delta x}{2}\frac{dr}{r} \\ &= -\mu\Delta z\frac{\partial}{\partial t}\int_{r_0}^{\Delta x} H_y(I,J,K)\frac{\Delta x}{2}\frac{dr}{r} \end{aligned} \tag{10.9}$$

which after evaluating the integrals and approximating the time derivatives as finite differences reduces to

$$\begin{aligned} &H_y^{n+1/2}(I,J,K) = H_y^{n-1/2}(I,J,K) \\ &+\frac{\Delta t}{\mu\Delta z}\left[E_x^n(I,J,K) - E_x^n(I,J,K+1)\right] \\ &+\frac{2\Delta t}{\mu\Delta x \ln\left(\dfrac{\Delta x}{r_0}\right)} E_z^n(I+1,J,K) \end{aligned} \tag{10.10}$$

Referring to Figure 10-3, for each Ez(I,J,K) component at the center of a thin wire, there are four associated H field components that must be computed at each time step using a form of (10.10). In addition to $H_y^{n+1/2}(I,J,K)$ of (10.10), the $H_y^{n+1/2}(I-1,J,K)$, $H_x^{n+1/2}(I,J,K)$, and $H_x^{n+1/2}(I,J-1,K)$ components must also be computed using the appropriate variation of (10.10). The electric field values are updated using the usual FDTD equations.

It is simple to implement this scheme after some consideration. Assigning a special ID array code to each of the affected H field components could be done, but this approach is cumbersome because flagging a particular H field location and vector direction, for example, identifying $H_y^{n+1/2}(I,J,K)$ as one of the affected H field values, does not provide the information necessary to evaluate (10.10). This is because this H field component may be adjacent to a wire centered on any of four electric field components, as can be seen from Figure 10-3. Thus, the ID array information would not only have to flag the particular H field components, but also include directional information as to which of the four orthogonal adjacent electric field components is centered inside a wire.

A simpler approach is to prescribe a special ID array value to the electric field locations at the center of a wire. These electric field locations are updated using the usual FDTD equations. After the H fields are updated throughout the FDTD space, all the FDTD cells are again checked to see which contain a wire, i.e., the E field locations at wire centers are found. For each of these the four orthogonal and adjacent H fields are updated using variations of (10.10) and the results written over the values obtained using the free space FDTD equations. This requires one extra storage location for each affected H field value to store the value of H at the previous time step, but since for most applications this involves only a very small portion of the total number of cells this does not create a serious impact on computer resources.

This approach is demonstrated in obtaining results for current flowing in a wire antenna with a nonlinear load and impedance of a thin wire monopole connected to a conducting box; results are given in Chapters 11 and 14, respectively.

10.4 LUMPED CIRCUIT ELEMENTS

Some applications of FDTD involve structures that contain lumped circuit elements small enough to be contained in one FDTD cell. An example is wire antennas, which may contain lumped circuit elements and sources in wire gaps. Consider such an antenna operating at 3 GHz. If the FDTD cells are taken as 1/10 wavelength cubes, these are 1 cm on a side, easily large enough to contain lumped circuit elements. The most commonly used computational method applied to these antennas, the thin-wire method of moments, makes provision for lumped circuit elements to be modeled in small gaps in the wire. In this section we show that lumped resistors and capacitors in one FDTD cell can be included very easily, with lumped inductors requiring somewhat more effort. Results obtained for wire antennas with lumped elements in one FDTD cell are given in Chapter 11 (for a circuit including a resistor, capacitor, and nonlinear diodes) and Chapter 14.

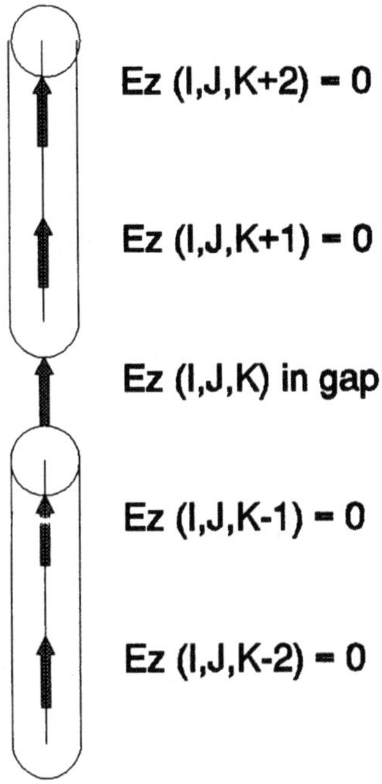

FIGURE 10-4. A section of a thin wire with a gap of length Δz modeled using FDTD field components.

A simple wire antenna geometry is shown in Figure 10-4. A line of electric field components located along the wire axis, in this case Ez components, are set to zero. If the wire radius is smaller than the FDTD cell cross-section the method described in the previous section may be used to modify the computation of the adjacent magnetic fields. However, the magnetic fields adjacent to the $E_z(I,J,K)$ field component used to model the gap in the wire will be calculated using the normal free space FDTD equations.

The simplest situation is a voltage source in the gap. In this case

$$E_z^n(I,J,K) = V_s(t)/\Delta z \tag{10.11}$$

where $V_s(t)$ is the specified source voltage. The current supplied by the source can be sampled by evaluating the line integral of the H fields encircling $E_z(I,J,K)$ in the gap as was done in (10.4) and (10.5) previously. The Fourier transforms of this voltage and current can be used to determine the antenna impedance and input power, as illustrated in Chapter 14.

It is also quite simple to model a lumped load consisting of a resistor and capacitor in parallel. To illustrate this we express (10.6) in a slightly different way as

$$\begin{aligned}
&\varepsilon\frac{\partial E_z(I,J,K)}{\partial t}+\sigma E_z(I,J,K)\\
&=\frac{H_x(I,J-1,K)-H_x(I,J,K)}{\Delta y}\\
&+\frac{H_y(I,J,K)-H_y(I-1,J,K)}{\Delta x}\\
&=(\nabla xH)_z
\end{aligned} \tag{10.12}$$

This simplifies the following equations by using $(\nabla xH)_z$ to represent the finite difference approximation to the curl operation. If the time derivative in (10.12) is approximated as a finite difference we obtain

$$\left(\nabla xH^{n+1/2}\right)_z=\varepsilon\frac{E_z^{n+1}-E_z^{n}}{\Delta t}+\sigma E_z^{n+1} \tag{10.13}$$

with the (I,J,K) location understood. As we have seen many times (10.13) is readily solved for E_z^{n+1} in terms of the previous time values of E^n and $H^{n+1/2}$, with the curl H term computed using spatial finite differences as in (10.12).

Rather than repeat this process let us consider the physical meanings of the terms. The curl H term gives the density of the total current (total in the sense of including both conduction and displacement current) flowing in the region surrounding the electric field component. The next term involving ε and the time derivative of E is the displacement current density flowing in this region in the z direction, and the σE^{n+1} term is the conduction current density in the z direction.

If we multiply (10.13) by $\Delta x\ \Delta y$ and insert some appropriate Δz factors we easily obtain the result

$$\Delta x\Delta y\left(\nabla xH^{n+1/2}\right)_z=\varepsilon\frac{\Delta x\Delta y}{\Delta z}\frac{\Delta z(E_z^{n+1}-E_z^{n})}{\Delta t}+\sigma\frac{\Delta x\Delta y}{\Delta z}\Delta zE_z^{n+1} \tag{10.14}$$

The first term is the total current flowing in the $\Delta x\ \Delta y$ area surrounding the Ez field component. The term $\varepsilon(\Delta x\Delta y)/(\Delta z)$ is readily identified as the capacitance of an FDTD cell for a z component of electric field, and $\sigma(\Delta x\Delta y)/(\Delta z)$ is the conductance. Furthermore, the product of Δz and Ez is the voltage in the z direction over the length of one FDTD cell. We can rewrite the above equation in terms of lumped elements as

$$\Delta x \Delta y \left(\nabla x H^{n+1/2} \right)_z = C \, \Delta z \, \frac{E_z^{n+1} - E_z^n}{\Delta t} + G \, \Delta z \, E_z^{n+1} \tag{10.15}$$

where now the first term is the total current flowing through the Δx by Δy area of the FDTD cell, C is the lumped "parallel plate" capacitance of the cell, and G is the lumped conductance of the cell in parallel with the capacitance, with $\Delta z \, E^{n+1}$ the voltage across the cell. In comparing (10.13) to (10.15), a lumped capacitance C is equivalent to setting an appropriate value of the permittivity ε of the cell based on the cell dimensions, and similarly for a lumped resistance, $R = 1/G$ and the conductivity σ of the cell. Thus, a lumped load that is a parallel combination of a capacitor and a conductance (resistance) can be modeled simply by setting the cell values of ε and σ appropriately. Equations (10.13) and (10.15) are interchangeable in terms of solving for E^{n+1}. However, one warning is that the cell permittivity cannot be set too low. FDTD cannot in general model materials with an epsilon that is too small (much less than free space) without becoming unstable unless extremely small time steps are used, much smaller than required by the Courant limit. If the conductance G (conductivity σ) is great enough so that the displacement current term involving C can be neglected, this term may be dropped. Indeed, if (10.13) or (10.15) is solved for E^{n+1} and G (or equivalently σ) is allowed to go to infinity, the correct result of $E^{n+1} = 0$ is obtained. However, making G (or σ) small enough so that the conduction current is of comparable magnitude with the displacement current (through the cell filled with free space), but nevertheless neglecting this displacement current term, will result in instabilities. A physical argument for this is that the capacitance of an FDTD cell cannot be made lower than the capacitance of the cell filled with free space by adding lumped elements in parallel with the cell capacitance.

While including a parallel combination of a capacitor and a resistor is simply equivalent to setting the permittivity and conductivity of the cell appropriately, an inductor requires modifying the stepping equations. A straightforward approach follows. Considering (10.15) again, if we add a lumped inductor in parallel with the capacitor and resistor (conductance) the resulting equation is

$$\begin{aligned} \Delta x \Delta y \left(\nabla x H^{n+1/2} \right)_z &= C \, \Delta z \, \frac{E_z^{n+1} - E_z^n}{\Delta t} \\ &+ G \, \Delta z \, E_z^{n+1} + \frac{\Delta z}{L} \int_0^{(n+1)\Delta t} E_z dt \end{aligned} \tag{10.16}$$

where L is the value of the lumped inductance in parallel with the C and G components. If we assume that Ez is constant over a time interval Δt, then the integration becomes a summation

$$\Delta x \Delta y \left(\nabla x H^{n+1/2} \right)_z = C\,\Delta z\, \frac{E_z^{n+1} - E_z^n}{\Delta t} + G\,\Delta z\, E_z^{n+1} + \frac{\Delta z \Delta t}{L} \left[E_z^{n+1} + \sum_{m=1}^{n} E_z^m \right] \tag{10.17}$$

which can be solved for the E_z^{n+1} in terms of the previous values of Ez and H. The result is

$$\begin{aligned} E_z^{n+1} &= \frac{C}{G\,\Delta t + C + \frac{(\Delta t)^2}{L}} E^n \\ &+ \frac{\Delta x \Delta y \Delta t}{\Delta z \left(G\,\Delta t + C + \frac{(\Delta t)^2}{L} \right)} \left(\nabla x H^{n+1/2} \right)_z \\ &- \frac{(\Delta t)^2}{L \left(G\,\Delta t + C + \frac{(\Delta t)^2}{L} \right)} \sum_{m=1}^{n} E_z^m \end{aligned} \tag{10.18}$$

Examining (10.18) we see that an auxiliary variable will be required to store the sum of the previous values of Ez. For an inductance L approaching 0, i.e., a short circuit, the correct result of Ez = 0 satisfies (10.18), but the last term will tend to be unstable, so that extremely small values of L should be avoided. For an inductance L = ∞, an open circuit, the inductance terms drop out of (10.18) as they should. Once again it should be remembered that the value of C should not be made less than the capacitance of the FDTD cell in the electric field component direction filled with free space.

With the basic approach illustrated above more complicated circuits including sources can be modeled, sometimes at the expense of auxiliary variables (and auxiliary difference equations if additional capacitors and/or inductors are included in the circuit). However, this capability allows us to include in the same FDTD calculation structures which are orders of magnitude different in geometrical structure (a coiled wire lumped element inductor in the gap of a wire antenna, for example) without requiring that the FDTD cells be made small enough to model the structures of the lumped elements. This extension is quite useful in antenna analysis, as will be illustrated in Chapters 11 and 14.

10.5 EXPANSION TECHNIQUES

Electromagnetic penetration through an aperture that then couples to a wire in the interior of an arbitrarily shaped conducting scatterer is an extremely complex problem. Generally, it cannot be solved analytically.[4] Analytic solutions have been found for some very restricted geometries;[5] however, these solutions are of limited value when applied to complex problems such as the response of internal aircraft cables when the aircraft is exposed to an external transient field. Numerical solutions have also been sought with some success. Typically,[6] a method of moments approach is used to find the external current and charge distribution on a scatterer with an aperture that is assumed to be open, and from the current and charge distributions, equivalent magnetic and electric dipoles can be obtained from Bethe hole theory. They represent the aperture excitation, which in turn drives the interior cavity where a wire may or may not be present. An evaluation of this type of numerical approach[7] has found that agreement between theory and experiment can range from good to poor. As might be expected, Bethe hole theory does not work well for large apertures.

An alternate numerical approach, well suited to complex systems such as aircraft, is to use FDTD with some simple modifications. Since FDTD recognizes only the presence of the scatterer and does not distinguish between interior and exterior regions, all that is needed is a sufficiently small cell size so that the aperture, interior structure, and cavity geometry can be modeled with reasonable accuracy. Computer resource limitations typically restrict the FDTD technique to relatively coarse cells, so that an accurate interior response prediction of a complex geometry cannot be made with a single FDTD run. This limitation is circumvented by an alternate numerical approach, which is called the expansion technique.

The FDTD code used in Chapter 4 to model the aircraft[8,9] divides the problem space into a $28 \times 28 \times 28$ mesh of cells. In order to keep the outer boundary far away to allow proper operation of the outer radiation boundary condition, only 20 cells were allocated to the aircraft model, even with the use of expanding mesh. For the aircraft length of 20 m this translates into a cell size 1 m in length and width, and 0.5 m in the other dimension. This size is well suited to external response predictions, but is much too coarse for modeling apertures and a realistic aircraft interior.

The expansion technique allows realistic interior response predictions to be made. The technique consists of making an initial computer run with a model of the entire aircraft. The electric fields, scattered from the aircraft and tangential to a sub-boundary, are stored on disk from this calculation (Figure 10-5). The portion of the aircraft inside the sub-boundary is then subdivided into smaller cells (typically fourfold or more) and the sub-boundary becomes the outer boundary for a second calculation. The same incident field used for the first calculation illuminates the subdivided portion of the aircraft on the second calculation (Figure 10-6). The same tangential E-field response as seen on the sub-boundary for the first run is imposed on the outer boundary of the second run. This operation is accomplished by reading the data back from the disk every 25 time steps. Because a finer grid is used on the second run, a finer time step in intermediate spatial and time points must be found by interpolation. The outer boundary then responds on the second run with a response nearly identical to the first-run response. This response is limited by the original cell size, i.e., the fields reaching the sub-boundary on the first run accurately model the

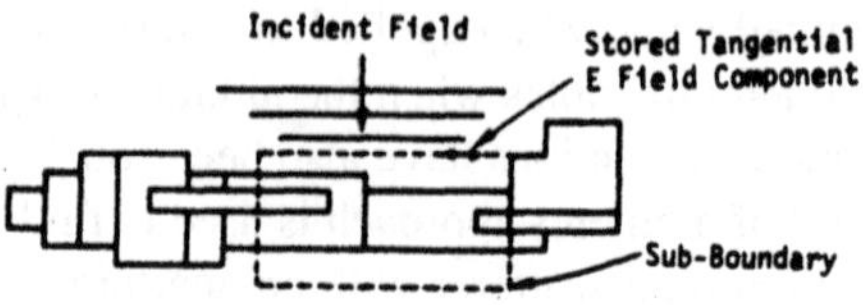

FIGURE 10-5. First run to obtain sub-boundary tangential E-fields.

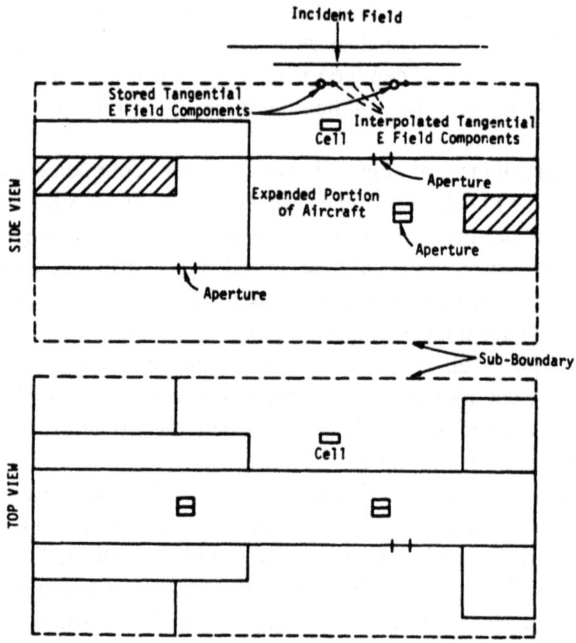

FIGURE 10-6. Second expanded run.

scattered fields up to a frequency limit f_{max} set by $f_{max} = 1/\lambda_{min}(\lambda_{min} \approx 4 \times$ cell size of the first run.

The result of this procedure is that on the second run, the missing portion of the aircraft still appears to be present, at least at low frequencies. Natural modes excited on the aircraft are preserved up to the frequency limit f_{max}. Those modes above f_{max} are not matched at the boundary on the second run. However, they are modes that are strongly damped.[10] Hence, the response at the aperture above f_{max} is primarily determined by the local fields incident on the aperture. So long as the aperture is at least a few diameters away from the outer boundary and so long as the aperture is small compared to the enclosed portion of the aircraft, error is minimized above f_{max}. Thus, apertures can be placed on the expanded portion of the aircraft for the second run and they will be excited by current and charge distributions associated with the excitation of the entire aircraft up to and usually well beyond f_{max}. The apertures so excited, in turn, couple energy into the interior of the aircraft, which because of the expansion, can be reasonably detailed.

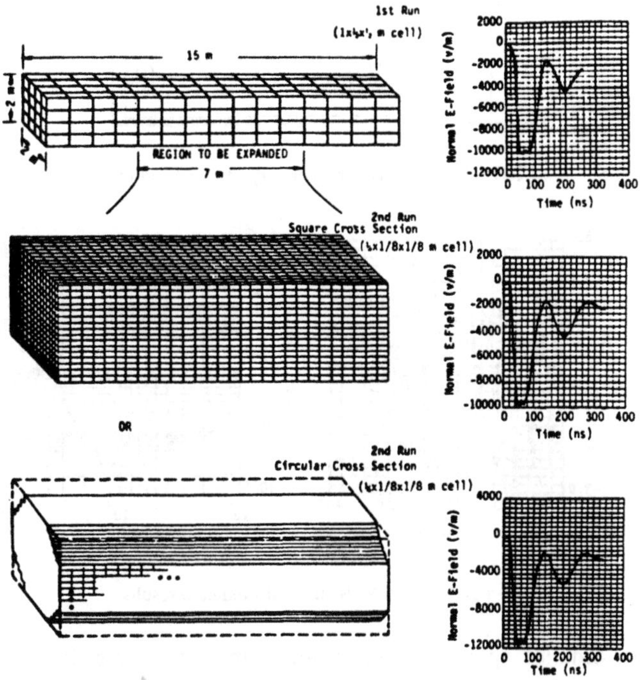

FIGURE 10-7. Different expansion geometries with typical exterior response predictions that demonstrate the stability of the expansion technique.

The expansion technique effectiveness is demonstrated on a simple cylinder geometry rather than a complex geometry, such as an aircraft. This allows examination of the technique using a well-understood geometry in terms of exterior response. A cylinder responds in much the same way as an aircraft fuselage when both are approximately the same size and both are excited by a transient wave with its electric field parallel to the major axis of the body. For convenience, a plane wave with double-exponential time dependence was used to illuminate the cylinder from above.

Stability of the technique is first examined using the exterior response of a cylinder. Solutions obtained in the second step should give answers nearly identical to the first run if the scattering geometry was unchanged, except for the finer gridding to more accurately model the scattering geometry. Starting with a 15 cell-long rectangular block model of the cylinder in a $28 \times 28 \times 28$ cell space in the first run, and going to a four-times-finer gridded rectangular block on the second run (or alternately, going to a nearly circular cross section) the expansion technique yields results (Figure 10-7) consistent with the geometries involved and confirms the stability of the technique.

A very simple interior geometry is then considered for the expanded portion of the cylinder (Figure 10-8). This rectangular-cross section region contained an interior wire terminated in 50-Ω resistors at internal bulkheads. Coupling to the outside was via a large (3×6 cell) asymmetrically located

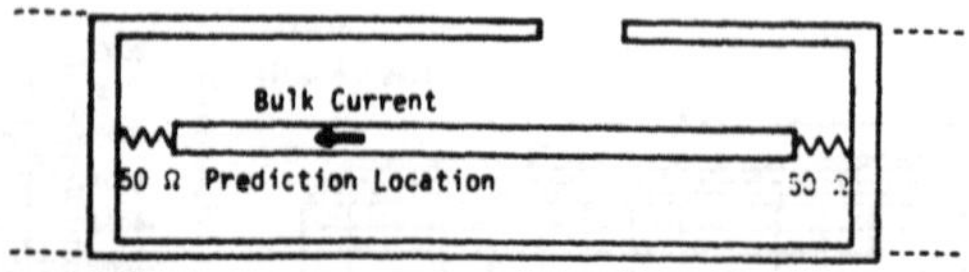

FIGURE 10-8. Interior geometry of expanded region.

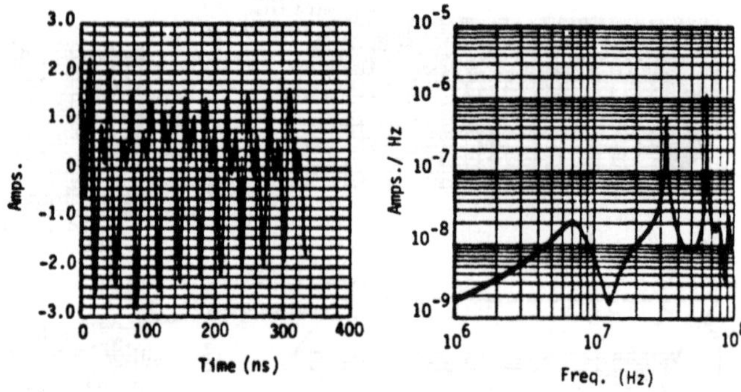

FIGURE 10-9. Sample of predicted results.

aperture. Smaller apertures and multiple apertures were also considered, along with different terminations, namely 0, 50, and ∞ Ω. A sample of the predicted results (Figure 10-9) for the first geometry examined (single, large aperture, and 50-Ω termination) reveals three prominent features in the frequency response seen in varying degrees in most of the predictions for the different geometries:

1. A low frequency response peak around the exterior fundamental resonance frequency (~10 MHz)
2. A mid-frequency response peak corresponding to the half-wavelength excitation of the interior cable that varies with the cable length and is more damped for the 50-Ω termination than for the 0- and ∞-Ω terminations
3. A high-frequency response peak at 65 MHz, presumably the second harmonic excitation of the interior cable

The transform was obtained using Prony's method[11] to find the poles and residues characterizing the data and, from them, the transform. This expedient was used because of the extreme truncation of the data record. The value of N, the number of poles sought, was varied from N = 16 to 36. The value of N selected for this analysis, N = 24, was the one that returned the lowest energy content when fitting the data. The energy content was determined using Equation 2 of Reference 11. The predictions were sampled at a rate of 75 MHz, so that a reasonable characterization of the poles below $\sim 0.7 \times 75$ MHz could be expected.[12]

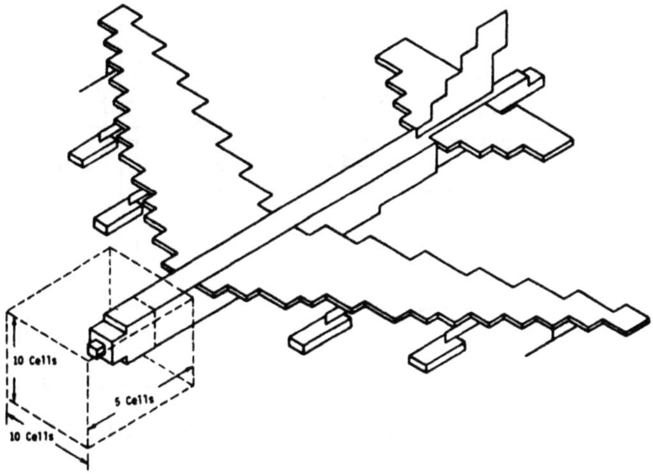

FIGURE 10-10. Aircraft model in a 50^3-cell space with sub-boundary indicated.

10.6 CODE REQUIREMENTS, LIMITATIONS, AND UTILITY

The technique has the following requirements:

1. Internal partitioning, such as bulkheads in the scatterer.
2. That the high-frequency noise picked up from the linear interpolation on the sub-boundary be filtered out. This step is most important for the interior response. The approach used for the cylinder was to apply Prony's method[11] to the data, and then remove the poles above 100 MHz.

The code is limited as follows:

1. To a frequency range of up to $f_{max}' = c/(4 \times \text{cell size of the second run})$. The exact value depends on how close to the sub-boundary the response is being sought. For interior responses with apertures away from the sub-boundary, the higher frequency limit may be approached.
2. By some slight waveform distortion between first and second runs, even when using the identical geometries for both runs. This is caused by dispersion effects that depend on cell size vs. wavelength. It is a slight effect for the expansion factors considered here.

The expansion technique economically increases the resolution of finite-difference solutions of the Maxwell equations, allowing interesting interior coupling problems to be addressed, thereby establishing its utility. An example of this utility is seen in Figures 10-10 and 10-11, in which an aircraft is first modeled in an $\sim 50^3$-cell space, and then the cockpit area is expanded, allowing significant interior modeling detail.

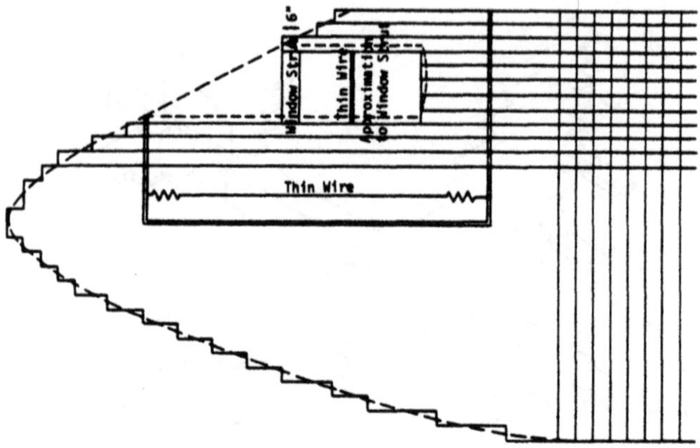

FIGURE 10-11. Expanded run showing cockpit area detail.

REFERENCES

1. **Umashankar, K., Taflove, A., and Beker, A.,** Calculation and experimental validation of induced currents on coupled wires in an arbitrary shaped cavity, *IEEE Trans. Ant. Prop.*, 35, 1248, 1987.
2. **Taflove, A. et al.,** Detailed FD-TD analysis of electromagnetic fields penetrating narrow slots and lapped joints in thick conducting screens, *IEEE Trans. Ant. Prop.* 36(2), 247, 1988.
3. **Holland, R. and Simpson, L.,** Finite-difference EMP coupling to thin struts and wires, *IEEE Trans. Electromag. Compat.*, 23, 89, 1981.
4. **Butler, C. M., Rahmat-Samii, Y., and Mittra, R.,** Electromagnetic penetration through apertures in conducting surfaces, *IEEE Trans. Electromag. Compat.*, 20, 1978.
5. **Seidel, D. B.,** Excitation of a Wire in a Rectangular Cavity, Interaction Application Memos, Air Force Weapons Laboratory Note Series, Memo 30, Baum, C., Ed., Washington, DC.
6. **Schuman, H. K.,** Circumferential Distribution of Scattering Current and Small Hole Coupling for Thin Finite Cylinders, Rome Air Development Center, RADC-TR-77-412, 1977.
7. **Buttingham, J. N., Deadrick, F. J., and Lager, D. L.,** An Experimental Study of the Use of Bethe Theory for Wires Behind Apertures, UCRL-52443, Lawrence Livermore National Laboratory, Livermore, CA, 1978.
8. **Kunz, K. S. and Lee, K. M.,** A three-dimensional finite-difference solution of the external response of an aircraft to a complex transient EM environment. I. The method and implementation, *IEEE Trans. Electromag. Compat.*, 20(2), 328, 1978.
9. **Kunz, K. S. and Lee, K. M.,** A three-dimensional finite-difference solution of the external response of an aircraft to a complex transient EM environment. II. Comparison of predictions and measurements, *IEEE Trans. Electromag. Compat.* 20(2), 333, 1978.
10. **Kunz, K. S. and Prewitt, J. F.,** Practical limitations to a natural mode characterization of electromagnetic transient response measurements, IEEE Trans. Ant. Prop., 28, 575, 1980.

11. **Van Blaricum, M. L. and Mittra, R.,** A technique for extracting the poles and residues of a system directly from its transient response, *IEEE Trans. Ant. Prop.*, 23, 777, 1975.
12. **Miller, E. K. and Lager, D. L.,** Information Extraction Using Prony's Method, UCRL-52329, Lawrence Livermore National Laboratory, Livermore, CA, 1977.

Chapter 11

NONLINEAR LOADS AND MATERIALS

11.1 INTRODUCTION

This chapter discusses the application of FDTD to problems that involve nonlinear circuit elements or materials. Dealing with nonlinearities is comparatively easy in the time domain as opposed to frequency domain calculations. Frequency domain approaches often involve iteration processes that themselves may involve transformation back and forth between frequency and time domains. Working directly in the time domain is often much simpler, because the parameters of the device or material may be modified as a function of the field strength in the FDTD cell as the computation progresses. In most situations this is not without some added computational effort, as the time step size often must be set much lower than normally required in order to avoid instabilities in the FDTD calculations.

In this chapter two approaches for dealing with nonlinear materials are demonstrated. In the first demonstration only one FDTD cell in the computational space is nonlinear. The example geometry is a wire antenna with a nonlinear diode load at the center illuminated by a pulsed plane wave. The difficulties in dealing with this geometry in a "brute force" fashion of changing the resistance of the diode as its voltage changes and reducing the time step size until the FDTD calculation is stable are shown, and an alternative approach that satisfies the nonlinear diode equation at each time step and allows the time step size to remain large is demonstrated. This demonstration utilizes the ability of FDTD to model lumped circuit elements in one FDTD cell, and it is assumed the reader is familiar with this, as discussed in Chapter 10.

If nonlinear materials are contained in only a relatively small number of the total FDTD cells in the problem space, then expending extra effort on these cells in order to maintain a large value of time step is justified. However, if the entire FDTD space is filled with nonlinear material it may be more efficient and is certainly simpler to change the material parameters as the field strength changes and to reduce the time step size until stability is reached. This approach is demonstrated in the final section of this chapter in an FDTD calculation of the low frequency attenuation of an incident plane way by a sheet of nonlinear magnetic material.

Application of nonlinear equations requires the use of the total field strength. The propagation calculation will be done directly in total fields. However, since the dipole problem involves an incident plane wave, we prefer to use the scattered field formulation for all cells except the cell that contains the nonlinear diode. Even in this cell scattered magnetic fields are used for determining the current through the cell, since the scattered magnetic fields due to the current flowing on the wire completely dominate the incident magnetic field in the vicinity of the wire. The total electric field (and

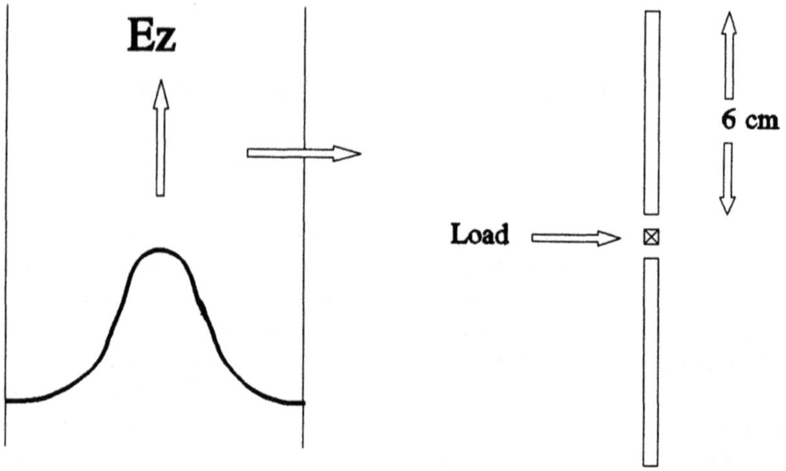

FIGURE 11-1. Gaussian pulsed plane wave incident on wire dipole with nonlinear diode load in center.

related diode voltage) is obtained by adding the incident to the scattered field at each time step. After the nonlinear diode equations are applied and the total electric field is determined, the scattered electric field is recovered by subtracting the incident field.

11.2 ANTENNA WITH NONLINEAR DIODE

In this section we discuss the capability of FDTD to include effects of nonlinear lumped circuit elements. Our demonstration geometry is a wire antenna with a nonlinear diode load connected across a central gap in the wire. Modeling a wire antenna including lumped loads was discussed in Chapter 10.

The first geometry considered is shown in Figure 11-1. The wire dipole is illuminated by a Gaussian pulsed plane wave. The pulse is as described in Section 3.4, but with $\beta = 64$ unless noted otherwise. The amplitude of the pulse will be varied so as to drive the diode into varying levels of nonlinear behavior so we can investigate the stability of the various nonlinear FDTD approaches. The FDTD space has cells that are 0.006-m cubes, and the problem space is $15 \times 15 \times 30$ cells, with the wire antenna in the z direction. The outer boundary is second order Mur. The time step will be at the Courant limit of 11.55 ps unless noted otherwise. Each dipole arm is ten cells long, with an additional center cell containing the nonlinear load. The wire diameter is 0.8118 mm. Because this diameter is much smaller than our FDTD cells, the sub-cell wire approximation of Chapter 10 is used except for the center cell containing the diode load. The lumped element circuit model for the cell at the center of the dipole for the initial demonstrations is shown in Figure 11-2. The nonlinear diode is in parallel with the free space capacitance C of the FDTD cell, as discussed in Chapter 10. The voltage across the diode

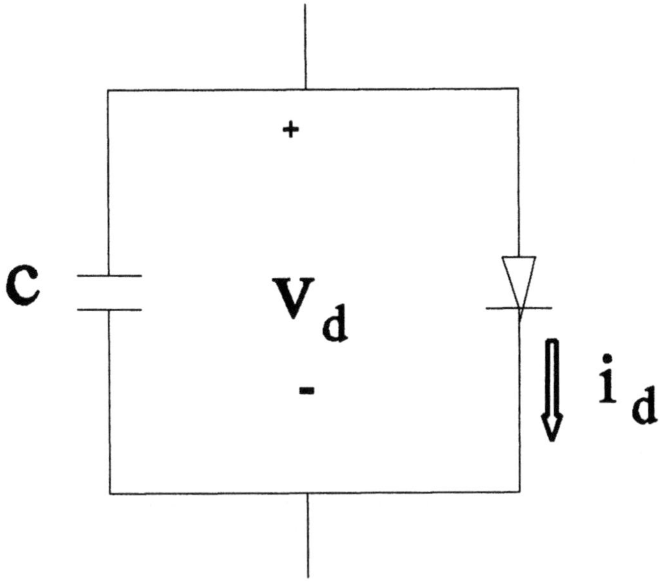

FIGURE 11-2. Equivalent circuit of nonlinear diode in parallel with FDTD cell capacitance.

is v_d, which is also the voltage across the FDTD cell, and i_d is the current through the diode.

The nonlinear relationship between the diode voltage and current is given by

$$i_d = 1.0 \times 10^{-8} v_d, \; v_d \le 0 \tag{11.1}$$

$$i_d = 2.9 \times 10^{-7} \left[\exp(15\, v_d) - 1\right], \; v_d \le 0 \tag{11.2}$$

This relationship is shown graphically in Figure 11-3 for positive diode voltages. For negative voltages the diode current is essentially zero on the scale of Figure 11-3. This particular diode relationship was taken from Reference 1 and is used later to calculate results to compare with results in (11.1).

We can expect to encounter difficulties if we neglect the FDTD free space cell capacitance $C = \varepsilon_0 (\Delta x \Delta y)/(\Delta z)$, as was discussed in Chapter 10. Ignoring this for the moment, let us consider how we may most simply include the nonlinear diode in our FDTD calculations. A simple way to model a nonlinear diode is as a voltage-controlled variable resistor, R. In Chapter 10 we saw that a lumped conductance $G = 1/R$ within an FDTD cell could be modeled by setting the conductivity of the cell as $\sigma = G(\Delta z)/(\Delta x \Delta y)$. This conductivity can be changed at each time step as a function of the voltage across the diode, with

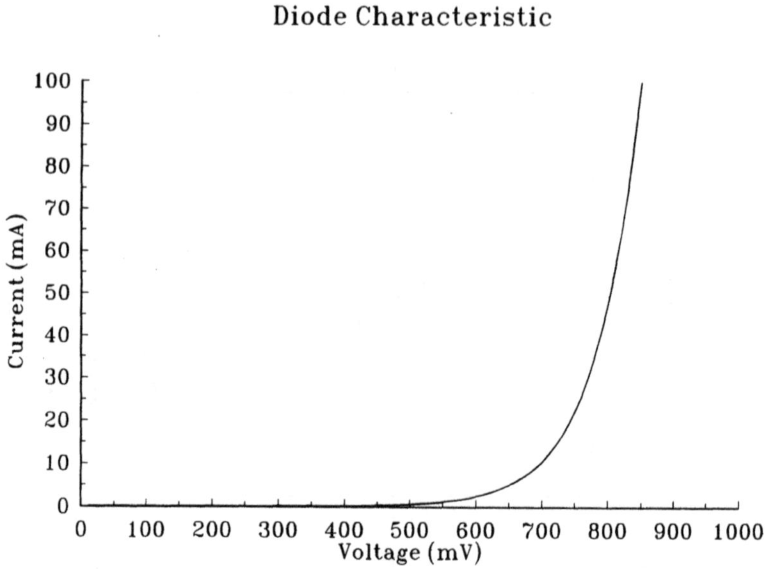

FIGURE 11-3. Current-voltage relationship for forward biased diode.

the latest value used to determine the next value of electric field directly from the current flowing through the cell. Explicitly we would determine the diode voltage at each time step from

$$v_d^n = E_z^n \Delta z \tag{11.3}$$

With this value of v_d^n we would find the current through the diode and the corresponding diode resistance from (11-1) and(11.2). We label this diode resistance R_d^n. If the capacitance of the FDTD cell is neglected, however, the diode current must also be equal to the line integral of the magnetic field around the cell, or

$$i_d^{n+\frac{1}{2}} = \Delta z \Delta y \left(\nabla x H^{n+\frac{1}{2}} \right)_z \tag{11.4}$$

as discussed in Chapter 10. If we then attempt to predict the next value of electric field from

$$E_z^{n+1} = R_d^n i_d^{n+\frac{1}{2}} / \Delta z \tag{11.5}$$

we find that for an incident Gaussian pulsed plane wave of 10 V/m peak amplitude our FDTD code produces overflow values of electric field almost immediately after the incident pulse excites the wire. A slight variation on the above scheme is to use the value of $i_d^{n+\frac{1}{2}}$ to determine $R_d^{n+1/2}$ from the diode equations (11.1) and (11.2) and then apply (11.5) to determine the next electric field value. Unfortunately, the same instability will result. The reason for this is easily seen. In either of these approaches there are two values for i_d, one determined from the surrounding magnetic fields using (11.4), the other determined from the diode equations (11.1) and (11.2). The approach described above makes no attempt to reconcile these two values, and the resulting electric field (diode voltage) jumps wildly between extreme values trying to produce an equality between them.

Having eliminated this approach, we now include the FDTD cell capacitance C of Figure 11-2 in our calculations. As above we use (11.3) to find the diode voltage, and the diode equations (11.1) and (11.2) to find the diode conductance $G_d^n = 1/R_d^n$. Then, using the free space value of the FDTD cell capacitance, we find from (7.15) that the next value of the electric field is given by

$$E_z^{n+1} = \frac{C}{C + G_d^n \Delta t} E_z^n + \frac{\Delta t \Delta x \Delta y}{C\Delta z + G_d^n \Delta t \Delta z} \left(\nabla x H^{n+\frac{1}{2}} \right)_z \tag{11.6}$$

This approach to modeling the nonlinear diode as a variable resistance will be stable as long as the $G_d^n \Delta t$ product remains relatively small as compared to the free space FDTD cell capacitance C. There are two conditions for which this will be true. One is when the conductance G of the diode is small, i.e., the diode is reverse biased or only very slightly forward biased. The other is when the time step size Δt is sufficiently small.

Before providing some examples of results obtained with this approach, which we will call the "variable resistor" method, let us consider another alternative. The difficulty with the variable resistor approach is that the diode current and voltage predicted by the FDTD update equations will not agree in general with the terminal relationships of the diode given in (11.1) and (11.2). But we can, however, ensure that all the equations applied to these quantities are consistent by solving the equations simultaneously at each time step. This is still an explicit approach, since future values depend only on past values. However, the explicit dependence is nonlinear.

To apply this approach to the lumped circuit of Figure 11-2 let us again consider the lumped element FDTD equation (7.15) repeated below for convenience as

$$\Delta x \Delta y \left(\nabla x H^{n+\frac{1}{2}} \right) = C\, \Delta z \frac{E_z^{n+1} - E_z^n}{\Delta t} + G\, \Delta z\, E_z^{n+1} \tag{11.7}$$

The first term is the total current flowing through the cell, the second is the displacement current flowing through the capacitor, and the third is the conduction current flowing through the cell conductance, in this case through the diode. Considering our circuit of Figure 11-2, the diode current is the conductance current, i.e., the total current less the displacement current. Making this identification and solving (11.7) for the conduction current through the diode, we obtain

$$i_d = \Delta x \Delta y \left(\nabla \times H^{n+\frac{1}{2}} \right)_z - \frac{C\,\Delta z}{\Delta t}\left(E_z^{n+1} - E_z^{n} \right) \tag{11.8}$$

We also need to determine the diode voltage. Because we are hoping to determine E_z^{n+1}, for maximum stability we should use the most recent value of electric field. Thus, instead of using (11.3) we use

$$v_d = E_z^{n+1} \Delta z \tag{11.9}$$

Equations (11.8) and (11.9) provide the values of diode voltage and current based on the FDTD equations. Equations (11.1) and (11.2) provide values of the diode voltage and current based on the diode characteristics. This system of nonlinear equations can be solved at each time step for E^{n+1} using a nonlinear equation solution method such as Newton-Raphson iteration. This value of E^{n+1} will satisfy both the FDTD equations and the nonlinear diode equations at each time step. The convergence is very fast since an initial guess for E_z^{n+1} of the previous value of E_z, E_z^n, provides the Newton-Raphson iteration with a good starting value. The computer time required for running the Newton-Raphson iteration for this demonstration is essentially identical with that for the variable resistance method, because only one FDTD component is involved. It should be emphasized that except for the magnetic fields surrounding the wire (which use a sub-cell modification to compensate for the thin wire) and the E_z field in the cell containing the nonlinear diode, all fields are updated using the normal FDTD scattered field formulation.

Now let us compare results for our test geometry using these two approaches. FDTD calculations of the total current (conduction plus displacement) flowing through the parallel combination of the FDTD cell capacitance and diode of Figure 11-2 are shown in Figure 11-4 for an incident Gaussian pulsed plane wave of peak amplitude 10 V/m. For comparison, results for the diode modeled as a short circuit ($E_z = 0$) and an open circuit in parallel with the FDTD cell capacitance C ($i_d = 0$) are also included. The Newton-Raphson iteration and the variable resistance approach provide nearly identical results that show the nonlinear behavior of the diode. However, comparing the current amplitude in Figure 11-4 to the diode characteristic shown in Figure 11-3 it is evident that the diode resistance remains high throughout the calculation.

To increase the nonlinear effect of the diode we repeat the calculation, but with the Gaussian pulse amplitude increased to 20 V/m. The total current through the dipole gap is shown in Figure 11-5. The peak diode

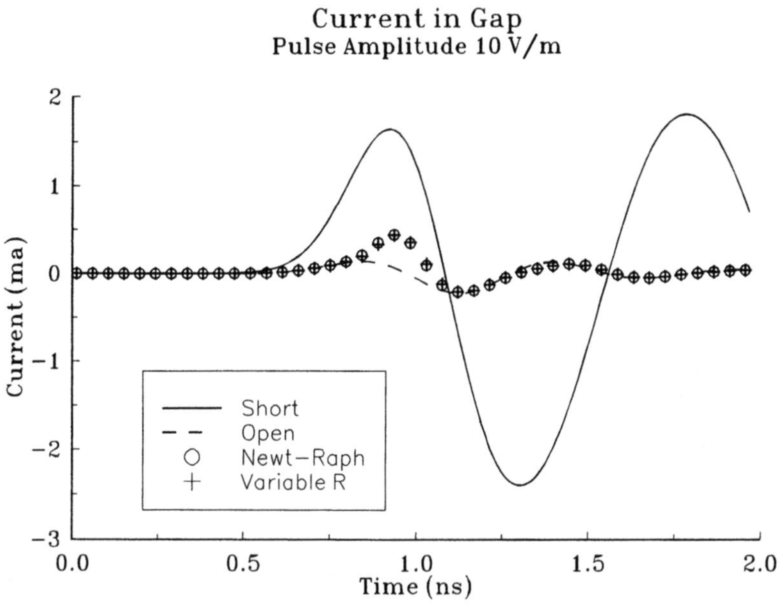

FIGURE 11-4. Transient total current flowing through the center of the dipole with incident 10 V/m Gaussian pulse, calculated for diode replaced with short and open circuits, and with diode modeled with Newton-Raphson iteration and as variable resistor.

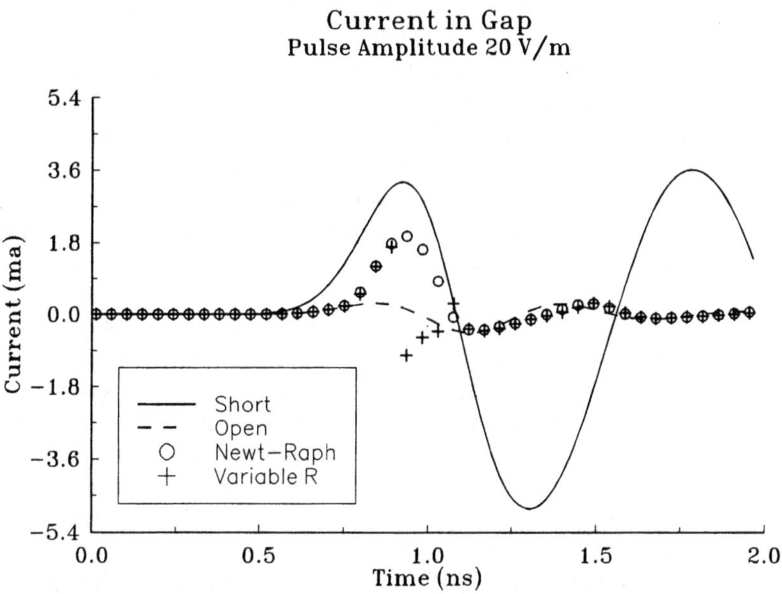

FIGURE 11-5. Transient total current flowing through the center of the dipole with incident 20 V/m Gaussian pulse, calculated for diode replaced with short and open circuits, and with diode modeled with Newton-Raphson iteration and as variable resistor.

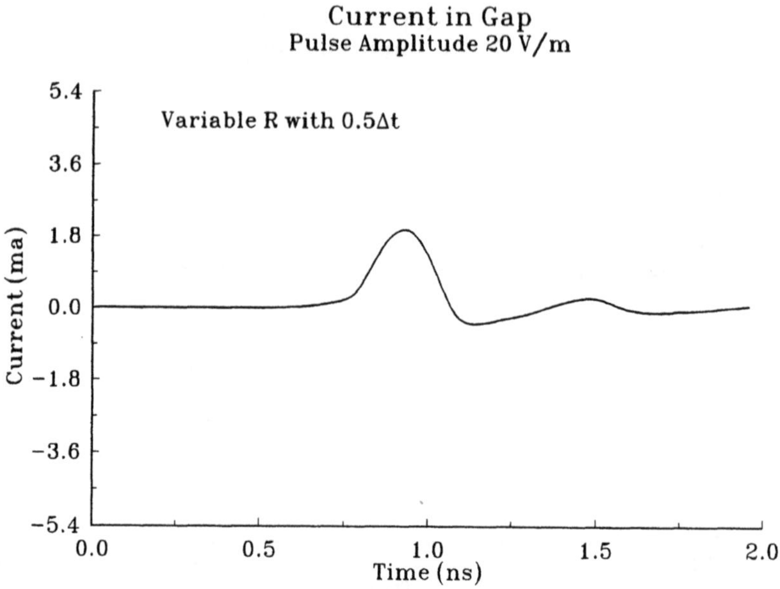

FIGURE 11-6. Transient total current flowing through the center of the dipole with incident 20 V/m Gaussian pulse, calculated with variable resistor approach, but with time step reduced by one half.

current is now large enough so that the diode is clearly exhibiting nonlinear behavior. The Newton-Raphson results are smooth and continuous, while the variable resistance results are discontinuous near the peak of the diode current. This calculation is repeated for the variable resistance method but with the time step reduced to half the Courant limit, and the pulse width β doubled to 128 to provide the same pulse duration. The results are shown in Figure 11-6. With the time step reduced the variable resistance method now provides results nearly identical to those produced by the Newton-Raphson method, although taking approximately twice the computer time to do so.

Let us stress these two methods even more by applying a 100 V/m peak pulsed plane wave, with the FDTD results for the gap current shown in Figure 11-7. The Newton-Raphson iterated results predict a gap current close to that for a short circuit when the diode is forward biased, and to that for an open circuit (in parallel with the cell capacitance) when the diode is reversed or only weakly forward biased. The variable resistance results, however, oscillate during the interval when the diode is forward biased. For this more nonlinear behavior, reducing the time step by a factor of one half (and correspondingly doubling β) improves the behavior of the variable resistance method, but does not entirely eliminate the oscillations, as shown in Figure 11-8. Presumably a further reduction in time step would eventually provide accurate results using this method.

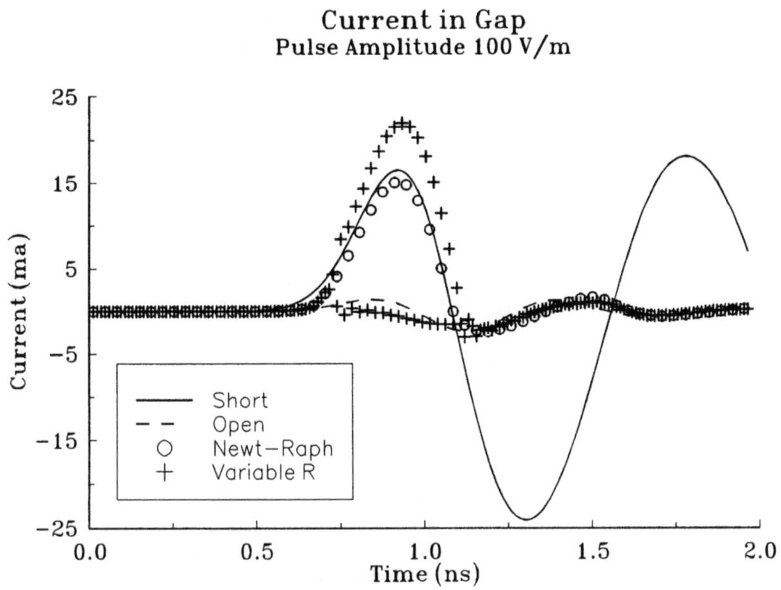

FIGURE 11-7. Transient total current flowing through the center of the dipole with incident 100 V/m Gaussian pulse, calculated for diode replaced with short and open circuits, and with diode modeled with Newton-Raphson iteration and as variable resistor.

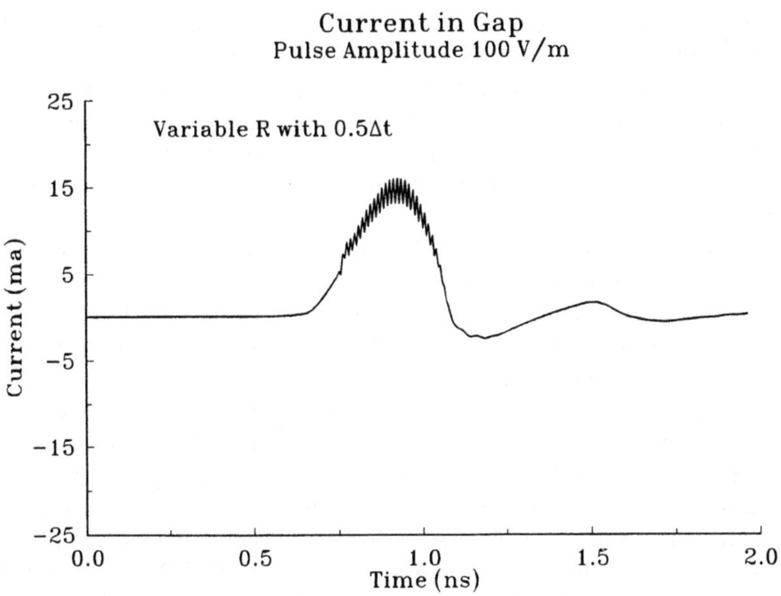

FIGURE 11-8. Transient total current flowing through the center of the dipole with incident 100 V/m Gaussian pulse, calculated with variable resistor approach, but with time step reduced by one half.

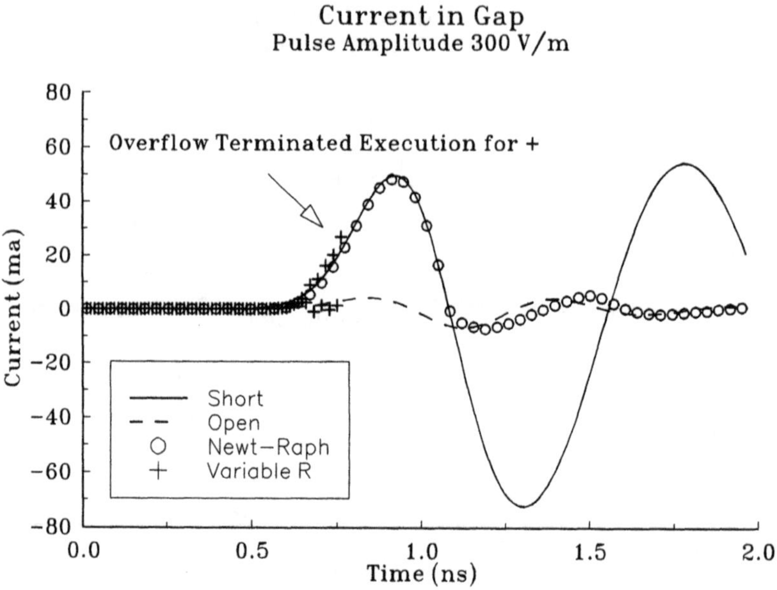

FIGURE 11-9. Transient total current flowing through the center of the dipole with incident 300 V/m Gaussian pulse, calculated for diode replaced with short and open circuits, and with diode modeled with Newton-Raphson iteration and as variable resistor.

For our final demonstration with this geometry the Gaussian pulse peak amplitude is increased to 300 V/m, with the results shown in Figure 11-9. The variable resistance method results oscillate and then overflow, terminating execution of the FDTD calculation. The Newton-Raphson iteration produces reasonable results, showing the diode behaving approximately like a short circuit when it is strongly forward biased. All results shown for the Newton-Raphson iteration shown here were calculated with the time step at the Courant limit.

While the results obtained using iteration to simultaneously satisfy both the FDTD and diode equations appear reasonable, accuracy has not been demonstrated. In order to do this we will compare FDTD results obtained using the iteration approach with published results for transient current through a wire antenna with a nonlinear diode load in the gap.

The specific example of interest is taken from Reference 1. The wire dipole still has a wire diameter of 0.8118 mm but now has arms 0.6 m long, as shown in Figure 11-10. The FDTD cell at the center of the wire again contains the lumped load. To approximate this geometry the only necessary adjustment of our previous example is to increase the number of cells used to model the wire dipole. Using the same 0.006 m cubical FDTD cells, 200 FDTD cells plus the gap cell were used for the dipole. Again, sub-cell modeling (described in Chapter 10) was used to adjust for the wire diameter being smaller than the FDTD cell size. The FDTD problem space was 39 × 39 × 240 cells, and was again terminated in second order Mur absorbing boundaries.

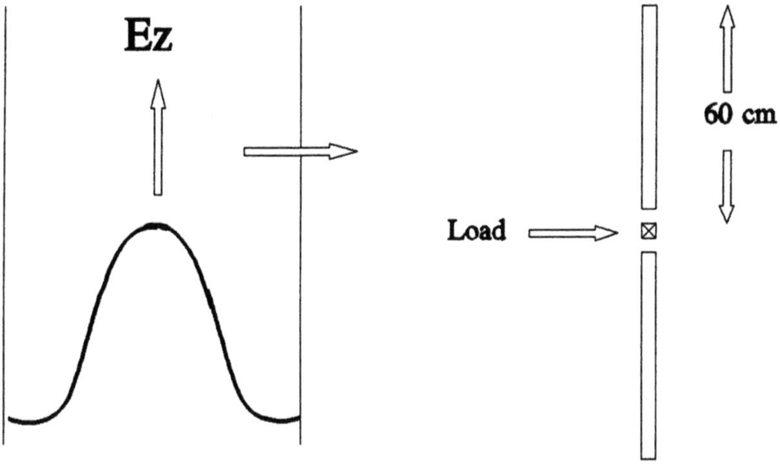

FIGURE 11-10. Gaussian pulsed plane wave incident on wire dipole with nonlinear diode load in center for comparison with results of Liu and Tesche.

As described in Reference 1, this dipole is loaded with two diodes in series with a 100-Ω resistor (the actual measurements were made using a single diode at the base of a monopole; image theory requires adding the additional diode and increasing the actual 50-Ω resistance to 100-Ω). The total diode junction capacitance of 0.5 pF must also be included in the model for accurate results at the frequencies contained in the pulse. The equivalent circuit used to approximate this lumped load is shown in Figure 11-11. The resistance R is now a fixed resistance representing the input resistance to an oscilloscope used to make the measurements. The capacitance C now includes both the FDTD free space cell capacitance and the diode junction capacitance. For our cell size the FDTD cell capacitance is only 0.053 pF, therefore, the diode junction capacitance dominates and the lumped capacitance in (11.9) was taken to be 0.5 pF.

Because the lumped circuit is different than in our previous example, the interaction must be slightly modified. This requires only substitution of

$$2v_d + i_d R = v_c = E_z^{n+1}\Delta z \qquad (11.10)$$

in place of (11.9) to determine the diode voltage, where R is the constant lumped resistance of the equivalent circuit in Figure 11-11.

The dipole considered in Reference 1 is excited by a pulsed plane wave. As shown in the reference, this plane wave has a peak electric field strength of approximately 390 V/m. For simplicity, this pulse has been approximated for our calculations by the same Gaussian pulse used in the previous examples in this section with β = 64, but with the peak amplitude set at 390 V/m. Our Courant limit time step of 11.55 ps provides a pulse duration of approximately 0.7 ns. The actual pulse shown in Reference 1 has approximately this duration,

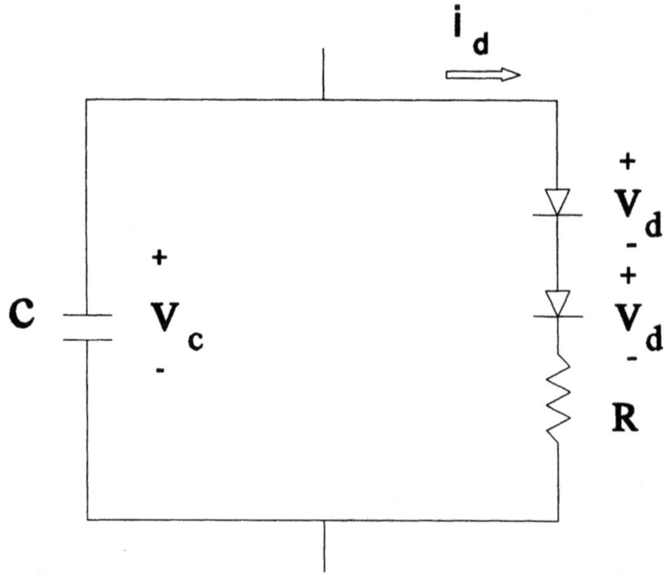

FIGURE 11-11. Equivalent circuit of nonlinear diodes in series with fixed resistor R and in parallel with capacitance C, the combination of the FDTD cell capacitance plus the diode junction capacitance, for comparison with results of Liu and Tesche.

but rings at a low amplitude out to about 3 ns. This has been neglected in the FDTD calculations.

The transient currents given in Reference 1 have a duration of approximately 14 ns. The corresponding FDTD calculation was 1300 time steps (allowing time for the incident pulse to reach the dipole). Three calculations of the current through the dipole load were made and compared to results calculated by Liu and Tesche.[1] As before, the total current flowing through the FDTD cell, determined by evaluating the curl of H around the cell containing the lumped load, is plotted. In the first calculation only the 100-Ω resistor (in parallel with the free space capacitance of the FDTD cell) was included. The results are shown in Figure 11-12. The agreement with the results of Liu and Tesche is quite good.

Next, the diodes were added in series with the 100-Ω resistor, and the FDTD cell capacitance C, shown in Figure 11-11, was set to 0.5 pF to model the combined diode junction capacitance. FDTD results for the two cases considered in Reference 1 were then calculated using the Newton-Raphson iteration approach to determine E_z in the antenna gap. These results are shown in Figures 11-13 and 11-14, and differ only in that the pulse initially forward biases the diode for the results in Figure 11-13, but reverse biases the diode in Figure 11-14. Comparing the peak current amplitudes to those in Figure 11-3 shows that the diode is highly nonlinear in these calculations. Some points taken from calculated results in Reference 1 are included in Figures 11-13 and 11-14 for comparison. The agreement between the FDTD results and both the calculated

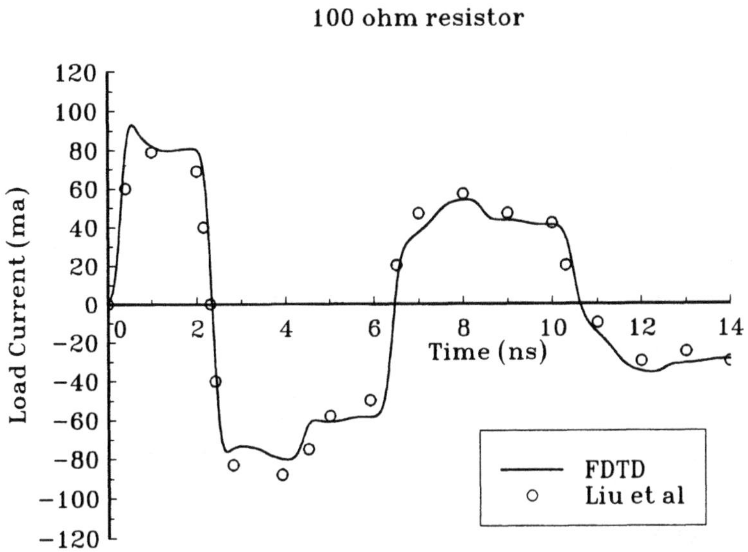

FIGURE 11-12. Transient current flowing through 100-W load and parallel FDTD cell capacitance at the terminals of the wire dipole calculated using FDTD and compared to calculated results of Liu and Tesche.

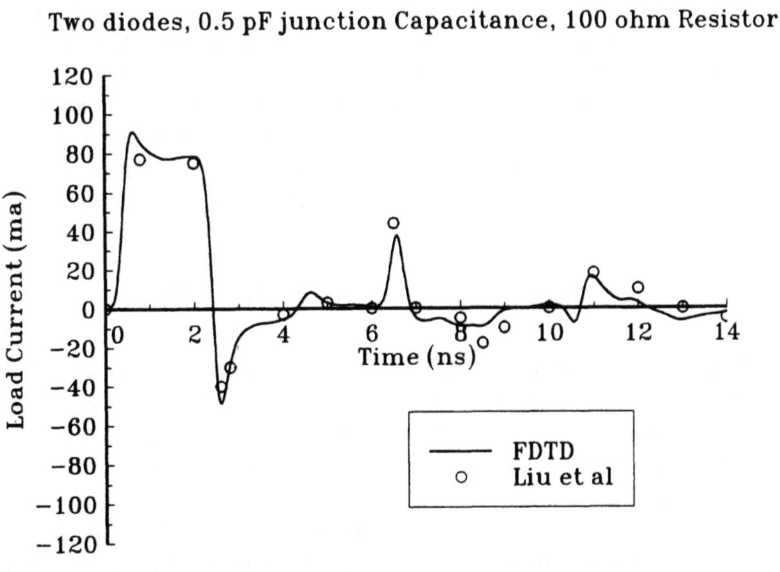

FIGURE 11-13. Total transient current flowing through equivalent circuit of Figure 11-11 located at the terminals of the wire dipole, calculated using FDTD and compared to calculated results of Liu and Tesche. Diode is initially forward conducting.

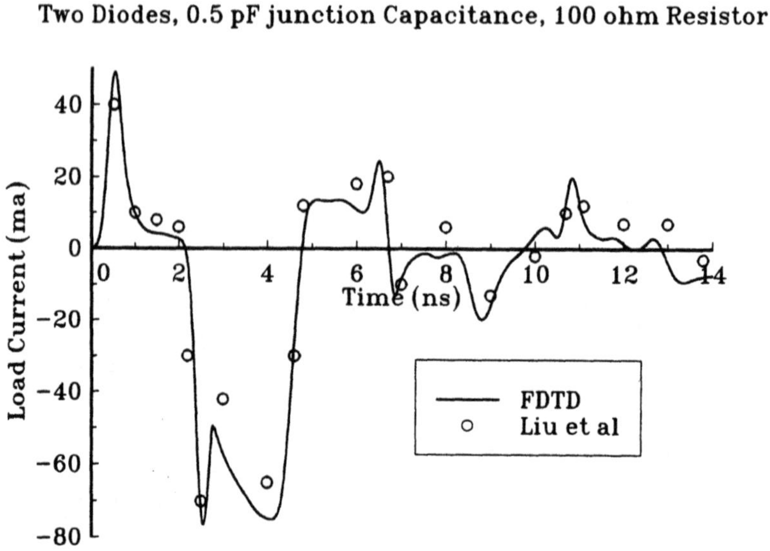

FIGURE 11-14. Total transient current flowing through equivalent circuit of Figure 11-11 located at the terminals of the wire dipole, calculated using FDTD and compared to calculated results of Liu and Tesche. Diode is initially reverse conducting.

and measured results in Reference 1 are excellent considering the different assumptions and approximations made in the analysis.

For problems in which only a relatively small number of cells contain nonlinear material or lumped circuit elements the iteration approach is highly advantageous. It allows us to calculate results for highly nonlinear situations without reducing the time step size. However, in problems in which all or most of the FDTD cells are filled with nonlinear materials, it is nearly as efficient and certainly simpler to reduce the time step size in order to maintain stability. In the next section an example of this type of calculation is given for shielding by a nonlinear magnetic sheet.

11.3 NONLINEAR MAGNETIC SHEET

In this section application of FDTD to propagation through a nonlinear magnetic material will be demonstrated. The FDTD calculations will require changing the permeability of the material at each time step based on the value of the magnetic field at the previous time step. Stability will be obtained by reducing the time step size well below the Courant limit. The frequencies of interest are 1 and 10 kHz. It is very unusual to apply FDTD at such low frequencies, but it is possible in this case because the conductivity of the material is so high that the time step is drastically reduced from typical free space values.

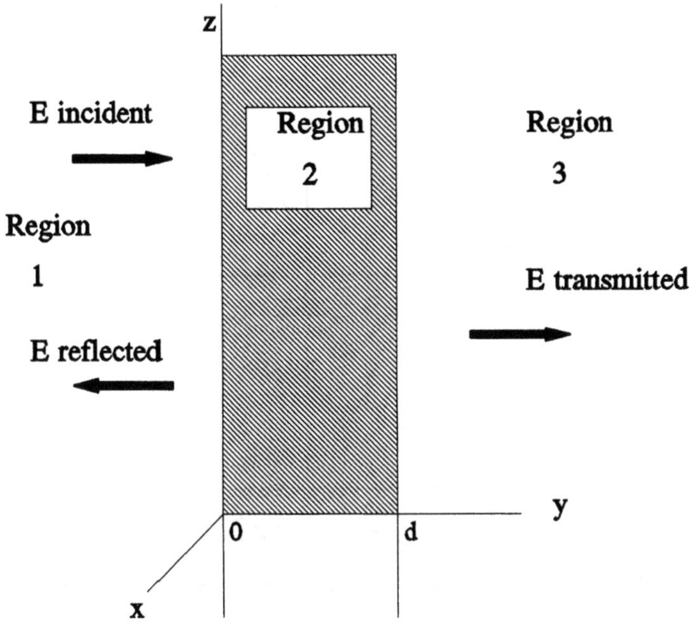

FIGURE 11-15. Infinite sheet of nonlinear magnetic material (Region 2) with free space on both sides.

The problem geometry is taken from Merewether[2] and is illustrated in Figure 11-15. A plane wave is incident from Region 1 on an infinite planar sheet of conducting saturable ferromagnetic material of thickness d. We would like to determine the field that is transmitted through the sheet. Merewether solved the problem using an implicit difference method. We demonstrate here the ability of FDTD to solve the problem explicitly.

The B-H characteristic of the nonlinear magnetic material is shown in Figure 11-16. The differential permeability is given by

$$d\mu(H) = \frac{\partial B}{\partial H} = \mu_m + B_s \exp\left(-|H|/H_c\right) H_c \tag{11.11}$$

where $\mu_m = 1.67 \times 10^{-4}$ H/m, $B_s = 1.53$ T, and $H_c = 120$ A/m. The conductivity of the material is 10^7 S/m, and the permittivity was taken as free space. The sheet thickness d is 1.26×10^{-4} m.

The electric and magnetic fields in the sheet were quantized to fit the geometry with $y = (j-1)\Delta y$ and $t = n\Delta t$. There are J magnetic field locations, with magnetic fields located at $y = 0$ ($j = 1$) through $y = d = J\Delta y$, so that magnetic fields are at the surfaces of the sheet. Electric fields are interleaved between the magnetic fields.

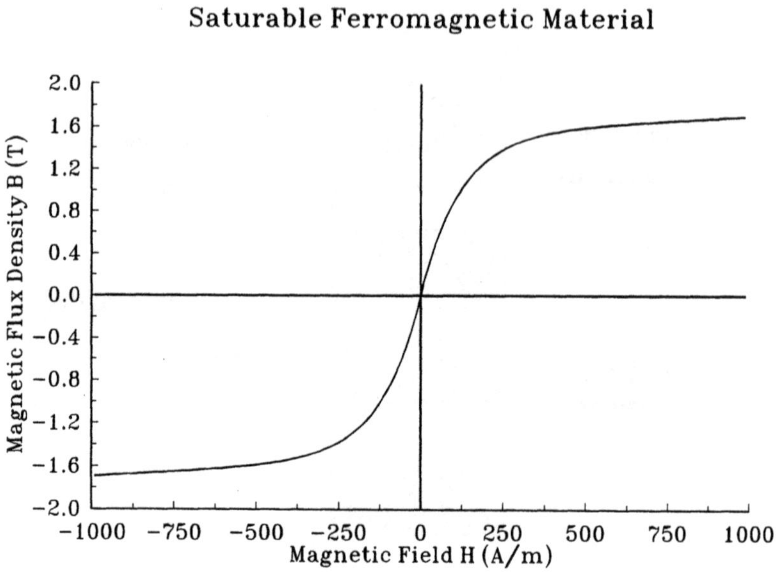

FIGURE 11-16. Magnetic characteristic of nonlinear magnetic material.

Let us first determine the FDTD equations. We assume only E_z, B_x, and H_x components. The update equation for the electric field is determined in the usual way from the Maxwell equation

$$\frac{\partial H_x}{\partial y} = \sigma E_z - \varepsilon_0 \frac{\partial E_z}{\partial t} \tag{11.12}$$

$$\begin{aligned} E_z^{n+\frac{1}{2}}\left(j+\tfrac{1}{2}\right) &= \frac{\varepsilon_0}{\varepsilon_0 + \sigma\Delta t} E_z^{n-\frac{1}{2}}\left(j+\tfrac{1}{2}\right) \\ &- \frac{\Delta t}{\left(\varepsilon_0 + \sigma\Delta t\right)\Delta y}\left[H_x^n(j+1) - H_x^n(j)\right] \end{aligned} \tag{11.13}$$

where in taking the differences we let $\sigma E_z \Rightarrow \sigma E_z^{n+\frac{1}{2}}$ to maximize stability for the high conductivity of the material.

The magnetic field update equation is affected by the nonlinearity of the material. Starting with the Maxwell equation and applying the chain rule for derivatives we obtain

$$\frac{\partial E_z}{\partial y} = -\frac{\partial B_x}{\partial t} = -\frac{\partial B_x}{\partial H_x}\frac{\partial H_x}{\partial t} = -d\mu\left(H_x\right)\frac{\partial H_x}{\partial t} \tag{11.14}$$

from which we easily obtain

$$H_x^{n+1}(j) = H_x^n(j) - \frac{\Delta t}{\Delta y d\mu\left(H_x^n(j)\right)}\left[E_z^{n+\frac{1}{2}}\left(j+\tfrac{1}{2}\right) - E_z^{n+\frac{1}{2}}\left(j-\tfrac{1}{2}\right)\right] \quad (11.15)$$

with the nonlinear differential permeability of (11.11) evaluated using the magnetic field from the previous time step.

Because the conductivity of the magnetic material is so high, the velocity of electromagnetic waves in the material will be extremely slow relative to the speed of light in free space. Thus, we will be able to use relatively large time steps, which is good because we want to make calculations at relatively low frequencies that compare with the results in Reference 2. While in previous 1-D demonstrations we calculated fields in the free space regions on either side of a slab of material, this must be avoided in this situation because the time steps required in the free space regions will be many orders of magnitude smaller than in the material. We, therefore, follow Merewether's approach and apply the boundary conditions at the two interfaces at y = 0 and y = d, so that all FDTD field components will be inside the material.

In order to accomplish this the fields in Regions 1 and 3 are separated into incident, reflected, and transmitted. In Region 1 we have

$$E_z(t,y) = E_{inc}(t - y/c) + E_{ref}(t + y/c) \quad (11.16)$$

$$H_x(t,y) = \frac{1}{\eta_0}\left[E_{inc}(t - y/c) + E_{ref}(t + y/c)\right] \quad (11.17)$$

and in Region 3

$$E_z(t,y) = E_{tra}(t - (y - d)/c) \quad (11.18)$$

$$H_x(t,y) = \frac{1}{\eta_0} E_{tra}(t - (y - d)/c) \quad (11.19)$$

where η_0 is the impedance of free space and c is the speed of light. If we multiply (11.17) by η_0 and add (11.16) we obtain

$$E_z(t,0) + \eta_0 H_x(t,0) = 2E_{inc}(t) \quad (11.20)$$

and performing a similar operation on (11.18) and (11.19) yields

$$E_z(t,d) - \eta_0 H_x(t,d) = 0 \tag{11.21}$$

Inside the magnetic material the conductivity is high, and so to a good approximation at the frequencies we will be concerned with the displacement current being much smaller than the conduction current. Thus, (11.12) can be approximated as

$$E_z = -\frac{1}{\sigma}\frac{\partial H_x}{\partial y} \tag{11.22}$$

Applying (11.22) to (11.20) and (11.21) yields the boundary conditions just inside the two surfaces

$$\eta_0 H_x(t,0) - \frac{1}{\sigma}\frac{\partial H_x}{\partial y}\bigg|_{y=0} = 2E_{inc}(t) \tag{11.23}$$

$$\eta_0 H_x(t,0) - \frac{1}{\sigma}\frac{\partial H_x}{\partial y}\bigg|_{y=d} = 0 \tag{11.24}$$

These two boundary conditions translate into one-sided second order accurate difference equations (11-2) and (11-3)

$$H_x^n(1) = \frac{1}{F}\left[2E_{inc}(n\Delta t) + \frac{1}{2\sigma\Delta y}\left(4H_x^n(2) - H_x^n(3)\right)\right] \tag{11.25}$$

$$H_x^n(J) = \frac{1}{2\sigma\Delta y F}\left[4H_x^n(J-1) - H_x^n(J-2)\right] \tag{11.26}$$

where

$$F = \eta_0 + \frac{3}{2\sigma\Delta y} \tag{11.27}$$

At each time step we use (11.25) and (11.26) to evaluate the magnetic fields on the surfaces after the interior magnetic fields have been evaluated using (11.15).

We are now ready to begin our FDTD calculations. From Merewether we obtain the expression for the incident field as

$$E_{inc}(t) = 1.2678 A e^{-f_0 t} \sin(2\pi f_0 t) \tag{11.28}$$

where A is the variable amplitude and f_0 is either 1 or 10 kHz.

Let us first present the results, and then discuss the FDTD parameters necessary to obtain them. For the first set the fundamental frequency f_o is 1 kHz, and the incident waveform is shown in Figure 11-17. Calculated transient values of the field transmitted through the nonlinear magnetic sheet are shown in Figure 11-18 for three different amplitudes of incident field and compared to the implicit results of Merewether with excellent agreement. For the second case f_0 is 10 kHz, and the incident waveform is shown in Figure 11-19. Calculated values of the transmitted field for three different amplitudes of incident field are shown in Figure 11-20, and again excellent agreement is obtained with the Merewether results.

Now let us examine the FDTD parameters needed to obtain these results. This is somewhat complicated because the material is both highly conductive and nonlinear. The usual approach is to determine the minimum wavelength in the material, set the cell size a fraction of that, determine the maximum velocity of propagation, and set the time step based on these according to the Courant limit. However, in this case determining the minimum wavelength and maximum velocity of propagation are complicated by the nonlinearity of the material. The permeability depends on the field strength, and the frequency spectrum is affected as well. Also, because the material is so highly conductive, we are near the limits of applicability of FDTD because the term that makes our differential equation hyperbolic is extremely small compared to the other terms.

Because our material is highly conductive, i.e., at the frequencies of interest $(\sigma)/(\omega\varepsilon) >> 1$, the equations for velocity of propagation and wavelength simplify to those for a good conductor and we can determine the maximum velocity and minimum wavelength as

$$v_{max} \sqrt{\frac{4\pi f_{max}}{\sigma \mu_{min}}} \tag{11.29}$$

$$\lambda_{min} \sqrt{\frac{4}{\sigma \mu_{max} f_{max}}} \tag{11.30}$$

As an estimate we will take f_{max} to be $6f_0$, as the spectrum of the incident waveform is below 1% of peak amplitude at frequencies above this. The maximum permeability occurs at H = 0 field strength and is given by

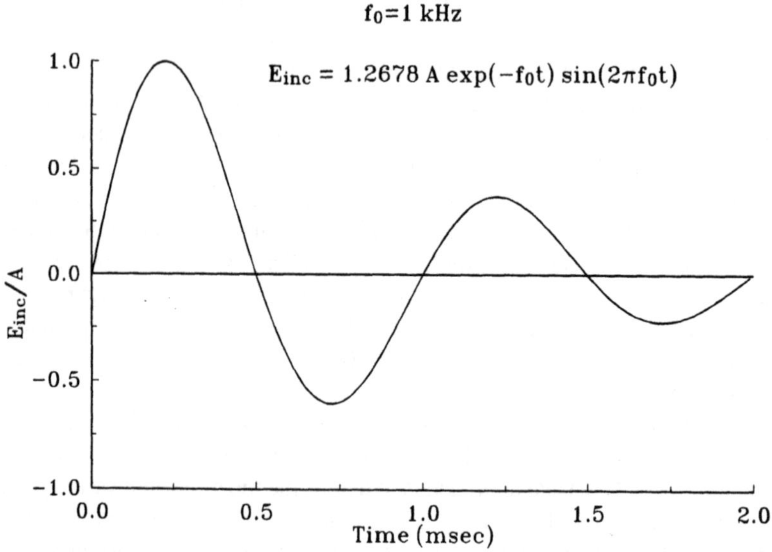

FIGURE 11-17. Normalized incident electric field for results in Figure 11-18; fundamental frequency is 1 kHz.

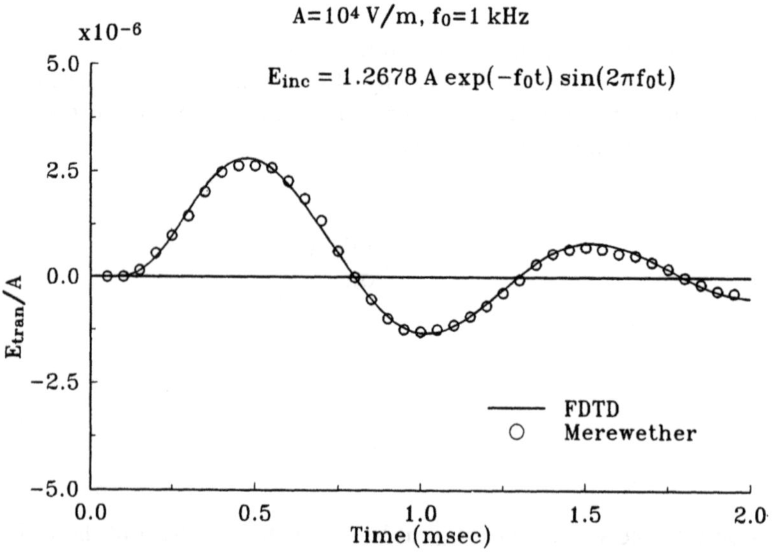

FIGURE 11-18a. Normalized transmitted electric field for incident waveform in Figure 11-17, with three different amplitudes (a–c).

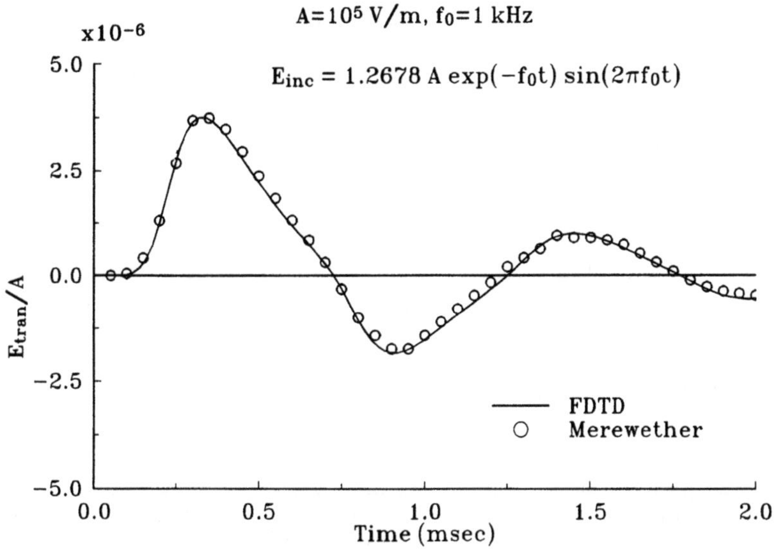

FIGURE 11-18b.

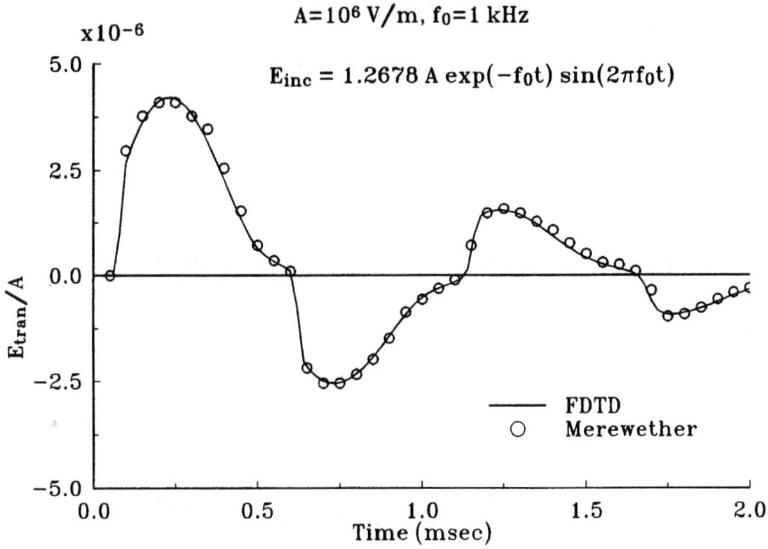

FIGURE 11-18c.

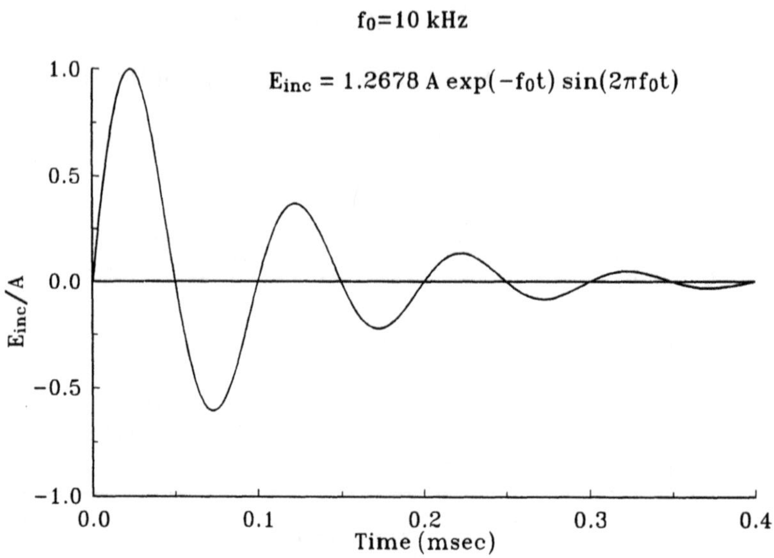

FIGURE 11-19. Normalized incident electric field for results in Figure 11-20; fundamental frequency is 10 kHz.

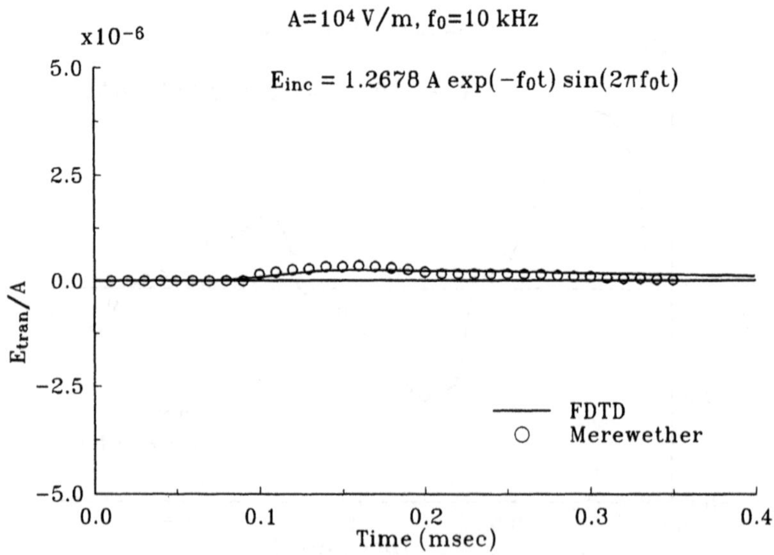

FIGURE 11-20a. Normalized transmitted electric field for incident waveform in Figure 11-19, with three different amplitudes (a–c).

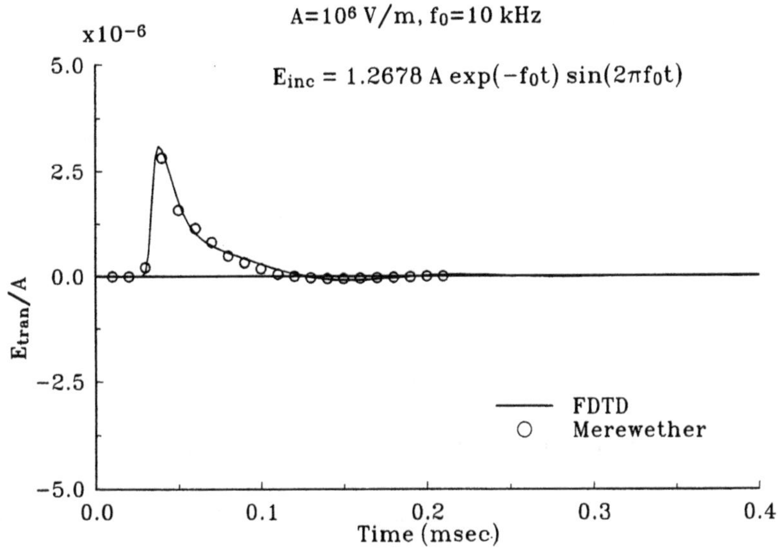

FIGURE 11-20b.

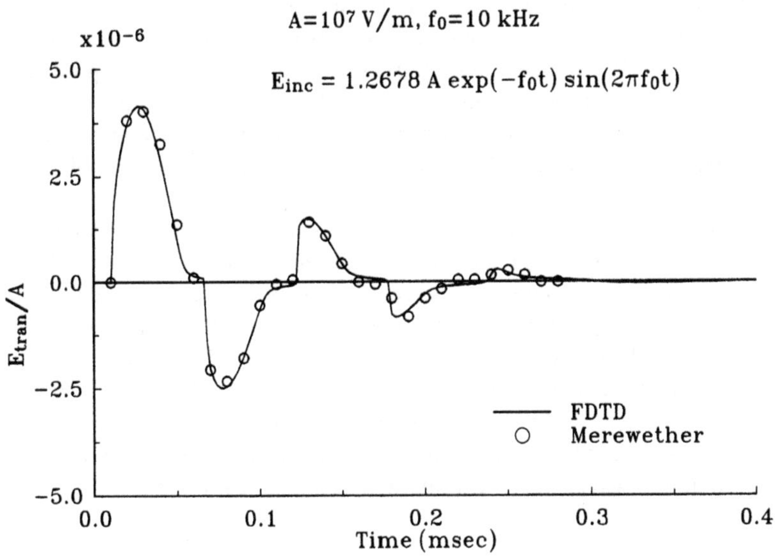

FIGURE 11-20c.

TABLE 11.1
FDTD Calculation Parameters for Results Shown in Listed Figures

Figure	Δy (μm)	$\lambda_{min}/\Delta y$	Δt (ns)	$\Delta y/(V_{max}\Delta t)$
18a	6.3	20.2	93.76	10
18b	6.3	20.2	93.76	10
18c	6.3	20.2	31.25	30
20a	2.0	20.1	94.12	10
20b	2.0	20.1	31.37	30
20c	2.0	20.1	31.37	30

$$\mu_{max} = d\mu(0) = \frac{B_s}{H_c} + \mu_m = 1.29 \times 10^{-2} \tag{11.31}$$

while the minimum permeability occurs at H = (∞) and is

$$\mu_{min} = d\mu(\infty) = \mu_m = 1.67 \times 10^{-4} \tag{12.1}$$

We usually like to have $\Delta y \leq (\lambda_{min})/(10)$, and Courant stability requires that $\Delta t \leq (\Delta y)/(V_{max})$. The parameters actually used to obtain the results in Figures 11-18 and 11-20 are summarized in Table 11.1. While a search for the precise maximum time step for which stability was obtained was not performed for each situation, values much larger than those shown in Table 11.1 would introduce oscillations and noise in the FDTD results. The parameter of 20 divisions per minimum wavelength is quite typical of FDTD calculations. Due to the combination of high conductivity and nonlinearity, however, the time step size must be reduced from 10 to 30 times less than that required by the Courant condition to maintain stability, with smaller time steps required as the material is driven more nonlinearly. Nevertheless, with only this decrease in the time step size relative to the Courant limit, the FDTD method is capable of producing accurate results for this very challenging combination of material parameters.

It is interesting to investigate the relative effects of the high conductivity and nonlinearity in forcing the time step size reduction relative to the Courant limit. In order to illustrate the relative importance some test calculations were made for various combinations of conductivity, permittivity, and nonlinearity. The test calculations were made for the excitation of (11.18) with $f_0 = 1$ kHz. The maximum frequency contained in the exciting waveform was again taken as $6f_0$, so that $\omega_{max} = 2\pi \cdot 6f_0$. Because not all calculations are made for high conductivities, general expressions were used to find the maximum velocity and minimum wavelength, with the wave number k given by

$$k(\omega,\mu) = \omega\sqrt{\frac{\mu\varepsilon}{2}}\sqrt{\sqrt{1+\left(\frac{\sigma}{\omega\varepsilon}\right)^2}+1} \tag{12.2}$$

TABLE 11.2
Time Step Reduction below Courant Limit (right column) Required for FDTD Stability as Conductivity and Permittivity are Changed for Linear Medium

σ (S/m)	ε_r	Δy (m)	$\lambda_{min}/\Delta y$	Δt (μs)	$\Delta y/(V_{max}\Delta t)$
0	Any	2400	20	8.3	1.0
10^{-6}	1	170	20	6.4	1.3
10^{-5}	1	630	20	3.6	2.3
10^{-4}	1	200	20	2.7	3.0
10^{-3}	1	65	20	2.6	3.2
10^{-3}	10	64	20	2.8	3.0
10^{-3}	100	63	20	3.6	2.3
10^{7}	1	0.0013	10	10.0	1.6
10^{7}	1	0.00065	20	2.6	3.2
10^{7}	1	0.00032	40	0.65	6.4

with

$$V_{max} = \omega_{max} / k(\omega_{max}, \mu_{min}) \tag{12.3}$$

$$\lambda_{min} = 2\pi / k(\omega_{max}, \mu_{max}) \tag{12.4}$$

and with μ_{min}, μ_{max} given by (11-31) and (11-32).

Let us first consider stability requirements for a linear medium. The time step size relative to the Courant stability limit is independent of the excitation amplitude A of (11.18) and of the permeability (for no magnetic losses). It does depend on the conductivity, and on permittivity for non-zero conductivity. This is illustrated in Table 11.2. The right-hand column indicates the factor below the Courant limit inside the material at which the time step must be set for stable calculations. From this table we see that only for zero conductivity can we actually compute with our time step at the Courant limit inside the material. As the conductivity increases the time step must be reduced, but with a plateau at relatively low conductivity values. This time step reduction also depends on the cell size, as shown in the data for the highest conductivity. For a given cell size (relative to the wavelength in the material) the time step reduction depends on the ratio of $(\sigma)/(\omega\varepsilon)$. Identical time step reduction factors in the right-hand column of Table 11.2 will be found to correspond to the same $(\sigma)/(\omega\varepsilon)$ ratio for the same cell size, that is, for the same $(\lambda_{min})/(\Delta y)$ ratio. This reduction in time step is not a consideration in typical FDTD calculations because most involve free space, and for all conductivity values considered here a time step at the Courant limit for free space will provide stable results.

Next let us consider the nonlinear medium with permeability as described previously and permittivity of free space, but with conductivity zero. Now the

TABLE 11.3
Time Step Reduction below Courant Limit (right column) Required for FDTD Stability as Excitation Amplitude is Changed, Nonlinear Magnetic Medium

s (S/m)	A	Δy (m)	$\lambda_{min}/\Delta y$	Δt (ns)	$\Delta y/(V_{max}\Delta t)$
0	1	25	20	950	1
0	10^3	25	20	950	1
0	10^6	25	20	240	4[a]
0	10^6	25	20	95	10[b]
0	10^6	25	20	32	30
0	10^9	25	20	38	25[a]
0	10^9	25	20	32	30
0	10^9	50	10	130	15

[a] Unstable.
[b] Marginally stable, but noisy; other entries stable and smooth.

stability depends on the amplitude A of (11.18). In Table 11.3 the reduction in the time step below the Courant limit required for stability is shown for various amplitudes. The boundary between complete stability and instability is not precise, as the FDTD results may not be so unstable as to overflow but yet still have noise and chatter over some parts of the computation. Thus, some entries in Table 11.3 have an indication of marginal stability. For small values of A no reduction in time step is required (a 1 in the right column), but as A is increased and the permeability goes through greater changes during the FDTD calculation the time step size must be reduced. Because Table 11.3 is for zero conductivity, this clearly indicates that a smaller time step is required by the nonlinearity of the material. This is the price paid for including nonlinearities by changing a parameter at each time step without simultaneously satisfying the relevant equations as described in the previous section.

Finally let us consider the nonlinear magnetic material with a conductivity of 10^7 S/m. Stability conditions for varying amplitudes of the excitation are shown in Table 11.4. Some of the entries in the right column are <1 because v_{max} is computed with μ_{min} while λ_{min} is computed from μ_{max}. Comparing Tables 11.3 and 11.4 it is clear that both the high conductivity and the nonlinearity of the material contribute to the necessity of reducing the time step size below the Courant limit.

To summarize, two approaches for including nonlinearities in FDTD calculations have been demonstrated. One of these allows the time step to remain at the Courant limit, but requires that a nonlinear equation be solved at each time step. The other simply changes the nonlinear parameter at each time step, but requires that the time step size be reduced below the Courant limit for stability. The former is most useful when only a relatively small number of FDTD cells contain nonlinear materials or devices, while the latter is simple to apply and is relatively efficient when most of the FDTD cells contain nonlinear materials.

TABLE 11.4
Time Step Reduction below Courant Limit (right column) Required for FDTD Stability as Excitation Amplitude is Changed, Conducting Nonlinear Magnetic Medium

σ (S/m)	A	Δy (μm)	$\lambda_{min}/\Delta y$	Δt (ns)	$\Delta y/(V_{max}\Delta\tau)$
10^7	10^{-3}	6.4	20	3200	0.3[a]
10^7	10^{-3}	6.4	20	2400	0.4
10^7	10^{-1}	6.4	20	3200	0.3
10^7	10^{-1}	6.4	20	2400	0.4[a]
10^7	1	6.4	20	47	20[a]
10^7	1	6.4	20	38	25[b]
10^7	1	6.4	20	32	30[b]
10^7	1	6.4	20	24	40
10^7	10^3	6.4	20	38	25[a]
10^7	10^3	6.4	20	32	30
10^7	10^3	13	10	150	13[a]
10^7	10^3	13	10	13	15

[a] Unstable.
[b] Marginally stable, but noisy; other entries stable and smooth.

REFERENCES

1. **Liu, T. K., Tesche, F. M., and Deadrick, F. J.,** Transient excitation of an antenna with a nonlinear load: numerical and experimental results, *IEEE Trans. Ant. Prop.*, 25, 539, 1977.
2. **Merewether, D. E.,** Electromagnetic pulse transmission through a thin sheet of saturable ferromagnetic material of infinite surface area, *IEEE Trans. Electromag. Compat.*, 11, 139, 1969.
3. **Strikwerda, J. C.,** *Finite Difference Schemes and Partial Differential Equations*, Wadsworth & Brooks/Cole, Pacific Grove, CA, 1979, 74.

Chapter 12

VISUALIZATION

12.1 INTRODUCTION

The FDTD technique, as well as other electromagnetic (EM) modeling techniques such as the method of moments, is numerically intensive, requiring significant computational resources for complex models. In addition to EM modeling techniques two broad areas stand out as numerically intensive — digital signal processing and scientific visualization. Our work on FDTD has been on computers optimized for scientific visualization applications, and we have found that this is an excellent hardware environment for FDTD. It is our general observation that computers optimized for visualization work well with FDTD and there is no reason to forgo visualization when working with FDTD on such a platform as the numerical resources needed to support one generally goes hand in hand with the resources to support the other. In fact, visualization should be considered whenever numerically intensive computer modeling is employed in whatever application.

The rationale for scientific visualization applied to FDTD can be made more specific by noting that it provides

- Increased confidence in the fidelity of the model when the model is rendered visually
- Effective diagnostics on field behavior, in particular the timing, magnitude, and source of scattering and coupling effects
- Physical insight as the evolution of the interaction is observed
- Intuition building as a multiplicity of interactions are viewed over time and conditions are varied such as the type of interaction object (geometry and material) and field configuration (angle of incidence and polarization, as examples)

The visualization discussion presented here is not meant to be definitive nor exhaustive, but rather suggestive of what visualization can do to help the EM modeler employing FDTD or for that matter other modeling techniques. Our discussion is based on our own experiences in developing visualization using graphical primitives on a Silicon Graphics workstation. These efforts can easily be matched or bettered by any modestly dedicated practitioner of EM modeling. We hope only to point out the advantages of scientific visualization for EM modeling and to encourage the reader.

12.2 TYPES OF VISUALIZATION

Visualization can be implemented in many forms with varying degrees of sophistication. In its simplest form it is static and monochromatic, a traditional radiation pattern (Figure 12-1), or a contour field plot (Figure 12-2). Color can be used to enhance the features as might be done with the contour plots. A further step is a 3-D plot of the geometry or fields, which is composed of 2-D slices. The slices can be orthogonal planes or more flexibly arbitrarily oriented slices. If the geometry is being rendered, it is desirable to have the option of hidden line removal. Rather sophisticated tools to perform these functions have evolved along with FDTD. One example is IMAGE 3.3, a mesh (the object in the gridded space)-verification tool[1] developed at Lawrence Livermore National Laboratory, Livermore, CA.

The next major step, and it is enormous, is to animate the visualization. Frequency domain codes are in essence static; time has been replaced by ω. A time domain code such as FDTD is, on the other hand, essentially dynamic and the time behavior of the fields is crucial. Animation is vital for a complete understanding of the behavior of the modeled system. A local monitor point can provide a time-dependent record, for instance, H_x (I_o,J_o,K_o) over N time steps. This is not animation, and the user must "fill in" the blanks. A plot of H_x over some region of space, typically a 2-D slice through the problem space, at successive times is the beginning of an animation if displayed as a succession of "stills" in time. If the computer employed is fast enough these "stills" can be displayed at a speed as slow as 6 to 10 frames per second (fps) for a possible animation when the "stills" are a time step between successive frames. Alternatively, individual frames that may take hours to compute can be assembled into frames on videotape or laser disc and played back at rates again as low as 6 to 30 fps or higher. This is true animation whether in color or black and white, or in 2- or 3-D. We have successfully animated waveguide field propagation in real time (~10 fps) in color over 28 × 28 2-D slices on a Silicon Graphics 4D220.[2] With the same machine, 2-D scattering, and coupling geometries up to a 100 × 100 cell display was possible.

This real-time animation capability raises two points. Point one is interactivity. A real-time animation display is a powerful "teaching" tool for a student and for an EM modeler. The student can quickly grasp powerful concepts and gain real physical insight. The modeler can quickly check the model and the physical process, and determine their significance. Visualization at this level of sophistication can in minutes (a few hundred to thousands of frames) impart the same understanding that might take weeks or months to obtain from conventional response plots.

The second point is that there is a cost to all this that will be looked at in more detail later in the chapter. For now it will be noted that the architecture of the Silicon Graphics machine is what is needed for effective visualization. There is separate hardware for the visualization activities, in effect an independent visualization CPU and the double buffering needed for a flicker-free display. Only recently have PCs begun to appear in this configuration and this development will hasten the adoption of more sophisticated visualization for FDTD applications on this category of computers.

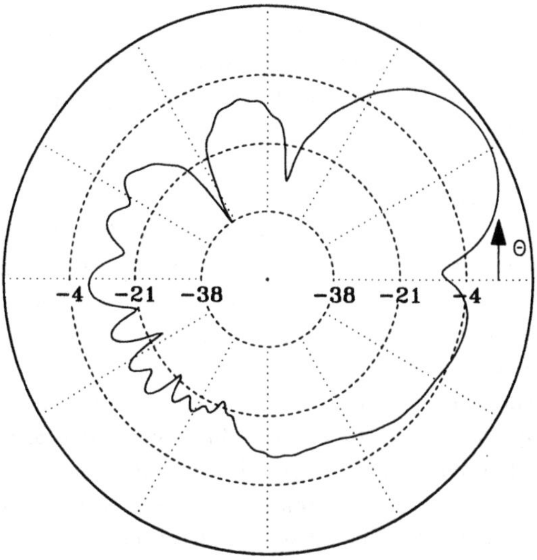

FIGURE 12-1. E-plane gain pattern of Eθ ρ = 0°, 30° shaped-end radiator.

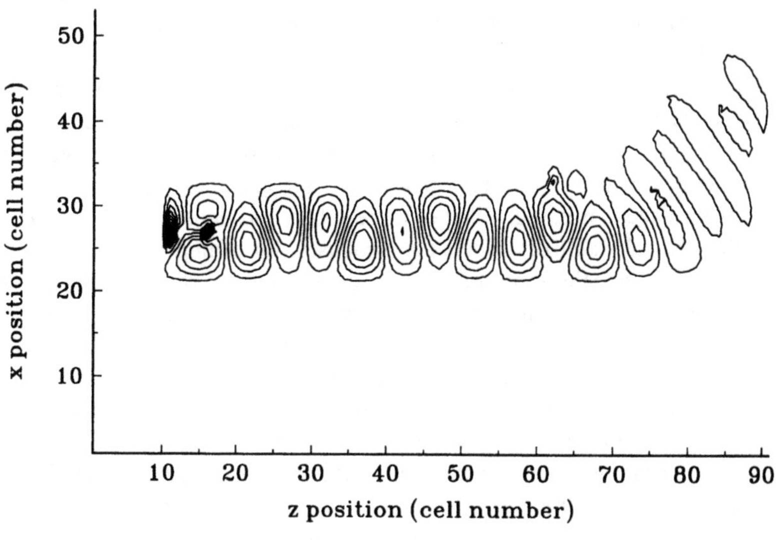

FIGURE 12-2. Contour plot of E_z for 30° shaped-end radiator x-z plane at y = 27.

12.3 EXAMPLES

Four specific rationales were given in the introduction for scientific visualization. Here, we shall give concrete examples of what was meant. We will treat the issues of model fidelity, model response, physical process insight, and intuition building.

12.3.1 MODEL FIDELITY

For a simple example of how model fidelity may be assured with visualization consider the cuts that can be made through a spherical scatterer to ensure sphericity. The use of different colors to denote different materials would allow the verification of the correct material modeling using exactly the same techniques used to verify the geometry.

Without visualization some glaring modeling errors are possible. Once when modeling an aircraft one of the authors left out a large section of the fuselage. This aerodynamically impossible aircraft still reflected electromagnetic energy and behaved in a seemingly reasonable fashion. The only obvious indication that something was amiss was the absence of a strong half-wavelength dipole resonance at the expected frequency based on the fuselage length. The error was discovered as a result of this observation, but the potential was there for not discovering the error or not discovering it quickly enough. Visualization makes events like this extremely rare.

12.3.2 MODEL RESPONSE

The minimum model response characterization is to select a modest number of monitor locations and to store the resulting response data from these locations. The aircraft response data of Chapter 4 is an example of this response characterization. Time response data are obtained directly from the code and the corresponding frequency domain response data are obtained through the simple expedient of applying a fast Fourier transform (FFT) to the time domain response data. This is sufficient to observe the major physical processes such as resonant behavior in the aircraft response and as just discussed, to note gross abnormalities in the response, which occur, for example, when part of the aircraft fuselage is missing. What of more subtle questions — the question of how the vertical stabilizer effects the responses, for example.

One approach to the above question is to remove the vertical stabilizer from the model and observe changes in the monitor location responses. A second approach is possible with scientific visualization, and this is to observe the fields scattering off the aircraft, and in particular the vertical stabilizer. The size and direction of the scattered as well as the total fields can be examined and their effects on adjoining portions of the geometry estimated.

Another example is a 2-D parallel plate waveguide operating in the TE mode. We can examine how a sinusoidal wave with E parallel to the plate surfaces scatterers from the leading edges of the plates and establish that this is the dominant scattering mechanism. For most other geometries there are other equally obvious dominant or at least important scattering mechanisms that are immediately identifiable.

12.3.3 PHYSICAL PROCESS INSIGHT

The example of the parallel plate waveguide and the dominant scattering from the leading edges of the two plates ideally illustrates how visualization provides the modeler insight into the physical processes that characterize the electromagnetic interaction being modeled. These processes are often categorized as scattering or coupling processes or mechanisms. The discussion presented here is based on an actual example of visualization presented on videotape in Reference 2.

For the parallel plate waveguide visualization allows a clear view of a cylindrical wavefront emanating from each leading edge of the waveguide in the scattered field display. Comparing two different geometries with two different plate spacings it is easily noted that with the wider spacing the interferences between the two scattered fields, one from the top plate and one from the bottom plate, is lessened and a significant total field appears within the plates. For the smaller plate spacing there is nearly complete cancellation due to interference and the total field between the plates is nearly zero.

The plate spacing was, of course, selected so as to be above and below cutoff for the frequency of the incident sinusoidal field for the large and small plate spacing. The wave penetrating the parallel plate waveguide structure with the wider spacing therefore comes as no surprise, as does the near-absence of penetrating fields in the parallel plate waveguide operating below cutoff. Even though the results may have been anticipated, the visualization clearly identifies the process and makes its effect obvious.

12.3.4 INTUITION BUILDING

As the modeler works with the visualization of increasingly different models, an ever-increasing set of experiences, similar to the parallel plate waveguide, are required. The modeler becomes aware of what are likely scatterer sources, how effective a shadow area may result in the region behind a plate, what portion of the frequency range of an incident pulse will penetrate an aperture, what does a bend in a waveguide do to the signal that encounters the bend, etc. These are not trivial insights.

Visualization literally makes the unseen world of electromagnetics visible. The designer of an anechoic chamber can see how the incident wave emanating from a horn and reflected off a reflecting dish (Figure 12-3) fills the bulk of the chamber with what is hoped to be a plane wave. The energy loss to the absorbing walls can be examined (Figure 12-4) to see where most of the absorption takes place and to decide where design changes may be advantageous. The figures shown here are static renditions of what was actually visualized of Silicon Graphics 4D 220 FDTD predictions in color on videotape.

With visualization intuition building is a natural process and it reinforces itself. The EM modeler develops this intuition which in turn lets the modeler

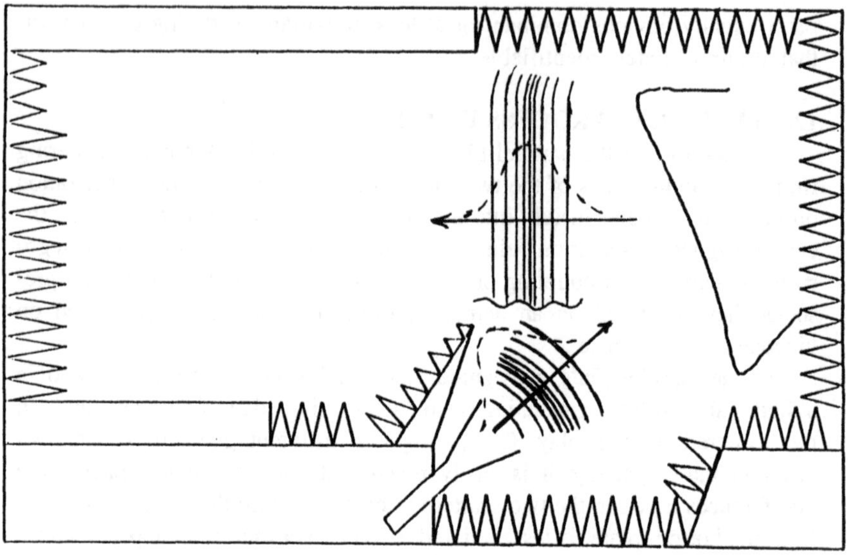

FIGURE 12-3. Idealized rendering of waveform "straightening".

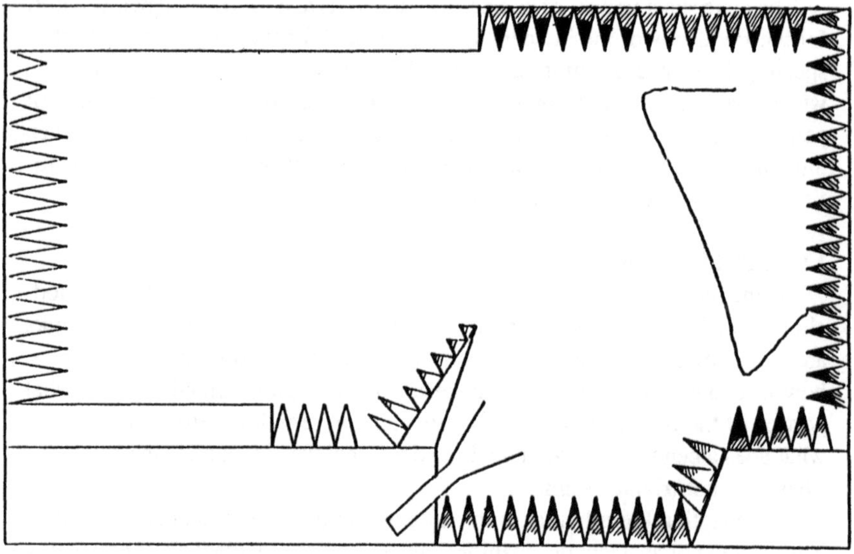

FIGURE 12-4. Idealized rendering of heat absorption.

anticipate what results may be obtained with a particular model and to tune the model to obtain optimum results before making any runs. Without visualization intuition building is a slow and laborious process. Visualization plays an important role in EM modeling, a role that will only grow as modeling becomes more sophisticated and the physical processes more complex.

12.4 RESOURCES AND COST OF VISUALIZATION

Having extolled the virtues of visualization we must now address the resources required and the costs. They are not insignificant when one addresses the most demanding form, animation. We shall assume color is desired over a wide range of colors mandating 24-bit color (16-meg colors) as opposed to 8-bit color (256 colors). Note that 8 bits provides a dynamic range of 256:1 if used for a linear gray-scale display and can only marginally convey amplitude information. We shall also assume at least 6 fps are required for animation. Ideally, 30 fps or more should be sought. If animation is not required then the cost of a single frame sets the visualization requirements, and these can be easily met in most cases.

A problem space $(NS)^3$ in size or nominally NS cells on a side may range from NS equal to around 30 and up to nearly 1000. A slice through the problem space will then be on the order of 30×30 cells up to 1000×1000 cells. If a cubical cell with square faces is represented by an array of pixels no less than 2×2 pixels, then on a screen with 1280×1024 pixel resolution a slice of up to 500×500 cells can be accommodated. We will take this as a reasonable current upper limit as it represents a problem space of a little over 10^8 cells, and we will use a 28×28 cell slice as the lower limit as this corresponds to our present EM courseware[2] visualization resolution.

The 500×500 cell slice would require 1000×1000 pixels to be transferred per frame or 1-meg pixel per frame. At 30 fps a 30-meg pixel/s transfer rate is required. At 3 bytes/pixel the transfer rate is nearly 100 Mb/s. Our visualization platform (Silicon Graphics 4D220) has a capability approximately 4 times this rate or 400 Mb/s. Graphic accelerator boards costing hundreds of dollars now offer approximately one third this capability on PCs. When the frame rate is dropped to 10 fps and only an array of 200×200 cells at 2×2 pixels per cell or 400×400 pixels is updated for each frame the data transfer rate drops to 400×400 pixels $\times$ 3 bytes/pixel $\times$ 10 updates per second $= 4.8 \times 10^6$ or ~5 Mb/s. This is well within the low end capability of visualization hardware. It would provide a slice through a $(200)^3$ cell or 8-meg cell problem space, which is a rather large problem space. For more powerful PCs with 100 Mb/s transfer rates no compromise is required and machines of this caliber are becoming common.

In the future matters only improve. Using the role of thumb that speed, memory, cost, etc. improve tenfold every 6 years, in 24 years machines will be available with CPUs that are roughly 10^4 times more powerful. The problem space that can then be modeled will be ten times larger than that of today because the modeling cost scales linearly with x, y, z, and t. A 2-D slice through the problem space will only increase in the number of pixels by the increase in two spatial dimensions or 100-fold. Having reached a performance level today capable of supporting visualization we will only find the requirements less demanding as a percentage of total resources required of

FDTD modeling and visualization. Thus, we expect visualization to be an inseparable adjunct to FDTD modeling.

REFERENCES

1. **McLead, R. R. and Allison, M. J.,** IMAGE 3.3, Tutorial Manual Ver. 1.5, UCRL-MA-104860, Lawrence Livermore National Laboratory, Livermore, CA, 1990.
2. **Kunz, K. S.,** Progress in Computational Electromagnetic Modeling for EM Courseware: A Personal Perspective at Penn State, presented at IEEE and URSI Joint Symp., Chicago, July 18–25, 1992.

PART 4: ADVANCED APPLICATIONS

Chapter 13

FAR ZONE SCATTERING

13.1 INTRODUCTION

Electromagnetic scattering has been an interesting problem since the earliest days of electromagnetic research. In the past, problems of interest included scattering from raindrops at light frequencies in order to explain the color of the sky and the origin of rainbows. Scattering from raindrops and ice crystals is still a topic of interest, but at microwave frequencies for applications such as radar meteorology. More recently the design of military aircraft with low scattering cross sections that make them less detectable by radar has renewed interest in this topic.

All components of an FDTD code required to make scattering calculations have been introduced in previous chapters, including the scattered field formulation itself, the outer radiation boundary condition (ORBC), and the far zone transformation. Should the scatterer contain frequency-dependent materials their effects can be included as well using the methods of Chapter 8. The only additional item needed is a suitable fast Fourier transform (FFT) computer code if frequency domain scattering results are desired. Let us consider these components as they apply to scattering calculations.

First, the scattered field formulation is extremely well suited to scattering calculations. Because the incident wave is specified analytically, it illuminates the scatterer accurately without being distorted by grid dispersion or other FDTD error sources. The scattered field, which is the field desired in scattering calculations, is directly computed. As this is the case, there is no need to divide the FDTD space into total and scattered field regions, as is the case when the total field formulation is used[1]. Launching the incident plane wave is also simpler, since it is specified analytically as described in Chapter 3.

The outer radiation boundary condition is extremely important in scattering calculations. This is especially true when trying to calculate scattering from targets with low level cross sections. The difficulty is that fields scattered from the target may partially reflect from the absorbing boundary and reilluminate the target. If, for example, a target shaped for low level scattering when illuminated from the incident field direction is reilluminated from another direction due to outer boundary reflection, the resulting scattered field may be much higher than the correct value. While most of the results presented in this chapter were computed using a second order Mur absorbing boundary, other absorbing boundaries with better performance, especially in the edge and corner regions, may be desirable for scattering applications.[2,3] Second order Mur will work very well if enough empty cells are used to separate the absorbing boundary from the scatterer, but the higher performance absorbing boundaries will work better if fewer free space cells are used. Moving the absorbing boundaries closer to the scatterer saves computer resources and allows larger targets to be considered with a given amount of computer memory.

Scattering cross sections are defined in the limit of the far zone of the scatterer, as the distance from the scatterer approaches infinity. This means that the near zone FDTD results must be transformed to the far zone. This can be done in the frequency domain,[1] but it is more efficient to use pulsed excitation and transform the transient fields to the far zone, especially if only backscatter or a few bistatic scattering directions are of interest. With a transient far zone transformation, scattering results for the entire band of frequencies for which the FDTD calculations are valid are produced from just one computation. Because the computation time for FDTD to reach steady state for sinusoidal and pulse excitation is approximately the same,[4] and because the FFT is extremely efficient, pulse excitation for scattering calculations with a transient far zone transformation is almost always preferred. In Chapter 7 an approach to transform the transient FDTD fields to the far zone was presented, and some examples given for scattering from simple shapes in 2- and 3-D were shown to illustrate the method. In this chapter we concern ourselves only with 3-D scatterers, and the 3-D far zone transformation of Chapter 7 is used.

While scattering calculations using the Yee cell can be quite accurate for many situations, for low level scattering from curved surfaces or flat surfaces that do not align with the grid, the results may be inaccurate. The error produced in approximating surfaces which do not align with the grid is commonly called "staircasing" error, because the stepped Yee cell approximation to the surface has the appearance of a flight of stairs (although the steps may be irregular). This difficulty can be solved by brute force by using extremely small cells, 1/20 to 1/50 of a wavelength or even smaller, but finite computer resources may prohibit this. Alternative FDTD algorithms for this situation are being actively researched,[5-7] but the present state of development requires the application of considerably more effort than the Yee cell algorithm discussed here. A special mesh for each object shape must be generated, and the mesh information used in the FDTD update equations. Later in the chapter a discussion of staircase errors is given.

Many of the methods described in previous chapters to deal with special materials such as frequency-dependent or nonlinear materials may be used to deal with special scatterers. Scatterers made of highly conductive materials or coated with thin layers of conductive material are difficult to model directly because the wavelength in such materials is relatively small as compared to free space. Thus, the FDTD cells must also be made quite small inside the conductive materials. A method to avoid this, commonly used in the frequency domain, is to model the boundary condition at the scatterer surface with an impedance. The time domain equivalent of this for both constant and frequency-dependent materials was presented in Chapter 9, and a result was given there for scattering from a lossy cylinder. Scattering results for a sphere composed of frequency-dependent material are given at the end of this chapter.

Because all of the tools needed for scattering calculations have been introduced in previous chapters, this chapter is not comprehensive. The next section presents the basic equations and approach for transient far zone scattering calculations using FDTD. The section following that gives a simple demonstration of the errors that can be encountered due to staircasing effects. Following that, the ease with which FDTD can calculate scattering from impedance sheets is presented. The importance of the absorbing boundary condition in producing accurate scattering results is then demonstrated, followed by an example of scattering from a frequency dependent material.

The chapter gives examples of scattering results obtained using FDTD. Also, several examples have already been shown in conjunction with the far zone transformation method in Chapter 7 and the surface impedance formulation of Chapter 9. The thrust of these examples is that scattered field formulation FDTD based on the Yee cell is extremely well suited to scattering calculations, within the limitations imposed by staircasing, and provided that the absorbing boundary is located sufficiently far from the scatterer to absorb the scattered fields.

13.2 FUNDAMENTALS

Given the tools already developed in previous chapters calculating scattering using FDTD is quite straightforward. First consider the basic definition of scattering cross section, reproduced here from Chapter 7. For 3-D we have

$$\sigma_{3D} = \lim_{R\to\infty}\left(4\pi R^2 \frac{\left|E^s_{3D}\right|^2}{\left|E^i\right|^2}\right) \tag{13.1}$$

where σ_{3D} is the scattering cross section in square meters, E^s_{3D} is the Fourier transform of the far zone electric fields E_θ or E_ϕ obtained using the transient far zone transformation of Chapter 7 (specifically from (7.7) or (7.8)), and E^i is the Fourier transform of the incident plane wave electric field. Because the far zone transform of Chapter 7 normalizes the far zone distance R to 1, the actual value of R used in evaluating (13.1) is 1 also. For phase-accurate results the incident field should be sampled at the same location as the time reference for the far zone transformation, specifically from the point where the $\bar{r}'$ vector of (7.9) and (7.10) starts, and the time shift of (7.16), which is applied to the far zone fields, must also be applied to the sample of the incident field E^i before the Fourier transformation is made. Also, E^i may be affected by the polarization of the incident field, i.e., it may be determined by the incident E_θ or E_ϕ or some combination of the two.

The corresponding 2-D scattering width is defined as

$$\sigma_{2D} = \lim_{\rho\to\infty}\left(2\pi\rho\frac{\left|E^{s}_{2D}\right|^2}{\left|E^{i}\right|^2}\right) \tag{13.2}$$

where E^{s}_{2D} is the Fourier transform of either E_z or E_ϕ of (7.23) or (7.24). The discussion of the variables E^i and ρ (R) given above for 3-D scatterers also applies for 2-D scattering calculations.

Examples of results obtained using scattered field FDTD with the transient far zone transformation and Mur second order absorbing boundaries are given in Chapter 7 for both flat plates in 3-D and circular cylinders in 2-D. While those results indicate the fundamental capability of scattered field FDTD to compute scattering cross sections, some additional discussions of this application follow. In the next section the errors that can be expected when Yee cell FDTD is applied to scatterers that have either curved surfaces or are not aligned with the grid are presented.

13.3 STAIRCASE ERRORS

Chapter 7 presented results for scattering from a flat plate using scattered field FDTD, and these were compared with the method of moments. The comparison indicated that FDTD was capable of producing accurate results for this geometry. The flat plate geometry considered then is shown in Figure 13-1, but with the computational space extended in size to 60 × 60 × 60. One reason that the FDTD results were good is that the flat plate naturally fits the Yee cell rectangular grid. To demonstrate what happens when the scatterer does not fit the grid we will use the geometry of Figure 13-2. Here, the same 29 × 29-cm square plate is being considered, but in this case it lies at a 45° angle with respect to the FDTD grid, and therefore must be approximated by a series of "steps" (not all the steps are shown in the figure for clarity). Figure 13-2 clearly shows the source of the term "staircase" used to describe the FDTD approximation to surfaces which do not conveniently fit the grid, and by association to describe the resulting errors.

Now let us investigate the error in approximating a plate which is actually flat by a staircased plate located at 45° with respect to the grid. To do this consider the sequence of Figures 13-3 to 13-6. In Figures 13-3 and 13-4 the scattering cross section of the flat plate is computed for normal incidence (with respect to the plate) for both the actual flat plate aligned with the FDTD grid (Figure 13-1) and the flat plate at 45° with the FDTD grid (Figure 13-2). Figures 13-3 and 13-4 indicate that for normal incidence the scattering from the staircased plate is almost identical to that from the grid-aligned flat plate. Consider next, however, the case of scattering from a flat plate with the incidence angle 85° from the plate normal in the $\phi = 0$ plane, as indicated in

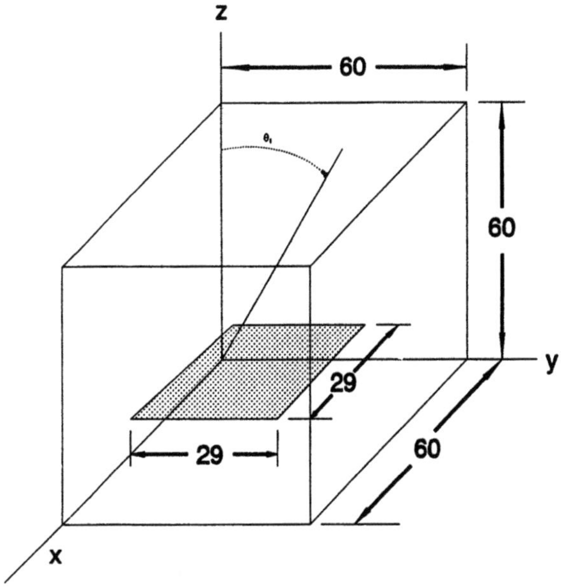

FIGURE 13-1. Flat plate 29 × 29 cm (1 cm cells) aligned in the FDTD grid.

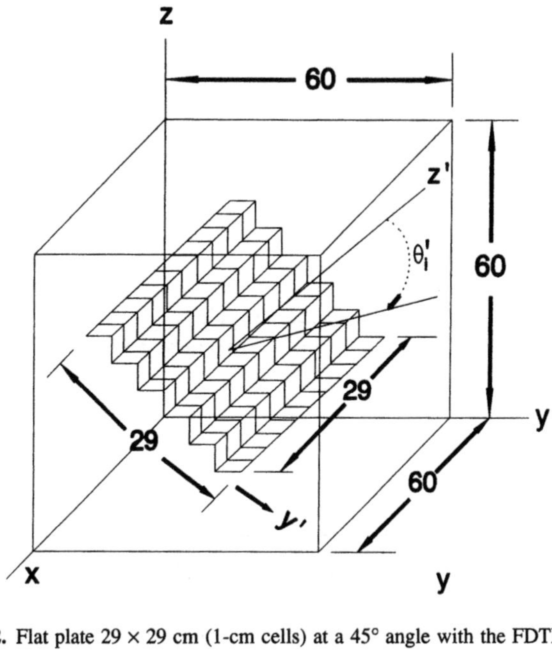

FIGURE 13-2. Flat plate 29 × 29 cm (1-cm cells) at a 45° angle with the FDTD grid showing "staircase" effect. Not all of the "steps" are shown.

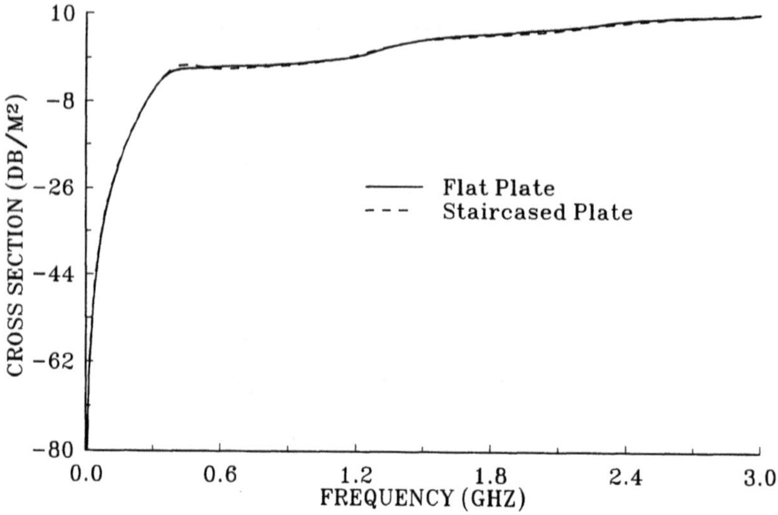

FIGURE 13-3. Copolarized backscatter for normal incidence, φ-polarized incident wave, for both flat and staircased plates.

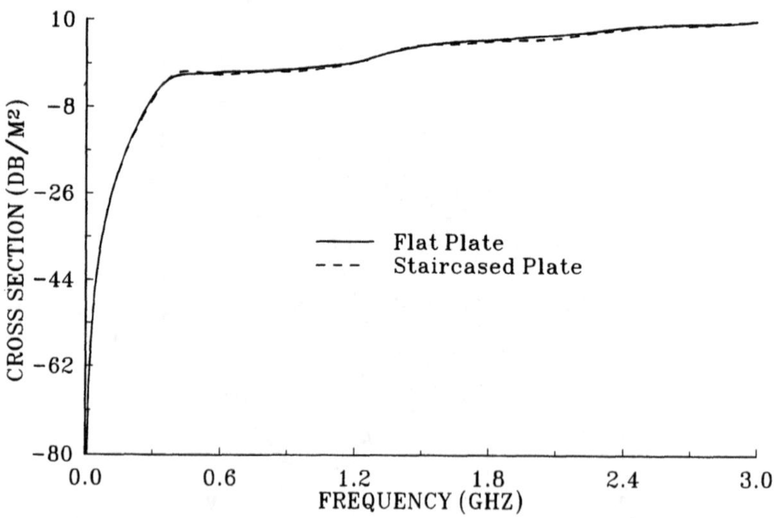

FIGURE 13-4. Copolarized backscatter for normal incidence, θ-polarized incident wave, for both flat and staircased plates.

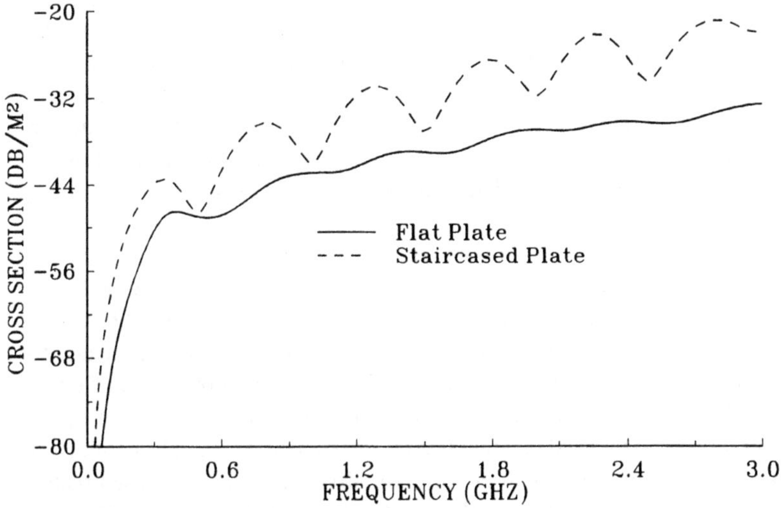

FIGURE 13-5. Copolarized backscatter for 85° incidence, ϕ-polarized incident wave, for both flat and staircased plates.

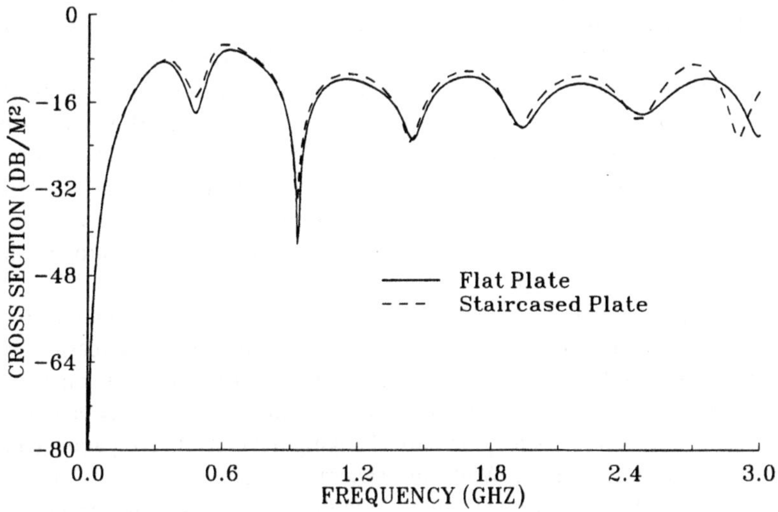

FIGURE 13-6. Copolarized backscatter for 85° incidence, θ-polarized incident wave, for both flat and staircased plates.

Figure 13-2. Even in this case, for ϕ-polarized incident field as shown in Figure 13-5, the flat and staircased approximation have very nearly the same scattering result. For the θ-polarized incident field results of Figure 13-6, the staircased plate produces very different scattering results than the flat, grid-aligned plate.

The physical explanation for this is quite simple. The ϕ-polarized incident wave has its electric field parallel to the plate surface, and relatively strong currents are induced on both the staircased and grid-aligned flat plate. In this situation, the incident E_ϕ- polarized wave is equivalent to an E_x component, and both the grid-aligned flat plate and the staircased plate directly interact with this field component.

On the other hand, the θ-polarized electric field is nearly perpendicular to the plate surface, and excites relatively weak currents on the flat plate. Indeed, for 90° incidence angle (relative to the plate) no currents would be excited on the flat grid-aligned plate at all, and the plate would not scatter, because for the one-cell-thick grid-aligned plate only E_x and E_y components are directly affected by the plate conductivity, and the incident field has only an E_z component. For the staircased plate, however, the E_θ incident field has both E_y and E_z field components, and both are directly affected by the staircased plate.

From this example we see that we can expect maximum staircase errors in situations where the electric field is perpendicular to the staircased scatterer surface. Does this mean that the Yee cell computer codes cannot be used for this situation? The answer is no. As the cell size is reduced the results from the staircased scatterer will eventually converge to the correct result, but very small cells (relative to the wavelength) may be required for accuracy.

Consider scattering from a sphere. In Figure 13-7 a sphere is approximated using 1-cm cubical Yee cells, with the staircasing of the sphere surface shown clearly. The incident field direction is $\theta = \phi = 22.5°$. The time domain far zone scattered field (with a Gaussian incident pulse) in Figure 13-8 clearly shows the pulse being scattered from the "staircase" as the ripple on the response. Nevertheless, the corresponding frequency domain backscatter result for the staircased perfectly conducting sphere, shown in Figure 13-9, is reasonably accurate. It could be improved by using smaller (and more) FDTD cells with a greater expenditure of computer resources. The results will be worse for incidence directions normal to the staircased flat surfaces, for example, from $\theta = 0$ and $\phi = 0$ or 90°, since these flat surfaces backscatter more strongly than the smooth sphere surface.

The results used thus far to illustrate staircasing errors have been the results for perfectly conducting scatters. For penetrable scatterers the staircasing error is generally less, because some of the energy is scattered from within the target where staircasing is not a problem. Also, the staircasing error may be reduced for penetrable scatterers by gradually changing the constitutive parameters at the surface of the target. While the sphere shown in Figure 13-7 may be composed of Yee cells with all edges set to the same material, this need not be the case. It would be more accurate to have the outermost edges set to differing constitutive parameters, depending on the portion of the edge actually within

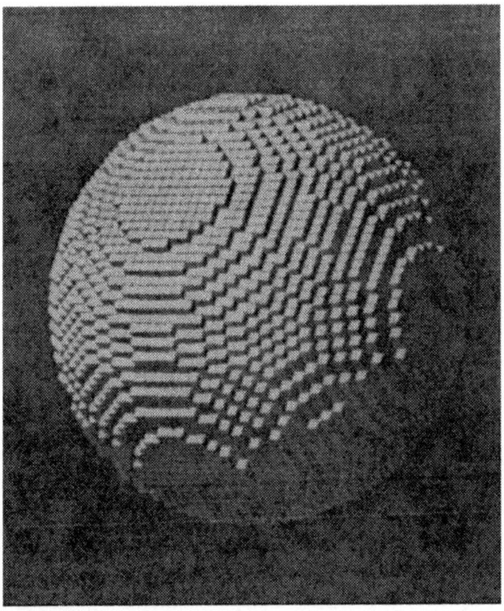

FIGURE 13-7. Staircased sphere in cubical Yee cells.

the sphere volume. Results using this approach will be more accurate than for an abrupt change in dielectric constant, with the improvement being greater for lower dielectric constants. An example of sphere scattering using a "fuzzy" surface approach is given at the end of this chapter for a sphere composed of frequency-dependent penetrable material.

If extremely accurate results are required for scattering from staircased targets with electric field polarization normal to the target surface, especially for high dielectric constants or perfectly conducting targets, there are two approaches to be considered. The first, as stated above, is the obvious approach of using smaller and therefore more FDTD cells. Accuracy can be checked by convergence as the cell size is reduced, much as can be done for method of moment solutions. The difficulty with this is that computer resources, in terms of both memory and processing time, may be exhausted before convergence is reached. The second approach is to abandon the cubical Yee cell, at least in the vicinity of the scatterer, in favor of alternative approaches.[5-7] At the present stage of research good results have been obtained for 2-D scatterers with several different approaches, while approaches valid in 3-D have been formulated but tested only on special geometries.

Applying these alternate approaches is quite complicated as compared to using cubical Yee cells. As explained in Chapter 3, "building" an object in Yee cells only requires setting the constitutive parameters at the appropriate field locations. The update equations for the electric and magnetic fields are the same for all field components (depending only on material properties). The field components are directly addressed in computer memory, i.e., the index on

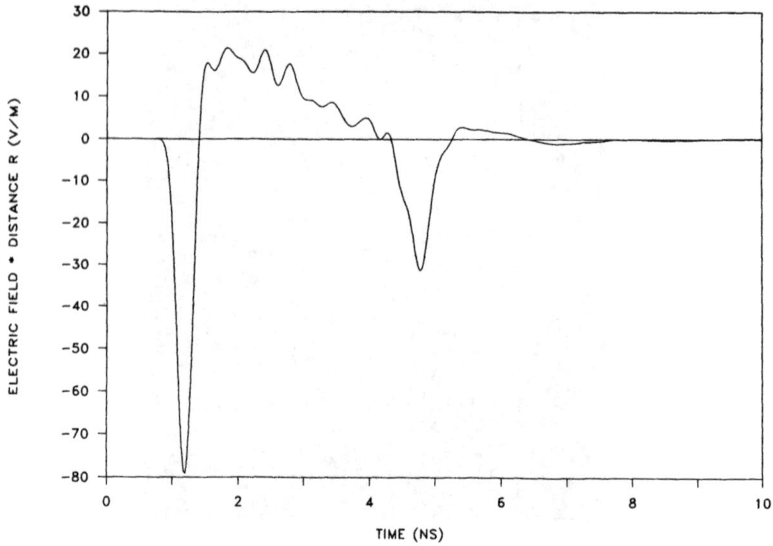

FIGURE 13-8. Transient backscattered far zone electric field for Gaussian pulse incident plane wave on staircased conducting sphere.

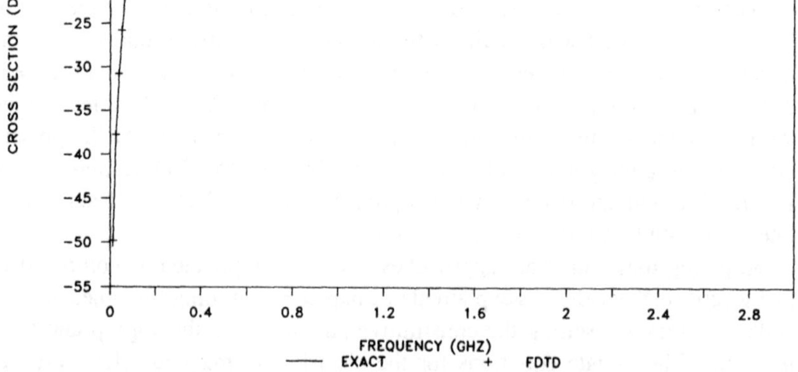

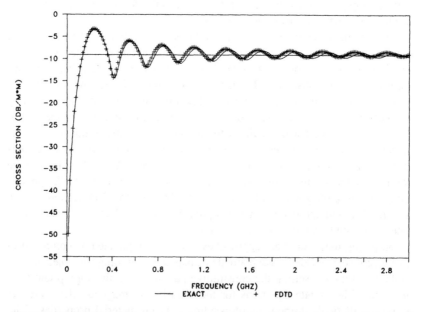

FIGURE 13-9. FDTD backscatter cross section for staircased conducting "sphere" compared with exact solution for smooth sphere.

the array (I,J,K) automatically determines the location of the fields in the space. With nonuniform cells, the cell mesh itself must first be generated to fit the surface of the scatterer. Then this mesh information must be incorporated into the FDTD update equations, usually requiring indirect addressing. This means that in order to find the location of an (I,J,K) field component one must use another set of array variables in which the locations are stored. In addition to the increased memory required to store both the field and grid information, nonuniform grids also require additional calculations to update the fields relative to Yee cell calculations. Thus, there is a trade-off between the two approaches, with the choice involving whether it is better to use larger and fewer nonuniform cells, which require more memory and calculation time for each cell (plus more human time to deal with the greater complexity), or smaller but more Yee cells, which require less memory and calculation time each (and less human time for programming due to the relative simplicity) for each cell update. There is no clear answer here. The choice depends on the geometry of the scatterer and its size in wavelengths, the polarization of the fields, the accuracy required, and the human and computer resources available for the calculation. At present, with the nonuniform cell FDTD still an active research area, Yee cells are preferred for most applications and are certainly more commonly used. However, in the future, with automated mesh generation and improved field update algorithms, the nonuniform cell methods may have wider use. Anyone interested in the application of nonuniform cell FDTD should read the literature[5-7] and consider the current state of development relative to the specific problem at hand.

13.4 IMPEDANCE SHEETS

Thin sheets of resistive or dielectric material are commonly encountered in radar cross section (RCS) scattering calculations. They are also used in waveguide and antenna components. Analysis of such sheets is simplified by using sheet impedances. In this section it is shown that sheet impedances can be modeled easily and accurately in FDTD calculations. The discussion and results of this section are taken from Reference 8.

Reference 9 reviews various approximate boundary conditions, including several for thin sheets and layers. These are applicable to sheets which are thin relative to the free space wavelength, so that they can be approximated by an electric current sheet. If the thin sheet is primarily conductive, the sheet impedance will be resistive, as is the case for resistance cards. A thin lossless dielectric sheet will have a purely reactive sheet impedance, while in general, the sheet impedance will be complex. These sheets are characterized by a discontinuity in the tangential magnetic field on either side of the sheet but no discontinuity in the tangential electric field. This continuity, or single valued behavior of the electric field, allows the sheet current to be expressed in terms of an impedance multiplying this electric field. These conditions imply that the effects of the sheet on the perpendicular electric field can be neglected. If this is not the case, more complicated models must be used, as discussed in Reference 10.

The sheet impedance can be defined in several ways. A convenient definition can be obtained by combining (3.3) and (3.5) of Reference 11

$$Y_s = \sigma T + j\omega\varepsilon_0(\varepsilon_r - 1)T \tag{13.3}$$

with

$$Z_s = 1/Y_s \tag{13.4}$$

where Y_s is the sheet admittance, Z_s is the sheet impedance, σ and ε_r are the conductivity and relative permittivity of the sheet material, T is the sheet thickness, and ε_0 is the free space permittivity.

Let us now consider how to incorporate this approximation into the FDTD method. The surface impedance approximation requires the impedance sheet to be thin as compared to the free space wavelength. In most FDTD calculations the FDTD cell size must be on the order of 1/10 wavelength or less for reasonably accurate results, so this condition is automatically met. Scattering from an infinitesimally thin perfectly conducting plate has been calculated by approximating the plate as being one FDTD cell thick with good results, as shown in Chapter 7. If it is assumed that the same approach can be applied to infinitesimally thin impedance sheets, then the plate thickness T in (13.4) becomes the thickness of the FDTD cell, and the conductivity and/or relative permittivity to be used in the FDTD calculations are adjusted in accordance with (13-4) to give the desired sheet impedance. Note that the FDTD cell dimension need not correspond to the thickness of the actual physical sheet. The FDTD cell thickness is used only to determine the conductivity and relative permittivity of the FDTD electric field location so that the desired sheet impedance is approximated. Note also that even if the wavelength in the material forming the impedance sheet is much smaller than a free space wavelength, the FDTD cell size need not be correspondingly reduced.

The approach is extremely simple, and its application will be demonstrated by examples. The first consists of calculating the far zone backscatter from a 29×29 cm flat plate of sheet impedance $Z_s = 500\ \Omega$. The FDTD calculations will use cubical Yee cells with 1-cm edges. Using T = 1 cm, the corresponding FDTD conductivity is $\sigma = 0.2$ S/m. The FDTD calculations shown in Figures 13-10 and 13-11 are made with the plate approximated by setting the conductivity to 0.2 S/m for x and y polarized electric field locations corresponding to single z dimension index over a range of x and y dimension indices to model the plate. The problem space size, orientation, and position of the plate, incident Gaussian pulse plane wave, and time step size are consistent with those in the flat plate examples of Chapter 7.

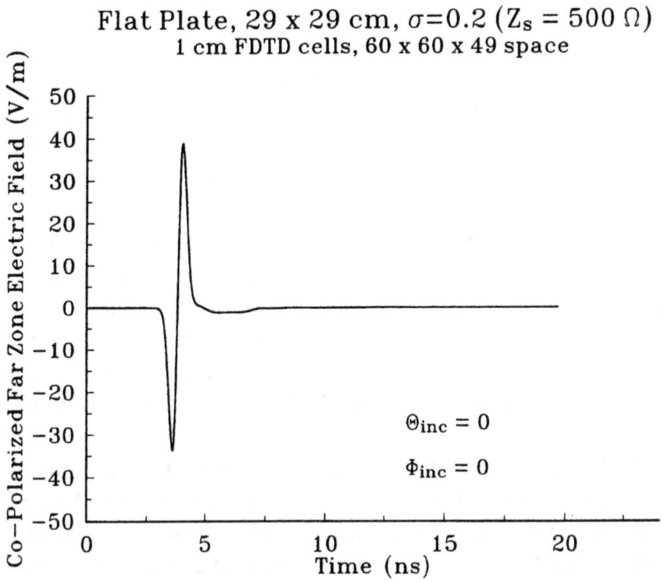

FIGURE 13-10. Copolarized backscatter far zone electric field vs. time for a 29 × 29 cm flat plate of sheet impedance 500 Ω, for a θ-polarized normally incident Gaussian pulse plane wave computed using FDTD.

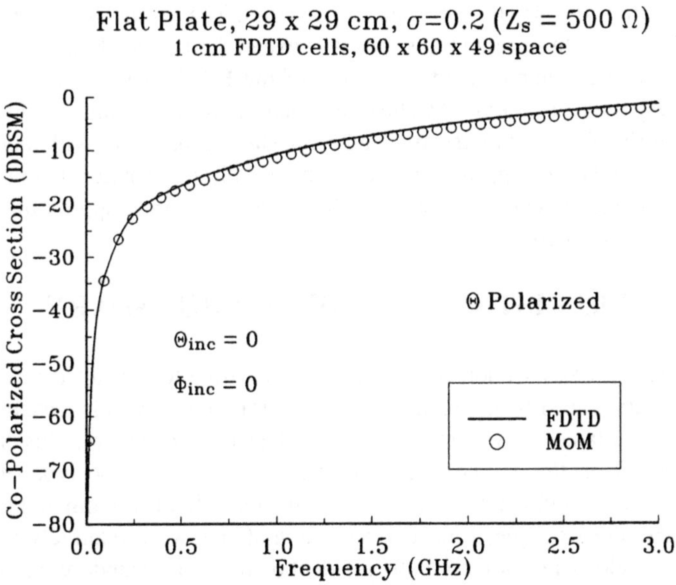

FIGURE 13-11. RCS for a 29 × 29 cm flat plate of sheet impedance 500 Ω, normal incidence, θ-polarized, obtained from FDTD results of Figure 1 and compared to the method of moments.[11]

Figure 13-10 shows the far zone backscattered electric field for a Gaussian pulsed plane wave normally incident on the plate. The RCS obtained from this far zone transient results is shown in Figure 13-11 and compared to results obtained using the method of moments.[11] The agreement is quite good. In Reference 8 results for this sheet impedance plate for non-normal incidence angles are given and also show good agreement with method of moments results.

In Figure 13-12 both FDTD and method of moments[11] results for scattering by a plate with a complex sheet impedance are shown. The sheet impedance is determined by applying (13.4) and (13.5) with conductivity 0.25 S/m, relative permittivity 3.0, and thickness 1 cm. The plane wave is a Gaussian pulse incident from $\theta = 45$, $\phi = 30°$, and ϕ-polarized. The FDTD results agree with the method of moments results for frequencies up to about 12 cells per wavelength.

The final result is for a plate with edge treatment. For this demonstration a 21 × 21 cm perfectly thin conducting plate is given a 4-cm border of sheet impedance $Z_s = 500\ \Omega$, resulting in a square plate 29 × 29 cm. This edged plate is modeled in FDTD by setting x and y polarized electric field locations for a single z dimension index as being either perfectly conducting for the central portion of the plate or with a conductivity of 0.2 S/m for the edges. The method of moments[11] calculations were made with a central perfectly conducting plate surrounded by four plates of sheet impedance Z_s=500 Ω attached to the central plate using overlap modes. Once again the plane wave is a Gaussian pulse incident from $\theta = 45$, $\phi = 30°$, and ϕ-polarized. The results are compared in Figure 13-13 with excellent agreement between the two methods.

This section demonstrated the ability of the FDTD method to easily and accurately model scattering by sheet impedances by comparing FDTD results for scattering from flat plates modeled using sheet impedances with method of moment results. The approach described here is directly applicable to the Yee cell, and demonstrated good accuracy for frequencies up to approximately 12 cells per wavelength.

13.5 DISTANCE TO OUTER BOUNDARY

Scattering cross-section results are very sensitive to unwanted reflections from the outer absorbing boundaries of the FDTD space. An adequate distance must be maintained between the scatterer and the outer boundary. This distance will depend on the geometry and material of the target, the type of outer boundary used, and the way in which the scattered fields illuminate the outer boundary. For example, Mur absorbing boundaries work relatively well for scattered fields which are normally incident on the outer boundary, not so well as the incidence angle moves away from normal, and are at their worst in the corners of the FDTD space. In Chapter 7 results for plate scattering with a 45° incidence angle were given, and the plate was moved out of the center of the FDTD space for these calculations. If the plate is located in the center of the space, the specular reflection from the plate will illuminate a corner of the

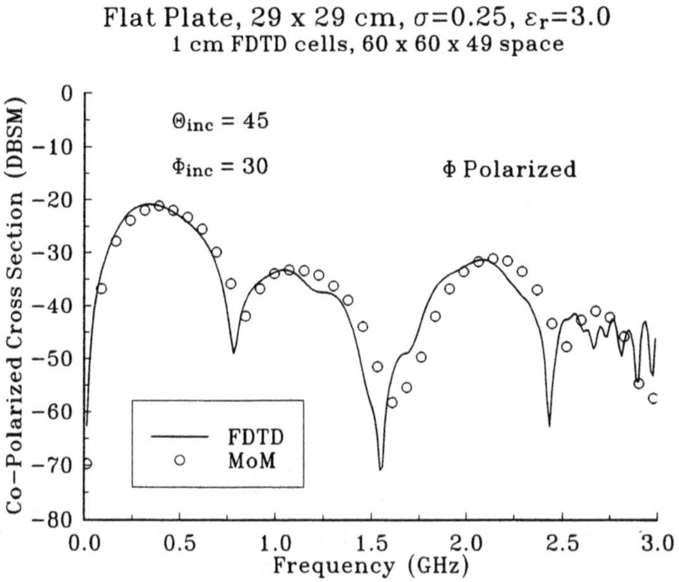

FIGURE 13-12. Copolarized RCS for a 29 × 29 cm flat plate of sheet impedance corresponding to conductivity of 0.25 S/m, relative permittivity of 3.0, and thickness 1 cm, for $\theta = 45$, $\phi = 30°$ ϕ-polarized incident plane wave obtained from Gaussian pulse FDTD results and compared to the method of moments.[11]

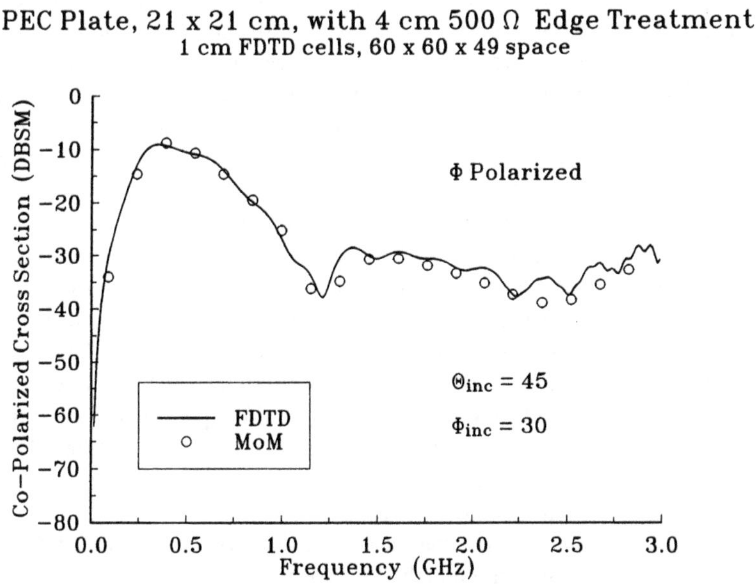

FIGURE 13-13. Copolarized RCS for a 21 × 21 cm perfectly conducting flat plate with a 4-cm 500 Ω edge treatment on all sides (total plate size 29 × 29 cm) for $\theta = 45$, $\phi = 30°$ ϕ-polarized incident plane wave obtained from Gaussian pulse FDTD results and compared to the method of moments.[11]

FDTD space and will be reflected back to the plate by the imperfect Mur absorbing boundary. From the plate it will again specularly reflect to the backscatter direction, resulting in significant error. This error is nearly eliminated by locating the plate so that the strong specular reflection does not illuminate the Mur absorbing boundary in the corner. Other absorbing boundaries, such as described in References 2 and 3, absorb better in the corner regions than does Mur's and require fewer cells between the scatterer and the outer boundary for the same absorption, but at the expense of additional complexity.

In the following sequence of results taken from Reference 11 the effect of the distance between the scatterer and the Mur second order absorber is illustrated. For the first example the scatterer is a square perfectly conducting rod with a 10:1 length to width ratio. In Figures 13-14 and 13-15 the rod is approximated using $100 \times 10 \times 10$ FDTD cells. The FDTD cells are 1/10 wavelength at the highest frequency shown. In these figures scattering results for end-on incidence are shown with 6 and 20 FDTD cells between the rod and the second order Mur absorbing boundaries, respectively, and the improvement in accuracy is striking. These problem spaces are $112 \times 22 \times 22$ or 54,200 cells and $140 \times 50 \times 50$ or 350,000 cells. Clearly, the need to allow so many cells between the scatterer and the absorbing boundary creates a severe impact on the computer resources needed.

Next we consider in Figures 13-16 and 13-17 the effects of reducing the FDTD cell size by half, so that the rod is approximated using $200 \times 20 \times 20$ cells with the cells size now 1/20 wavelength at the highest frequency. Again, backscatter for end-on incidence is shown, with the second order Mur absorbing boundaries separated from the scatterer by 8 and 20 cells, respectively. The improvement obtained by increasing the separation of the Mur absorber from the scatterer is significant, especially at the higher frequencies. The FDTD problem spaces used for these two figures are $216 \times 36 \times 36$ or 280,000 cells and $240 \times 60 \times 60$ or 864,000 cells. The results in Figure 13-15 appear to be of an accuracy similar to those in Figure 13-17, however, even though Figure 13-17 has FDTD cells half the size and required nearly 2.5 times as much computer memory. Let us consider the reason for this. The results in Figures 13-15 and 13-17 both use the same number of cells between the rod and the absorbing boundary, but since the cells used in Figure 13-17 are half the size, the Mur absorber is actually closer to the scatter in this calculation than it is in Figure 13-15. One would expect that if the number of cells between the rod and the absorbing boundary were increased to 40 with the smaller cells used, maintaining the same distance to the absorbing boundaries, the results would be even more accurate.

From this example we see that basing the decision on how much distance to place between the scatterer and the Mur absorber on the number of cells may not be entirely valid, but considering the actual distance is important as well. This example also illustrates that the allocation of computer resources to provide smaller cells to more accurately model the interactions with the scat-

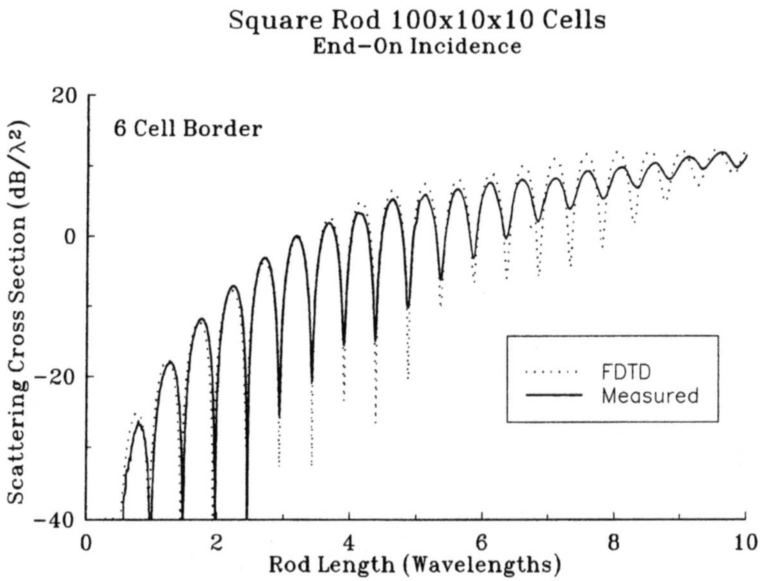

FIGURE 13-14. FDTD calculation of end-on backscatter for 100 × 10 × 10 cell square rod with 6 cells between the rod and the absorbing boundaries compared to measurements.

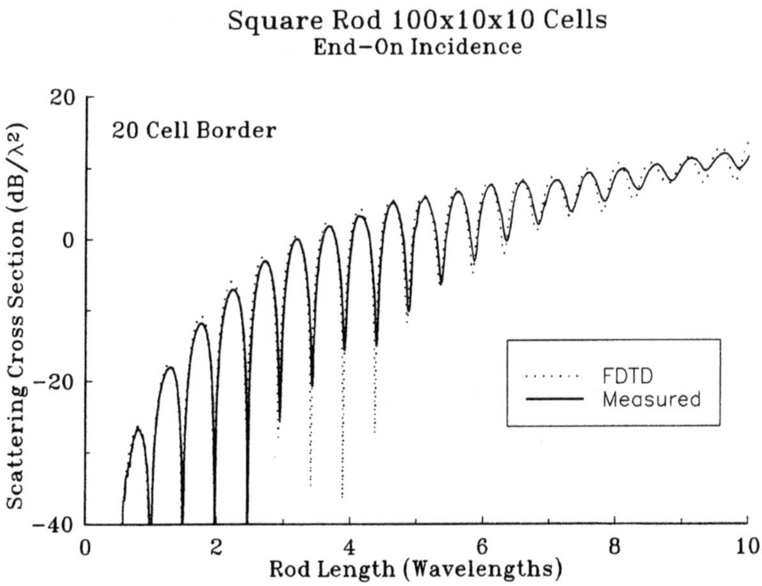

FIGURE 13-15. FDTD calculation of end-on backscatter for 100 × 10 × 10 cell square rod with 20 cells between the rod and the absorbing boundaries compared to measurements.

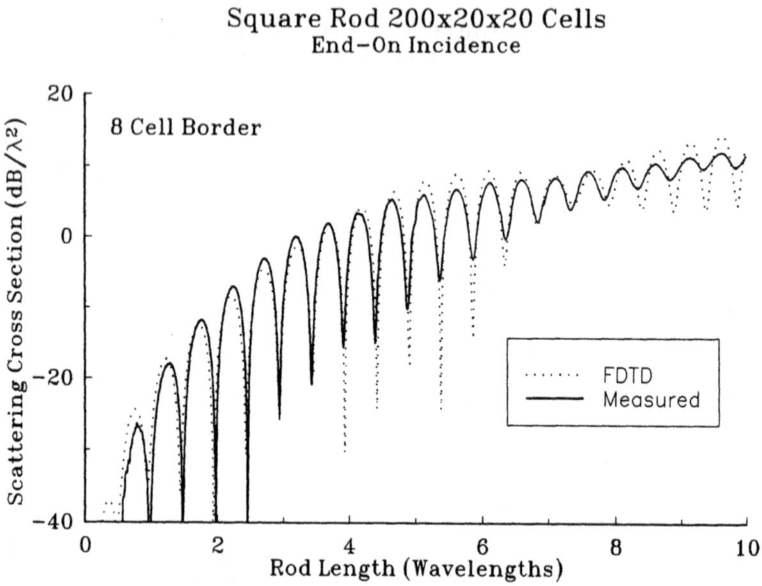

FIGURE 13-16. FDTD calculation of end-on backscatter for 200 × 20 × 20 cell square rod with 8 cells between the rod and the absorbing boundaries compared to measurements.

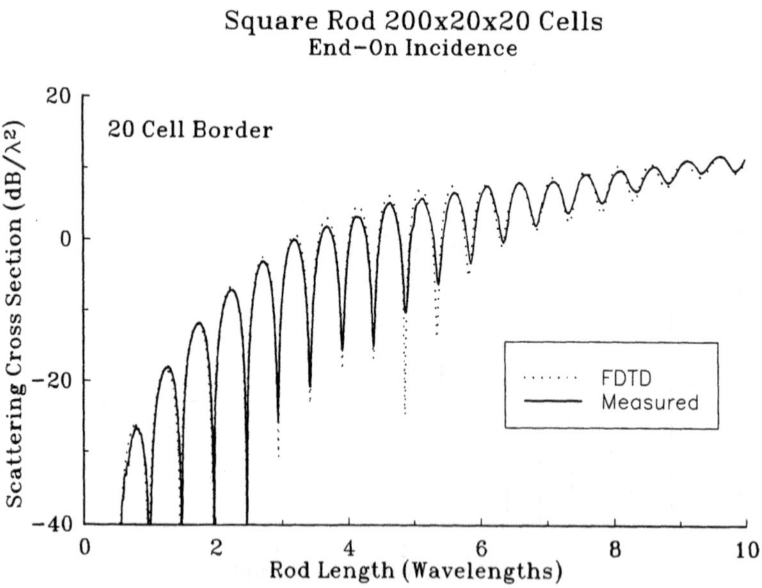

FIGURE 13-17. FDTD calculation of end-on backscatter for 200 × 20 × 20 cell square rod with 20 cells between the rod and the absorbing boundaries compared to measurements.

terer, as opposed to using larger cells to locate the absorbing boundary as far from the scatterer as possible, is important in obtaining the most accurate results possible with the computer resources available. Ideally, one would like to make convergence tests on both variables, cell size, and distance between the scatterer and the absorbing boundaries, and observe convergence to the same result.

This example also points out the tremendous savings in computer storage (and, of course, computer execution time) that will result from using absorbing boundaries[2,3] that reduce the reflections from the outer boundary without requiring the distance needed by the Mur formulation used here.

13.6 FREQUENCY-DEPENDENT MATERIALS

For the final example in this chapter FDTD scattering results for a frequency-dependent material will be given. The material permittivity is described by one second order Lorentz pole with parameters corresponding to (8.37) and (8.38) of $\varepsilon_\infty = 1$, $\gamma_1 = 5.01 \times 10^{16}$, $\alpha_1 = 3.99 \times 10^{16}$, and $\beta_1 = 0.28 \times 10^{16}$. The complex permittivity of the material is shown in Figure 13-18.

To illustrate the application of the frequency-dependent scattered field FDTD formulation of Section 8.7 in 3-D, the backscatter cross section for a sphere made of this material is calculated. The FDTD space is $71 \times 71 \times 71$ cells, using Liao's[2] second order absorbing boundary. A smooth cosine pulsed plane wave with pulse width $\beta = 24$ excites the sphere. The time step for both FDTD calculations was set at $0.45 \times$ the Courant limit. This was required for stability due to the relative permittivity being <1. The angle of incidence is $\theta = 70$, $\phi = 20$. The sphere radius is 15 nm or 25 FDTD cells.

The outer surface of the sphere is "fuzzy", as discussed in Section 13-3. The specific approach used was to determine the distance from the center of the sphere to the two endpoints of each Yee cell electric field component. If both endpoints lie within the sphere volume, constitutive parameters are set for the Lorentz material. If both endpoints are outside the parameters are free space. If one endpoint is inside the sphere volume and one is outside, the multiplying constants for the FDTD update equations are modified in proportion. For example, consider that for a certain E field component 40% of the corresponding Yee cell edge is inside the sphere volume and 60% is outside. First, compute the multiplying constants in (8.61) for the Lorentz material and for free space. Then, for the dispersive calculation for this particular component use material parameters in (8.61) which are the sum of 0.4 times the dispersive constants for the Lorentz material plus 0.6 times the constants for free space. For the calculations shown here about 1000 different sets of constants were used for the various electric field components on the sphere surface.

The far zone transient scattering was Fourier transformed and used to produce the frequency domain backscatter. The result is shown in Figure 13-19. Compare this with the exact solution and with an FDTD calculation which does not include the frequency dependence of the material. For the nondispersive

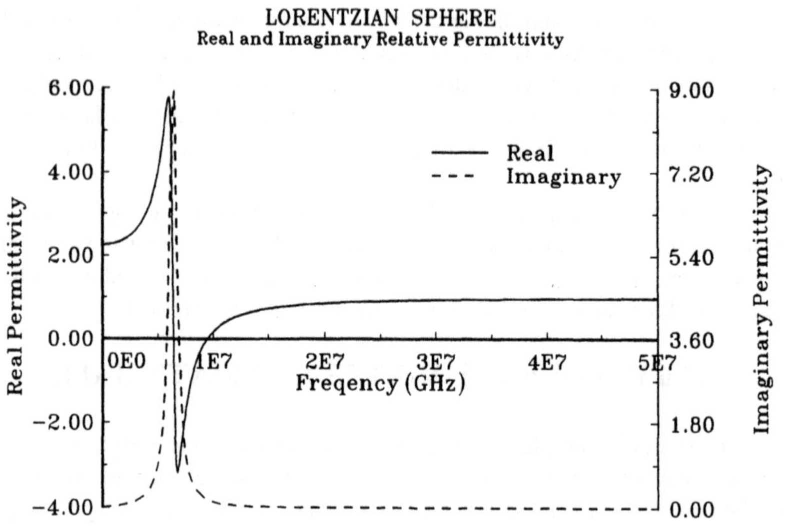

FIGURE 13-18. Complex permittivity for second order Lorentz material used for sphere scattering calculation.

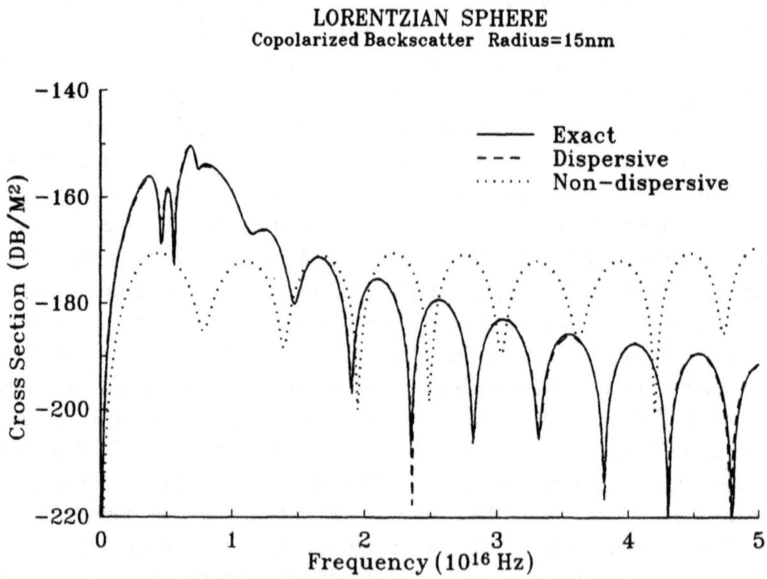

FIGURE 13-19. FDTD dispersive material scattering for sphere composed of Lorentz material of Figure 13-18 compared to non-frequency-dependent FDTD calculation and with exact solution. FDTD results obtained from single transient calculation and FFT.

FDTD calculation the permittivity and conductivity was chosen to match the Lorentz material at a frequency of 1.7×10^{16} Hz. The frequency-dependent calculation required approximately 5 h on a workstation with a 7-MFLOPS computations speed, about twice the computer time for the non-frequency-dependent calculation. The accuracy of the fuzzy dispersive calculation is excellent. The results also clearly illustrate the advantage of using the scattered field frequency-dependent approach to obtain wide bandwidth scattering from frequency-dependent materials.

REFERENCES

1. **Taflove, A. and Umashankar, K.,** Review of FD-TD numerical modelling of electromagnetic wave scattering and radar cross section, *Proc. IEEE,* 77, 682, 1989.
2. **Liao, Z. et al.,** A transmitting boundary for transient wave analysis, *Sci. Sin.* (Ser. A), 27(10), 1063, 1984.
3. **Mei, K. and Fang, J.,** Superabsorption: a method to improve absorbing boundary conditions, *IEEE Trans. Ant. Prop.,* 40(9), 1001, 1992.
4. **Furse, C. et al.,** Improvements to the finite-difference time-domain method for calculating the radar cross section of a perfectly conducting target, *IEEE Trans. Microwave Theory Tech.,* 38(7), 919, 1990.
5. **Madsen, N., and Ziolkowski, R.,** Numerical solution of Maxwell's equations in the time domain using irregular nonorthogonal grids, *Wave Motion,* 10, 583, 1988.
6. **Fusco, M.,** FDTD algorithm in curvilinear coordinates, *IEEE Trans. Ant. Prop.,* 38(1), 76, 1990.
7. **Holland, R., Cable, V., and Wilson, L.,** Finite-volume time-domain (FVTD) techniques for EM Scattering, *IEEE Trans. Electromag. Comp.,* 33(4), 281, 1991.
8. **Luebbers R. and Kunz, K.,** FDTD Modeling of thin impedance sheets, *IEEE Trans. Ant. Prop.,* 40(3), 349, 1992.
9. **Senior, T. B. A.,** Approximate boundary conditions, *IEEE Trans. Ant. Prop.,* 29, 826, 1981.
10. **Maloney, J. and Smith, G.,** The efficient modeling of thin material sheets in the Finite-Difference Time-Domain (FDTD) method, *IEEE Trans. Ant. Prop.,* 40(3), 323, 1992.
11. **Newman, E.,** A User's Manual for the Electromagnetic Surface Patch Code: ESP Version IV, The Ohio State University Research Foundation, ElectroScience Laboratory, Department of Electrical Engineering, Columbus, OH, 1988.
12. **Trueman, C., Kubina, S., Luebbers, R., Kunz, K., Mishra, S., and Larose, C.,** RCS of cubes, strips, rods and cylinders by FDTD, *Proc. Annu. Symp. Applied Computational Electromagnetics Society,* Monterey, CA, March 1992.

Chapter 14

ANTENNAS

14.1 INTRODUCTION

In this chapter application of the FDTD method to antennas is discussed, and several fundamental example calculations are presented. FDTD calculation of the basic antenna parameters, including self- and mutual impedance and admittance, gain, efficiency, and radiation patterns is demonstrated for several different antenna geometries. The FDTD results are compared both with calculations made with the method of moments (MoM) and with measurements.

While extremely accurate results for impedance for relatively simple 2-D antenna geometries can be obtained if FDTD is used to model the feed region in detail,[1] the approach taken here is to use only the usual Yee cell FDTD field components to model each source. This corresponds to the usual way thin wire MoM computer codes model approximately a source as a thin gap in the wire or as a frill of magnetic current. In this discussion the term "source" is used in a rather restrictive sense, meaning the spatial location(s) in which the antenna problem is being excited, i.e., where a source is supplying energy to the electromagnetic calculations. It is not to say that an antenna feed region or subsystem whose geometry affects the antenna performance cannot be modeled using the approach here. For example, later in this chapter we consider a shaped-end waveguide antenna which is fed by a metallic probe located near the closed end of the waveguide. This probe is shaped and located so as to excite a particular waveguide mode. The metallic probe is modeled using a number of FDTD field components. However, the probe is driven by a single applied electric field in one FDTD cell, located where the probe contacts the waveguide wall. The advantages of using the FDTD field components to excite the antenna problem are that relatively large, 3-D antenna geometries can be considered, and that calculations can be made using parallelepiped FDTD cells.

In addition to sources, lumped loads can also be modeled within one FDTD cell, much as is done in thin-wire MoM calculations, where lumped loads are also located in infinitesimal wire gaps. The general topic of modeling wires with lumped sources and loads was considered in Chapter 10, and FDTD antenna calculations including lumped loads within a single FDTD cell are illustrated later in this chapter. On the other hand, extended regions of lossy dielectric or magnetic materials also can be included easily in FDTD calculations, while this is relatively difficult using the MoM. Further, aperture antennas can be more readily accomodated using FDTD than with the MoM.

Application of FDTD to antennas has occurred recently relative to other applications such as shielding and radar cross-section (RCS). This is somewhat surprising, since the geometrical and material generality of FDTD suggests that it might have significant application to antenna analysis, especially in situations in which other structures, especially those that are electromagnetically penetrable, are nearby. Also, when FDTD is applied to radiating antenna calculations it loses one of its disadvantages relative to the MoM in some other applications that require results at multiple far zone angles. For example, in scattering applications the MoM produces results for different plane wave incidence angles efficiently from a single impedance matrix, while FDTD requires a complete recalculation for each different incidence angle. However, for antenna radiation problems, FDTD can produce far zone fields in any number of different directions efficiently during one computation. Because FDTD also provides wide frequency band results with pulse excitation, it is extremely efficient in antenna applications, since from one FDTD computation results for impedance and radiation patterns over a wide frequency band can be obtained.

However, one reason FDTD has lagged in antenna applications is that the MoM can provide results for small, relatively simple antennas with much less computer time and memory required than can FDTD. This is because the MoM finds only the currents flowing on the wire or conducting surface, while FDTD must calculate the fields in the entire computational region. This region must contain enough cells to allow 15 to 20 cells between the antenna and the absorbing boundaries, and if the antenna is small and geometrically simple the overhead involved with computing fields in all the surrounding free space cells makes FDTD much less efficient than MoM. It is only for relatively large antennas, or antennas with geometries and/or materials that are not easily included in MoM formulations, that FDTD becomes a competitive method, and these situations require fairly powerful computers that have become generally available only recently.

The FDTD calculations described in this chapter are converted to the frequency domain for comparison with frequency domain methods, such as the MoM. However, it should be kept in mind that FDTD is capable of computing transient far zone radiation for antennas excited by nonsinusoidal sources using the transient far zone transformation of Chapter 4. These transient calculations can be done more efficiently for most antenna geometries using FDTD than by applying frequency domain methods. They also allow efficient determination of wide bandwidth gain, as will be shown later in the chapter.

Because the antenna problems considered here are excited by sources located in FDTD cells within the problem space, there is no analytically specified incident field and the total field form of FDTD is used.

Situations do occur where antenna parameters can be determined from scattering calculations. For example, in Reference 2 the impedance of a dipole antenna was determined by performing two scattered field calculations with a pulsed incident plane wave. In one the open circuit voltage in the dipole gap was computed, and the second computed the short circuit current. By applying Fourier transforms and fundamental circuit theory, the dipole impedance was determined. However, for most situations exciting the antenna by a local source is preferred. Indeed, the dipole impedance determined using two scattered field calculations in Reference 2 can be determined with only one total field FDTD calculation using a voltage source in the dipole gap, as will be demonstrated in the next section of this chapter.

The remainder of this chapter is divided into three sections. In Section 14.2 the fundamentals of applying FDTD to antennas will be presented and demonstrated using the simple geometry of a pair of parallel wire dipole antennas. This antenna geometry can be easily modeled using the MoM, which will be utilized to provide comparison results. In Section 14.3 a more complicated geometry of a wire monopole attached to a conducting box will be considered. For this example, FDTD is capable of producing accurate results for wide band impedance and gain with much less computer time than would be required by a series of MoM calculations to cover the frequency band. Finally, in Section 14.4 FDTD results for a shaped-end waveguide antenna will be calculated using FDTD, with gain and radiation pattern results compared with measurements. This example illustrates the capability of FDTD with parallelepiped cells to accurately predict radiation from an antenna with curved surfaces.

14.2 IMPEDANCE, EFFICIENCY, AND GAIN

In this section calculation of self- and mutual antenna impedance, efficiency, and gain will be demonstrated. So that comparison data can be obtained easily, a thin wire antenna geometry consisting of a pair of parallel wires will be used so that the MoM can be easily applied. Modeling wires in FDTD was considered in Chapter 10, and we will use the approach described there. The wire geometry is shown in Figure 14-1. Two wire dipoles of 57 and 43 cm length are parallel and separated by 10.5 cm. Both are center fed, and are symmetrically positioned. For purposes of defining voltages and currents for our mutual impedance calculations the longer antenna is labeled as antenna 1 with voltage and current V_1 and I_1 (not to be confused with the I spatial index in the Yee notation) at the center cell, similarly for the shorter antenna considered as antenna 2. The problem space is 61 × 51 × 80 cells, with the cell dimensions $\Delta x = \Delta y = 0.5$ cm, $\Delta z = 1.0$ cm. Making the two transverse dimensions smaller results in a greater length to diameter ratio, so that a thin-wire MoM code may be used to provide comparison results over a wider band of frequencies. Thinner wires may be modeled in FDTD using sub-cell methods described in Chapter 10, as will be demonstrated in the next section of this chapter.

For the FDTD calculations the longer dipole is fed at the center with a Gaussian pulse of 1.0 V maximum amplitude so that the electric field in the gap of antenna 1 is specified as

$$E_z^n(I,J,K) = -V_1(n\Delta t)/\Delta z \tag{14.1}$$

where $V_1(t)$ is a Gaussian source voltage given by

$$V_1(t) = 1.0e^{-\alpha(t-\beta\Delta t)^2} \tag{14.2}$$

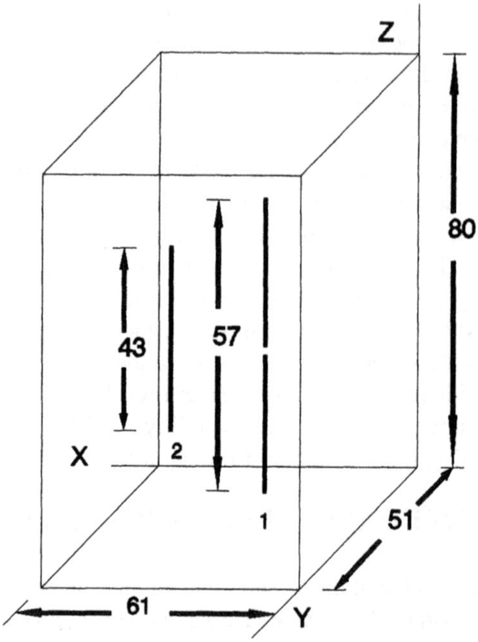

FIGURE 14-1. FDTD problem space for demonstrating calculation of impedance, efficiency, and gain. Dimensions are given in FDTD cells, with cell size 0.5 × 0.5 × 1 cm.

with $\beta = 32$, $\alpha = (4(\beta\Delta t))^2$ and the pulse truncated for $t < 0$ and $t > 2\beta\Delta t$, as described in Chapter 3. The time step was set at the Courant limit of 11.11 ps.

The currents $I_1(t)$ and $I_2(t)$ flowing through the centers of their respective dipoles can be sampled by taking the discrete approximation of the line integral of the H_x and H_y components encircling this electric field, so that $I_1(t)$ is obtained from

$$\begin{aligned} I_1(n\Delta t) = {} & \left[H_x^{n+\frac{1}{2}}(I, J-1, K) - H_x^{n+\frac{1}{2}}(I, J, K)\right]\Delta x \\ & + \left[H_y^{n+\frac{1}{2}}(I, J, K) - H_y^{n+\frac{1}{2}}(I-1, J, K)\right]\Delta y \end{aligned} \tag{14.3}$$

with the values of spatial indices I,J,K corresponding to the location of the $E_z(I,J,K)$ electric field along the wire axis at the antenna location through which the current I_1 is to be determined; the method is similar for $I_2(t)$. During the progress of the FDTD calculations these currents are saved for each time step. The FDTD calculations are continued until all transients are dissipated, so that the Fourier transform yields the steady-state frequency domain response of the antenna. The $(\Delta t)/2$ time offset between $V_1(t)$ and $I_1(t)$ due to the time offset between electric and magnetic fields in FDTD calculations can be neglected because it is a small fraction of the period of the waveform, even at the highest frequency considered.

Along with the applied Gaussian voltage pulse the currents are Fourier transformed to the frequency domain. With one antenna fed by a voltage source it is simpler to determine admittances than impedances, and the same information is available. (Keep in mind that $Z_{11} \neq 1/Y_{11}$; rather the following equations must be solved simultaneously for V_1 and V_2 in terms of I_1 and I_2 and impedances substituted for admittance expressions which result.) From the admittance parameter equations

$$I_1(\omega) = V_1(\omega)Y_{11} + V_2(\omega)Y_{12} \tag{14.4}$$

$$I_2(\omega) = V_1(\omega)Y_{21} + V_2(\omega)Y_{22} \tag{14.5}$$

with $V_1(\omega)$ the frequency domain driven dipole voltage and V_2 zero for the passive antenna, we easily obtain the self-admittance of dipole 1 and the mutual admittance (since $Y_{12} = Y_{21}$) between the dipoles by dividing the appropriate complex Fourier transforms of $V_1(\omega)$, $I_1(\omega)$, and $I_2(\omega)$.

For comparison, MoM results were obtained using the Electromagnetic Surface Patch Version 4(ESP4)[3] computer code. The wire radius for the MoM calculations was taken as 0.281 cm, providing the same cross-sectional area as the 0.5-cm square FDTD cells. While the FDTD calculations should be valid up to approximately 1.5 GHz based on having 20 FDTD cells per wavelength, the thin wire approximation for the MoM code becomes questionable at approximately 1 GHz, and this was taken as the upper frequency limit for comparison of results. The FDTD calculations were continued for 8192 time steps to be sure that all transients had dissipated.

Figure 14-2 shows the Gaussian pulse voltage applied to the one cell gap at the center of the longer, driven dipole. Figures 14-3 and 14-4 show the current flowing in the center cell of the driven and passive dipole, respectively. All are plotted on the same time scale, corresponding to about 4500 of the 8192 total time steps. Figures 14-5 to 14-7 show the magnitude of the Fourier transforms of the voltage and current results of Figures 14-2 to 14-4. The current results indicate the complicated frequency domain behavior of the coupled dipole system.

The self-admittance was obtained by dividing the complex Fourier transform of the driven dipole current by that of the Gaussian voltage pulse feeding the dipole at each frequency. The results are shown in magnitude and phase in Figures 14-8 and 14-9 and compared with ESP4 MoM results. Considering the differences in how the feed region is modeled (a 1-cm gap in the FDTD calculations vs. an infinitesimal gap in ESP4) the agreement is quite good.

The mutual admittance was obtained in a similar manner, dividing the complex Fourier transform of the passive dipole current by that of the Gaussian pulse. The results are shown in Figures 14-10 and 14-11. Again, the agreement is quite good considering the different approximations and assumptions made in the FDTD approach relative to the ESP4 computer code, especially in modeling the feed region.

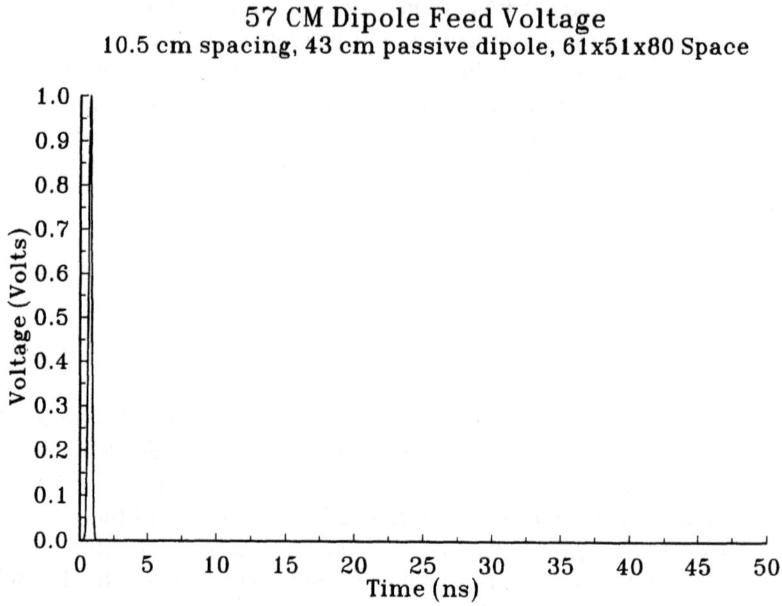

FIGURE 14-2. Gaussian voltage pulse across center cell of antenna 1.

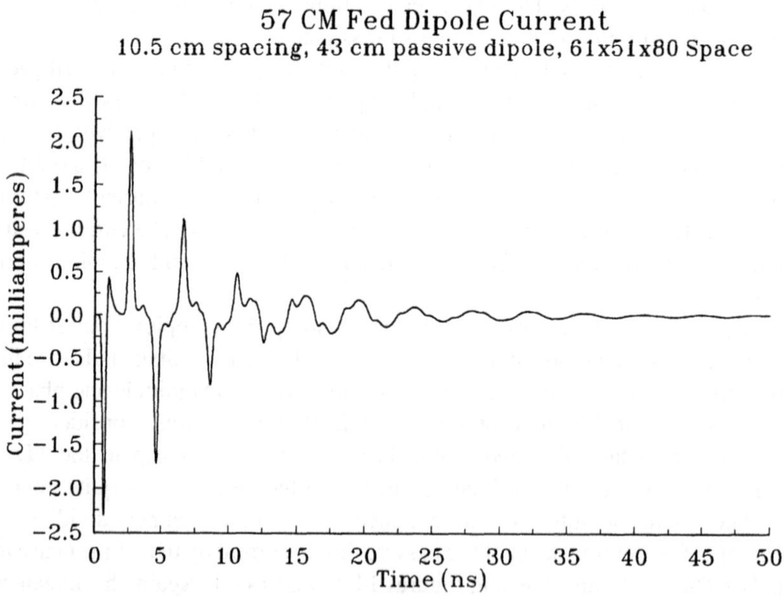

FIGURE 14-3. Calculated current flowing through source in gap at center of antenna 1 excited by Gaussian voltage source.

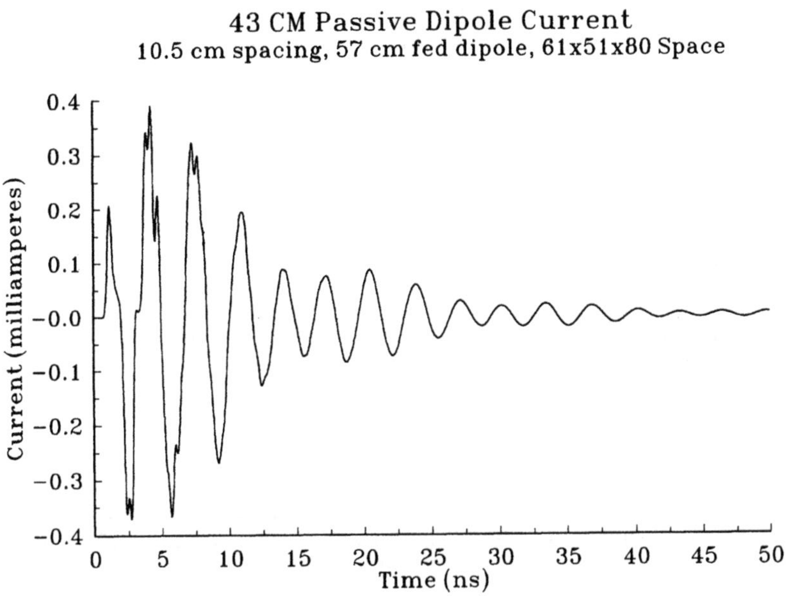

FIGURE 14-4. Calculated current flowing through center of antenna 2 with antenna 1 excited by Gaussian voltage source.

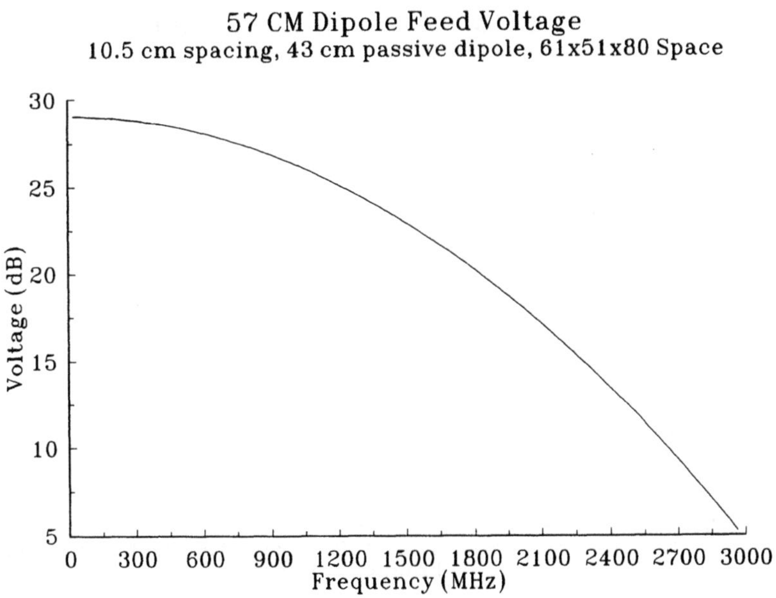

FIGURE 14-5. Magnitude of the discrete Fourier transform of the Gaussian pulse voltage across the center cell of antenna 1.

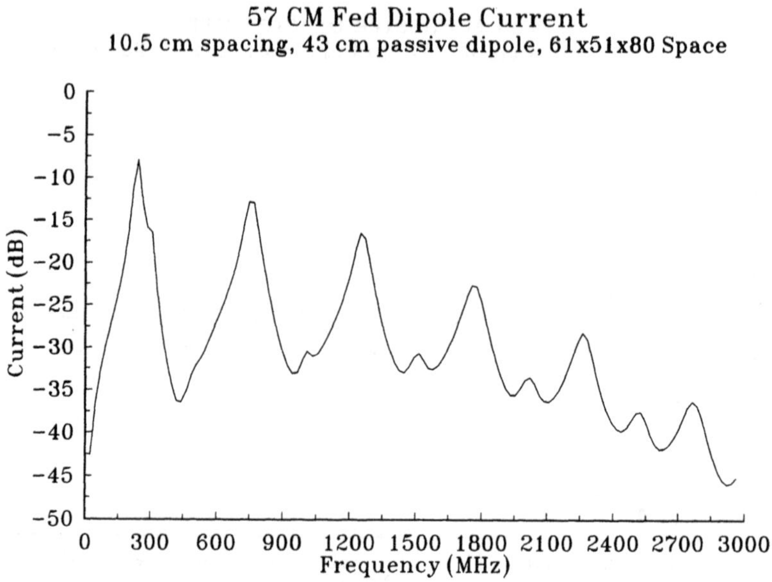

FIGURE 14-6. Magnitude of the discrete Fourier transform of the current flowing through the center cell of antenna 1.

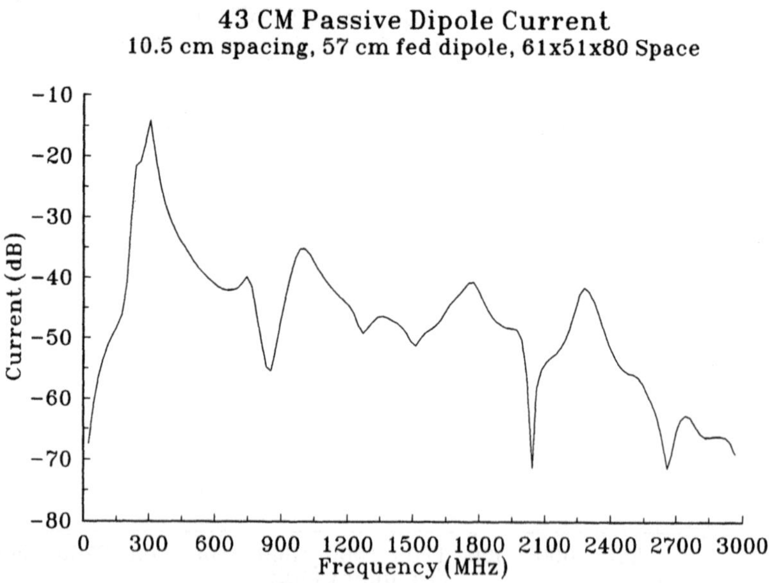

FIGURE 14-7. Magnitude of the discrete Fourier transform of the current flowing through the center cell of antenna 2.

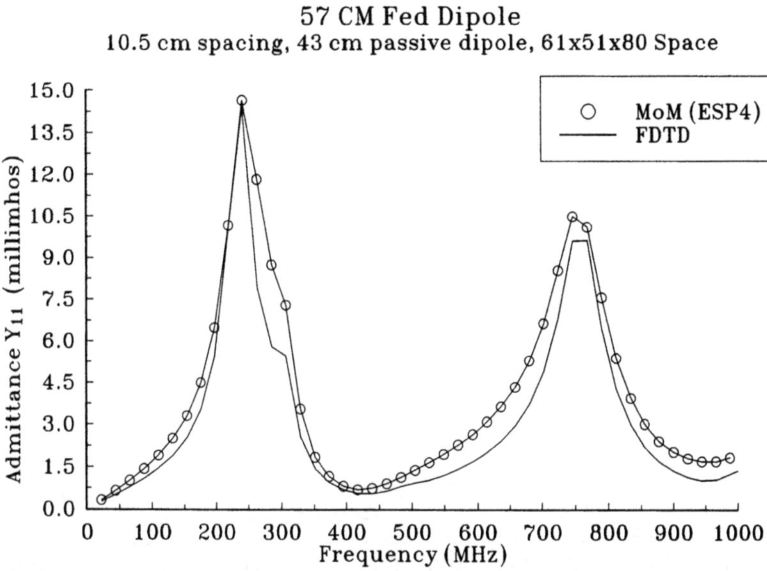

FIGURE 14-8. Self-admittance magnitude for antenna 1 calculated using FDTD and compared to the MoM.

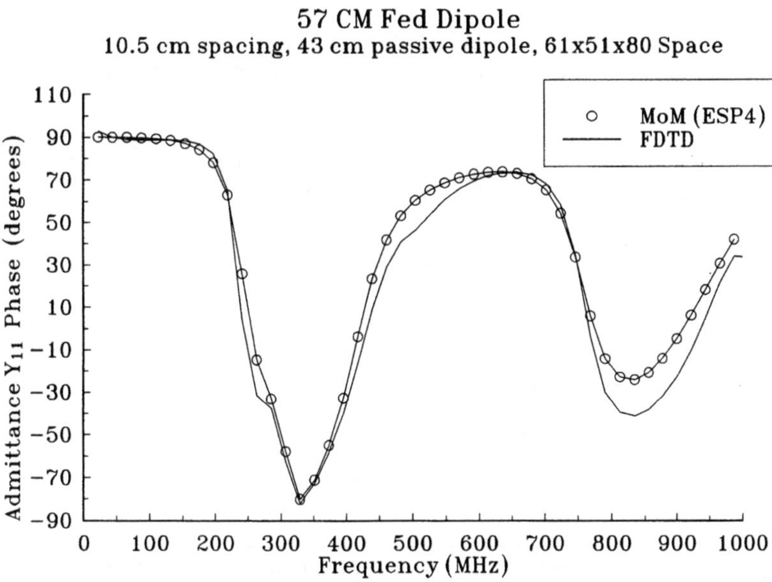

FIGURE 14-9. Self-admittance phase for antenna 1 calculated using FDTD and compared to the MoM.

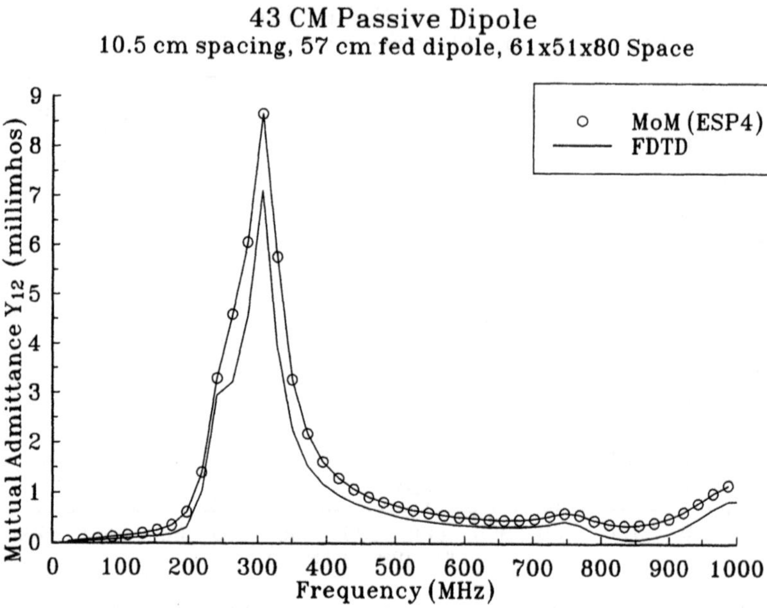

FIGURE 14-10. Mutual admittance magnitude calculated using FDTD and compared to the MoM.

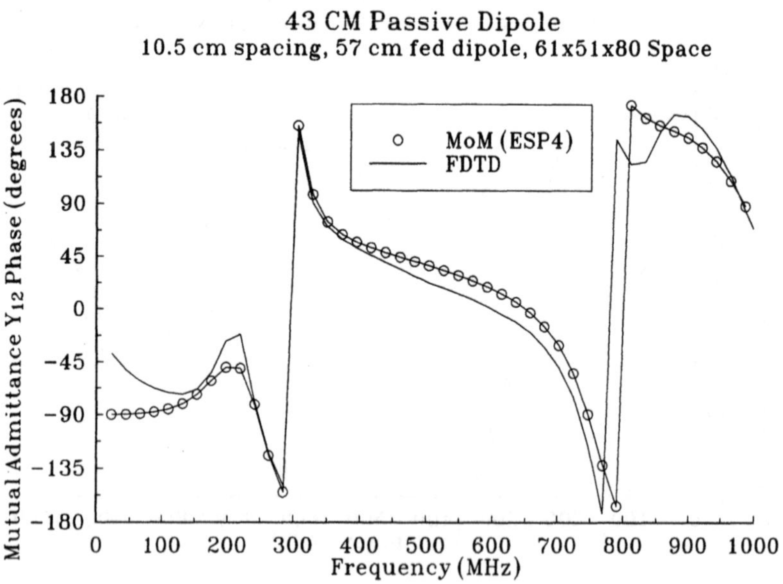

FIGURE 14-11. Mutual admittance phase calculated using FDTD and compared to the MoM.

Having demonstrated the capability of FDTD to calculate antenna self- and mutual admittance (and therefore impedance), we now proceed to deal with efficiency and gain. Again, we want to use pulse excitation so that results for a wide frequency band can be obtained from one FDTD computation. The approach which we will use for this is simple and straightforward.

Let us consider a typical situation in which an antenna is fed with one voltage source modeled as an electric field $E_z^n(I,J,K)$ with corresponding voltage $V_1(t)$ across the cell at the antenna feed gap, and that this source supplies a time domain current $I_1(t)$, just as described in (14.1) to (14.3) for the above mutual impedance example. If after all transients are dissipated the time domain results for these two quantities are Fourier transformed, the equivalent steady-state input power is given quite simply by

$$P_{in}(\omega) = \mathrm{Re}\left[V_1(\omega) I_1^*(\omega)\right] \tag{14.6}$$

at each frequency.

Dissipated power is also computed quite simply. Consider that an FDTD electric field component $Ez(I',J',K')$ is in a region with conductivity σ. If we assume that the electric field is uniform within a single FDTD cell, then the equivalent steady-state power dissipated in this region is given by

$$\begin{aligned} P_{diss} &= \iiint \sigma \left|E_z(\omega)\right|^2 dv \\ &= \sigma \left|E_z(\omega)\right|^2 \Delta x \Delta y \Delta z \\ &= \frac{\sigma \Delta x \Delta y}{\Delta z} \left|E_z(\omega) \Delta z\right|^2 \\ &= G \left|V_z(\omega)\right|^2 \end{aligned} \tag{14.7}$$

where $E_z(\omega)$ is the Fourier transform of $E_z(I',J',K')$. Furthermore, we also see from (14.7) that we can equivalently determine the dissipated power by considering the FDTD cell to contain a lumped conductance G with a voltage V_z across the cell in the z direction. Thus, (14.7) indicates that a lumped resistance $R = 1/G$ in a wire gap can be approximated by adjusting the conductivity σ appropriately. If many FDTD cell locations contain dissipative materials the computation in (14.7) is repeated for each such cell in each field component direction, with the total power dissipated given by the sum.

To determine the antenna gain, the far zone electric field in the desired direction must be determined. Using the approach given in Chapter 7 this can be done for the transient far zone fields. Since the far zone electric field is computed so that the $1/r$ amplitude factor and the r/c time delay are suppressed, where r is the far zone distance and c the speed of light, the antenna gain relative to a lossless isotropic antenna in the θ, ϕ direction is given by

$$\text{Gain}(\theta,\phi) = \frac{\left|E_F(\omega,\theta,\phi)\right|^2 / \eta_0}{P_{in} / 4\pi} \tag{14.8}$$

where $E_F(\omega,\theta,\phi)$ is the Fourier transform of the transient far zone time domain electric field radiated in the Θ,ϕ direction and η_0 is the impedance of free space.

We again will use the geometry shown in Figure 14-1 for demonstration. However, while the shorter dipole was a continuous conducting wire for convenient calculation of mutual admittance, here a 50-Ω resistor will be located in a one-cell long gap in its center so that considering the pair of dipoles as a single antenna fed at the center of the longer dipole, the antenna array will have an efficiency of < 100%.

The FDTD computations were made using the same parameters as for the previous example. The 50-Ω resistor was approximated by setting the conductivity associated with the Ez field at the center of the shorter dipole to 8 S/m. This value was obtained using (14.7) for the cell dimensions given with a conductance G = 1/50. Calculations again were run for 8192 time steps to allow for complete dissipation of transients, and again only the first 4500 are shown on the time domain plots which follow in order to increase clarity. During the progress of the FDTD calculations the z-directed current flowing through the feed gap at the center of the longer dipole and the z-directed electric fields at the center of both dipoles were saved for each time step along with the far zone transient electric field.

The MoM results for efficiency and gain were again obtained using the ESP4[3] computer code using the same wire radius of 0.281 cm.

The time domain results for the current in the wire gap fed by the Gaussian pulse and for the voltage across the 50-Ω resistor in the center of the shorter wire are shown in Figures 14-12 and 14-13. The complex Fourier transforms of these are used in (14-6) and (14-7) to determine the steady-state frequency domain input and dissipated powers, shown in Figures 14-14 and 14-15.

The input power peaks when the longer wire is resonant, while the level of the dissipated power depends on the interactions between the two wires in a more complicated way. The antenna efficiency is determined from the input and dissipated powers in the usual way as

$$\text{Efficiency} = \frac{P_{in} - P_{diss}}{P_{in}} \tag{14.9}$$

and the result is shown in Figure 14-16 and compared with results from ESP4.[3] The agreement is quite good, except at the lower frequencies. The FDTD results are questionable here because the antenna is very short electrically and

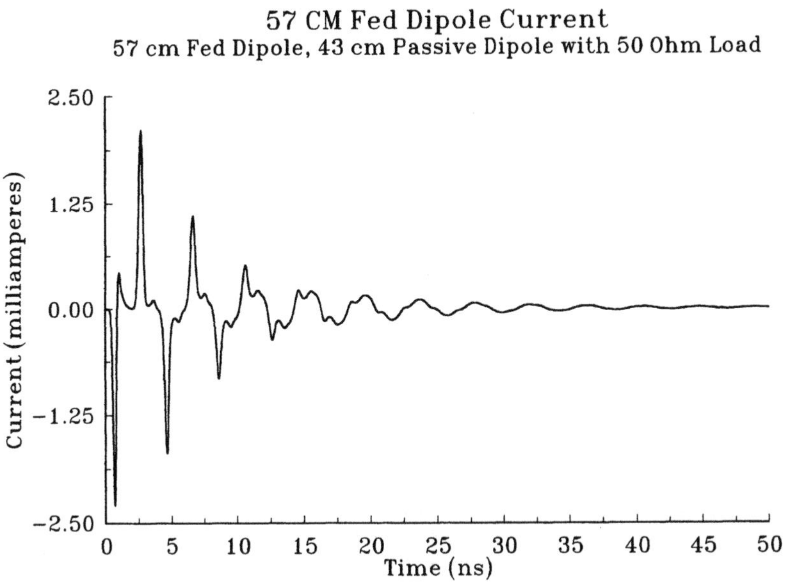

FIGURE 14-12. Time domain current flowing through the voltage source in the gap of wire 1 with a 50-Ω resistive load at the center of wire 2.

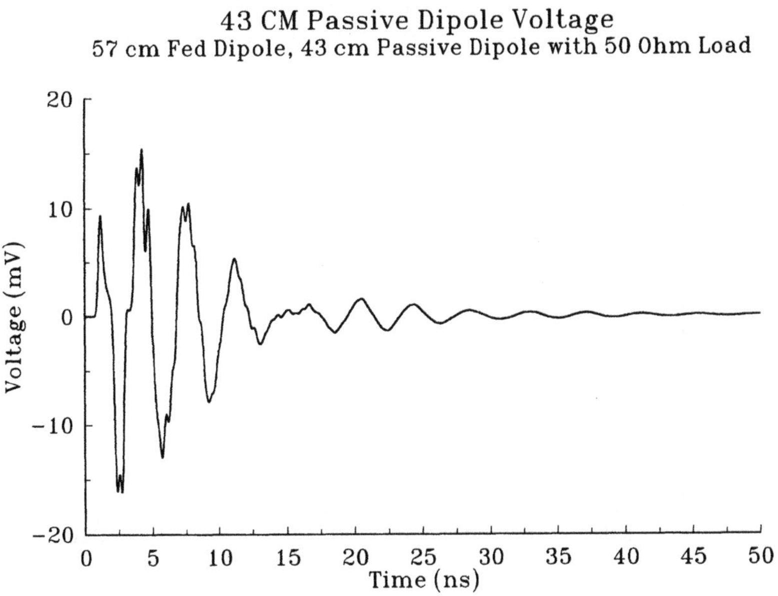

FIGURE 14-13. Time domain voltage across the 50-Ω resistor at the center of wire 2.

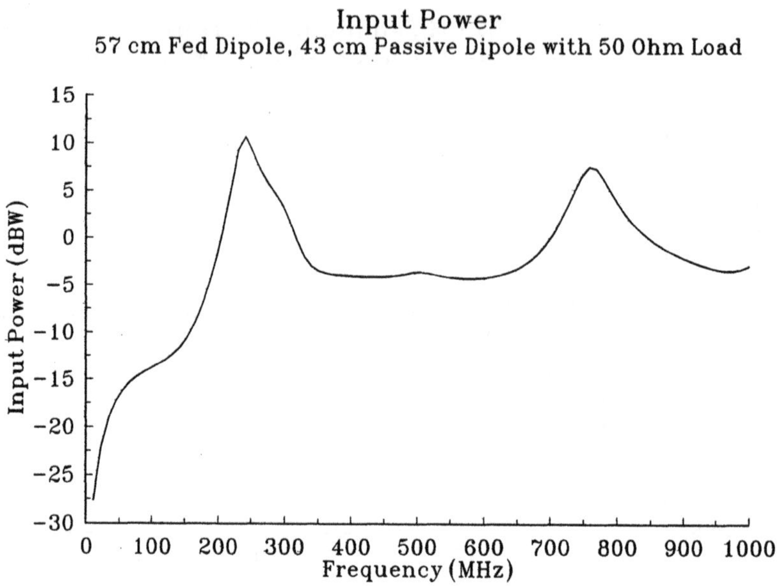

FIGURE 14-14. Input power vs. frequency from the complex Fourier transforms of the transient FDTD input voltage and current in the gap of the longer wire.

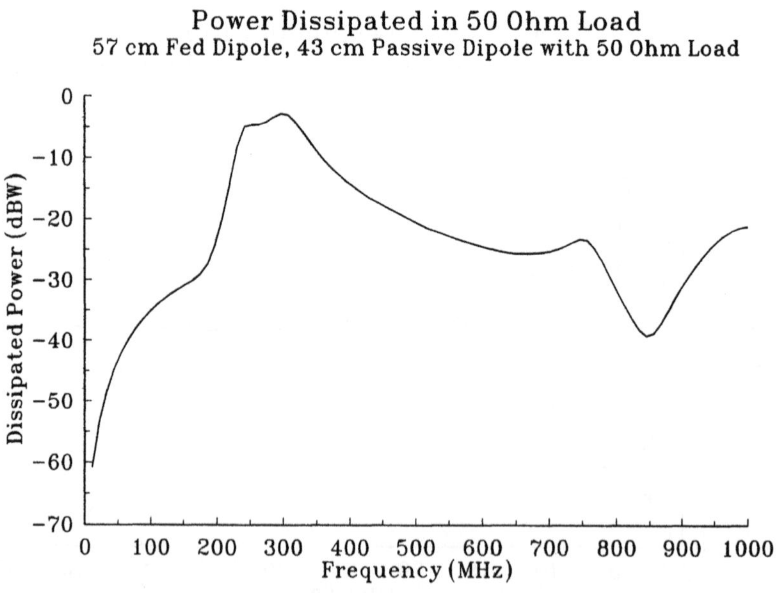

FIGURE 14-15. Power dissipated in the 50-Ω resistor from the complex Fourier transform of the transient FDTD voltage.

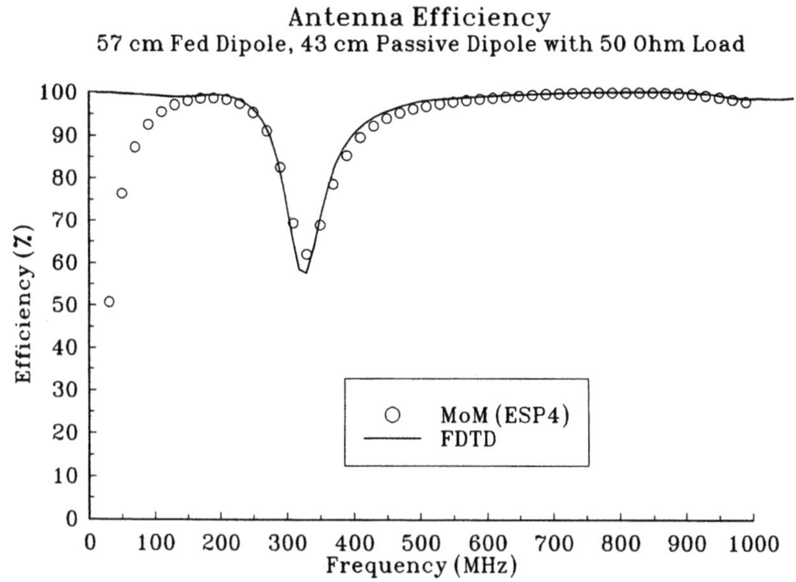

FIGURE 14-16. Wide band efficiency computed from the transient FDTD results and compared with the MoM for the two wire antenna with a 50-Ω load at the center of antenna 2.

the voltage and current are very nearly 90° out of phase. A very small error in the computed phase of the fed dipole current will result in a relatively large error in the value of P_{in}. This is also a problem, although presumably to a lesser extent, with the MoM results, since at 10 MHz ESP4 predicted an efficiency of –46% (this value is not shown in Figure 14-16), with the negative sign indicating more power dissipated than supplied.

Proceeding now to determine the absolute gain, Figure 14-17 shows the transient far zone E_θ electric field in the $\theta = 90$, $\phi = 0$ direction such that the shorter wire is acting somewhat as a director element for the longer fed wire. The Fourier transform of this is shown in Figure 14-18, with the peaks corresponding roughly to the resonances of the fed dipole. The absolute gain with respect to a lossless isotropic antenna is computed using (14.8) and the results are compared with the MoM in Figure 14-19. Except at the lower frequencies (where the input power level is extremely low and difficult for FDTD to accurately determine), the results are quite reasonable, especially considering the fundamentally different approaches of the two methods in approximating the lumped sources and loads.

In this section an approach for calculating the fundamental antenna parameters of self- and mutual impedance (admittance), efficiency, and gain using FDTD has been formulated and demonstrated on a simple wire antenna geometry. This geometry was chosen for convenience in obtaining validating results using the MoM. For this wire geometry the MoM is simple to apply, and takes much less computer time and memory than does FDTD. However, in the next two sections of this chapter FDTD is applied to other antenna geometries for

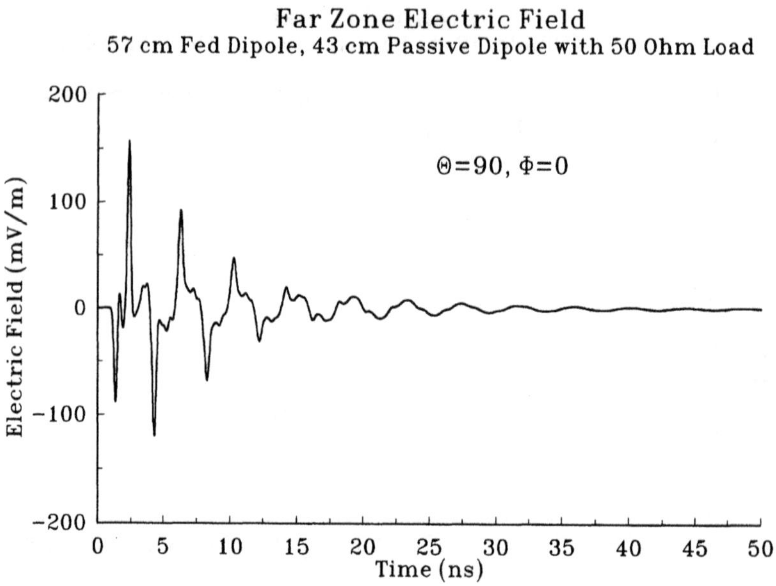

FIGURE 14-17. Time domain far zone radiated electric field normalized to a unit radial distance.

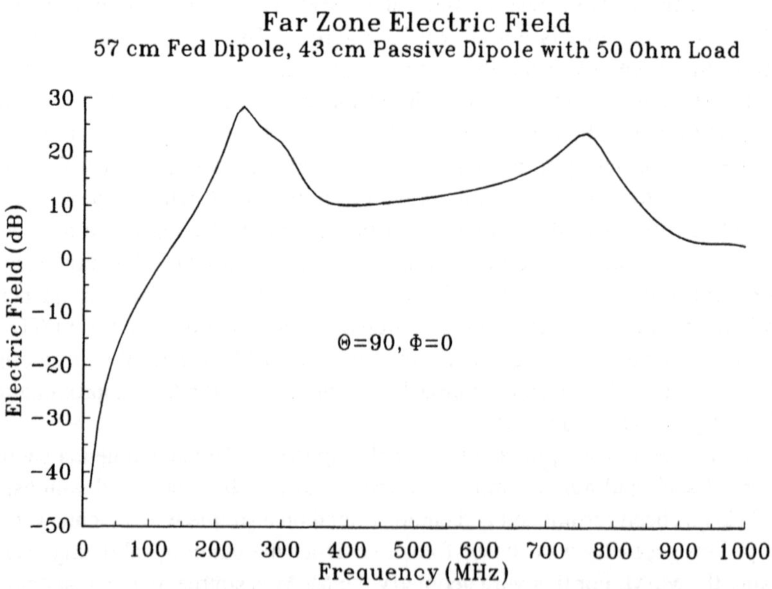

FIGURE 14-18. Fourier transform magnitude of the transient far zone radiated electric field.

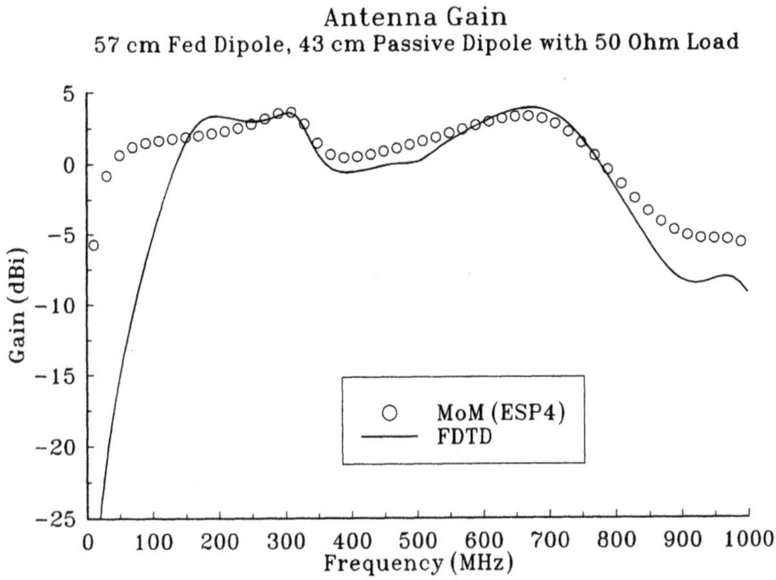

FIGURE 14-19. Absolute gain with respect to isotropic of the two wire antenna with 50-Ω load at the center of antenna 2 computed from the transient FDTD results and compared to the MoM.

TABLE 14-1
Dimensions of Monopole on Box shown in Figure 14-20

a = 60 mm		b = 10 mm
h = 50 mm		r = 0.5 mm
	c = 50, 130, or 200 mm	
	w = 10 or 30 mm	

Note: The monopole wire radius is r.

which it produces accurate results and requires less computer time than the MoM. The first of these is a wire monopole on a conducting box, considered in the next section.

14.3 MONOPOLE ANTENNA ON A CONDUCTING BOX

The geometry for our example is shown in Figure 14-20. A wire monopole antenna is connected to a conducting box and fed at the junction between the monopole and the box. The antenna connection is centered in the top of the box in the y dimension, but may be offset in the x dimension. The geometry is an approximation to a small hand-held radio unit with attached antenna. It is desired to calculate a radiation pattern at a frequency of 1.5 GHz, and the input impedance and gain over a band of frequencies. Three different size boxes are considered, with the dimensions given in Table 14-1.

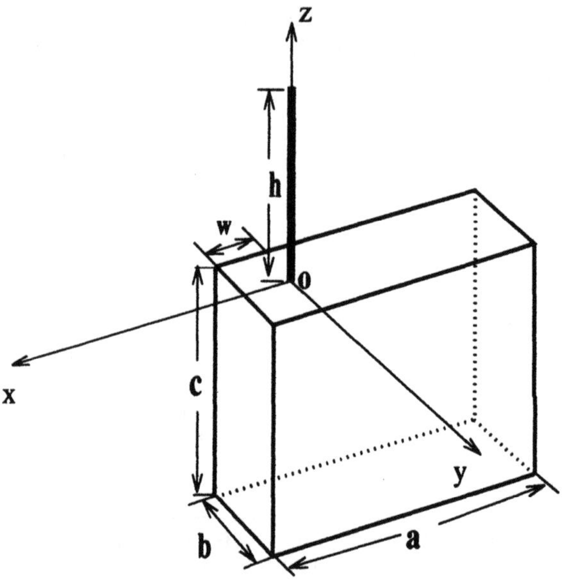

FIGURE 14-20. Geometry of monopole antenna on conducting box. Dimensions are given in Table 14.1.

At the highest frequency of interest, 6 GHz, the boxes are approximately 3, 8, and 12 square wavelengths in surface area. The monopole is centered in the x dimension (w = 30 mm) for radiation patterns and gain, while both centered and offset locations (w = 10 mm) are considered for impedance.

Both first and second order Mur absorbing boundaries acting on electric fields are used, with second order applied except when noted otherwise. It was found that the low frequency impedance results were very sensitive to the size of the problem space, i.e., the number of cells between the antenna and the Mur absorbing boundaries. A relatively large border of cells between the antenna and the absorbing boundaries was used in the impedance and gain calculations, as reflections from the absorbing boundaries reduced the accuracy, especially at lower frequencies. Making the FDTD space much smaller reduced the accuracy of the impedance at low frequencies due to reflections from the absorbing boundary.

The wire diameter is much smaller than the FDTD cells used, and using smaller cells in order to more accurately model the wire diameter would be a tremendous waste of computer resources. On the other hand, impedance results are quite sensitive to the wire radius. In order to include the effects of the relatively small wire radius, the magnetic fields surrounding the electric field components along the wire monopole axis were calculated as described in Chapter 10, except for the magnetic fields surrounding the $E_z(I,J,K)$ electric field component at the antenna base when that component was used as the driving source. It was found that attempts to calculate these magnetic fields using the sub-cell method produced less accurate impedance results.

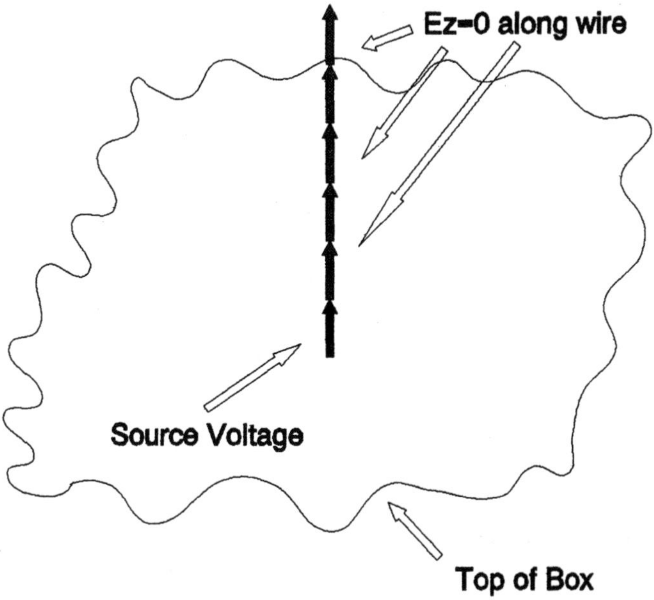

FIGURE 14-21. FDTD monopole antenna excited using the electric field component on the wire axis at the base of the wire. Corresponds to MoM gap feed.

Now let us discuss how the antenna was fed. Two methods were used. The first of these is illustrated in Figure 14-21, and corresponds to the approach used in the previous section for a wire dipole. The conducting box is modeled in the usual way by setting FDTD electric field components to zero over the surface of the box. These Yee cell field components are not indicated in Figure 14-21; however, the E_z field components on the wire monopole axis are indicated by arrows. These E_z components are set to zero, except for the component at the base of the monopole. This electric field component was used as the voltage source driving the antenna. If we let $V_1(t)$ be the source voltage, then the $E_z(I,J,K)$ electric field at the base of the monopole is given by (14.1). This method for driving the antenna corresponds to the MoM gap source, except that usually in MoM calculations the gap is infinitesimally thin, while the FDTD gap is one spatial interval (Δz in this case) in length.

The other approach investigated for feeding the monopole antenna is shown in Figure 14-22. In this method the E_z field components along the wire axis are all set to zero, including the one just above the top of the box. The four electric field components on the surface of the box going radially from the monopole axis are, however, driven. Since the $E_z(I,J,K)$ component at the base of the antenna is now set to zero, the thin wire approach of Chapter 10 will be used to calculate the magnetic fields surrounding all of the E_z electric fields along the wire axis, including this one.

The method used to calculate these magnetic fields assumes a $1/\rho$ variation of the E_x and E_y components adjacent to the wire, where ρ is the radial distance from the wire. The radial field dependence of the coaxial cable that this source

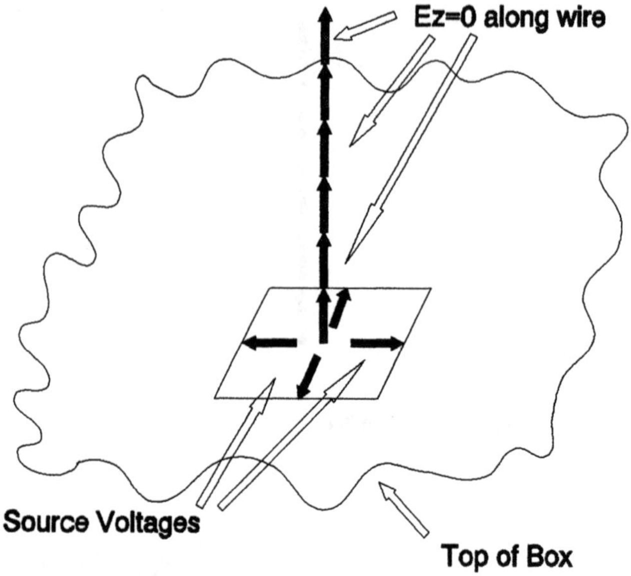

FIGURE 14-22. FDTD monopole antenna excited using the electric field components on the top of the box at the base of the wire. Corresponds to MoM magnetic frill feed.

is supposed to approximate is also $1/\rho$. Thus the E_x and E_y source components on the top of the box at the base of the monopole as shown by arrows in Figure 14-22 are driven as

$$E_x^n(I,J,K) = -E_x^n(I-1,J,K) = \frac{-V_1(n\Delta t)}{\ln\left(\frac{\Delta x}{r}\right)} \cdot \frac{2}{\Delta x} \tag{14.10}$$

$$E_y^n(I,J,K) = -E_y^n(I,J-1,K) = \frac{-V_1(n\Delta t)}{\ln\left(\frac{\Delta y}{r}\right)} \cdot \frac{2}{\Delta y} \tag{14.11}$$

This approach corresponds to the magnetic frill method for exciting wires in the MoM, since one can consider these impressed electric fields to be equivalent to a magnetic current source circulating around the wire.

In the following calculations the approach shown in Figure 14-22 and described in (14.1) is used for all calculations except where explicitly noted otherwise.

Now let us consider the FDTD parameters for the various calculations to follow. First let us consider the single frequency radiation patterns. The FDTD cells were sized at $\Delta x = \Delta y = \Delta z = 5$ mm, corresponding to 40 cells per wavelength at 1.5 GHz. This relatively small cell size is required by the size of the box rather than the necessity of having so many cells per wavelength, since for this cell size the "b" dimension of the box is only two cells.

The time step is chosen slightly below the Courant limit for all calculations at $\Delta t = \Delta z/(2c)$.

For the smallest box an FDTD space of $72 \times 62 \times 70$ cells was used and far zone fields were calculated at 9° increments (21 directions, making use of the problem symmetry). For the c = 130 mm box the space was $72 \times 62 \times 86$ cells, and far zone fields were calculated in 6° increments. For the largest box with c = 200 mm, the problem space was $72 \times 62 \times 100$ cells and patterns were calculated in 4° increments.

The radiation patterns were desired in the $\phi = 0$ (x-z) plane. The far zone fields were calculated using the 3-D transient far zone transformation of Chapter 7. The far zone fields in all pattern directions were calculated during 1 FDTD computation. Once steady-state conditions are reached the magnitude of the electric field is easily determined at each far zone angle.

Since radiation patterns are desired at only one frequency, for these calculations the source voltage $V_1(t)$ was specified as a 1.5-GHz sinusoidal time variation.

Next we consider calculating the input impedance of the monopole. Since impedance results over a wide bandwidth are desired the antenna was excited with a Gaussian pulse source voltage $V_1(t)$ rather than the sinusoidal time variation used for the single frequency radiation patterns. The Gaussian pulse width is approximately 0.04 ns. Since higher frequencies are involved, and since impedance requires greater accuracy than radiation pattern calculations, the cell sizes were reduced from those used for the pattern calculations. For the smallest box (c = 50 mm) 1.67 mm cubical cells were used, corresponding to approximately 30 cells per wavelength at the highest frequency. The problem space for this box was $130 \times 90 \times 180$ cells. For impedance calculations for the intermediate size box (c = 130 mm) a cell size of 2.5 mm was used, since for this box too much computer memory was required with the 1.67 mm cells. This larger cell size corresponds to 20 cells per wavelength at 6 GHz. With the larger cells the problem space for the larger box could be reduced to $100 \times 90 \times 150$, fewer than required for the smaller box. The impedance results for the larger cell size appeared to be just as accurate as with the smaller cell size, as will be shown. The larger cell size allowed the absorbing outer boundary to be located farther from the antenna with fewer FDTD cells required.

The transient current $I_1(t)$ through the base of the monopole was sampled at each time step by calculating the discrete line integral of the magnetic field around the $E_z(I,J,K)$ electric field located at the monopole base using (14.3).

After all transients have dissipated, the resulting transient current $I_1(t)$ and the Gaussian excitation voltage $V_1(t)$ were Fourier transformed. The complex transforms were divided at each frequency to obtain the input impedance of the monopole.

For calculating the wideband gain the input power to the antenna is needed. The equivalent steady-state input power is obtained at each frequency from the complex Fourier transforms of $V_1(t)$ and $I_1(t)$ from (14.6), and the absolute gain with respect to isotropic from (14.8).

All computations were performed on an NEC SX-2N computer. The measurements were made in an indoor anechoic chamber.[4] The monopole was fed with a coaxial cable terminated in an SMA type 219T miniature coaxial connector mounted flush with the top surface of the box.

Radiation patterns were calculated for all three box sizes, c = 50, 130, and 200 mm, with the monopole centered on the top of the box (w = 30 mm). The radiation patterns are shown in Figures 14-23 to 14-25, and are compared to measurements and to MoM calculations made using a code developed specifically for this geometry.[5] The computer time required by the FDTD calculations varied from 7 min of CPU time for the radiation pattern for the smallest box to 20 min for the largest box. Some of this time increase was due to calculating the radiation patterns in finer angle increments for the larger boxes. The MoM computer code required about 9 min of computer time for each radiation pattern calculation since the same number of modes was used for all three box sizes.

Next consider the impedance calculations. The transient time domain current through the base of the monopole due to the Gaussian source voltage computed using FDTD is shown in Figure 14-26 for the smallest box size (c = 50 mm) with the monopole centered on the top of the box (w = 30 mm). This calculation required approximately 8 min of CPU time, about the same as for the radiation pattern calculation despite the smaller cell size and greater number of cells. This is because no far zone fields were computed while the impedance calculations were being made. The impedance obtained by dividing the Fourier transforms source voltage $V_1(t)$ and the monopole base current $I_1(t)$ of Figure 14-26 is shown in Figure 14-27 and compared to measurements.

The results shown in Figure 14-28 use the same cell size, time step, and problem space size as in Figure 14-27 and provide two different comparisons. First, FDTD results are shown for both methods of feeding the monopole, the E_z feed of Figure 14-21 and the $E_{x,y}$ feed of Figure 14-22. The FDTD results are changed slightly, with the $E_{x,y}$ feed perhaps being more accurate at the lower frequencies and but less accurate at the higher frequencies. This slight change is similar to that observed in MoM thin-wire calculations, where the gap and magnetic frill feeds typically yield similar but slightly different results. The second comparison involves the fact that the results of Figure 14-28 were obtained using first order Mur absorbing boundaries. Comparing the E_z fed results in Figure 14-28 to those obtained using second order Mur in Figure 14-27, it can be seen that the only effect is a slight change in the real part of the impedance at the lowest frequencies. The second order Mur results are slightly more accurate. However, first order Mur performs quite well due to the large number of cells between the antenna and the absorbing boundary.

Returning to the E_z wire axis feed approach of Figure 14-21 and (14.1) to excite the antenna, impedance results are shown in Figure 14-29 for the monopole centered in the top of the middle sized (c = 130 mm) box, and in Figure 14-30 for the smallest box (c = 50 mm) but with the monopole offset in the x direction from the center of the box top (w = 10 mm). The agreement with measurements is again quite good for both cases.

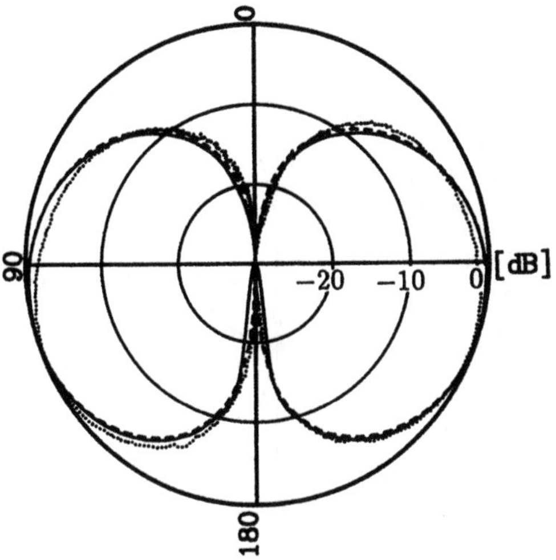

The patterns. c=50 mm.

— FDTD, - - - MoM, ... Expt

FIGURE 14-23. Calculated and measured radiation pattern at 1.5 GHz for monopole antenna centered (w = 30 mm) on smallest box (c = 50 mm).

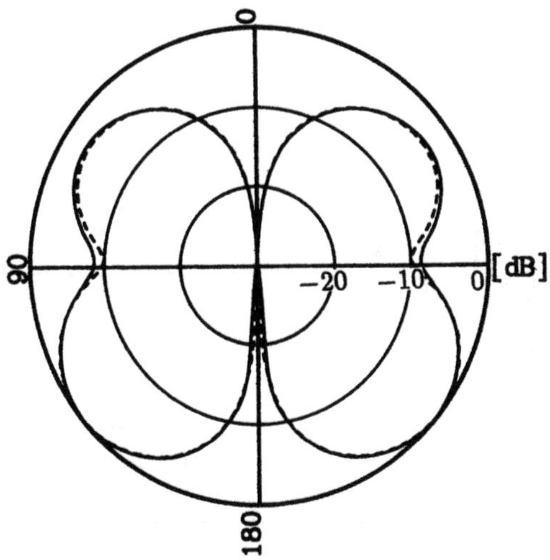

The patterns. c=130 mm.

— FDTD, - - - MoM

FIGURE 14-24. Calculated and measured radiation pattern at 1.5 GHz for monopole antenna centered (w = 30 mm) on intermediate size box (c = 130 mm).

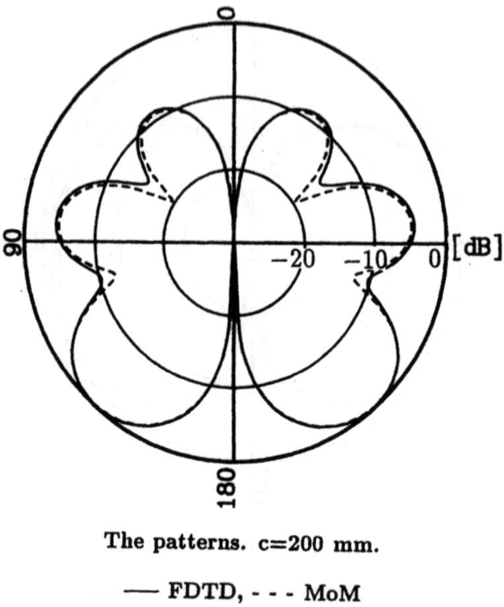

FIGURE 14-25. Calculated and measured radiation pattern at 1.5 GHz for monopole antenna centered (w = 30 mm) on largest box (c = 200 mm).

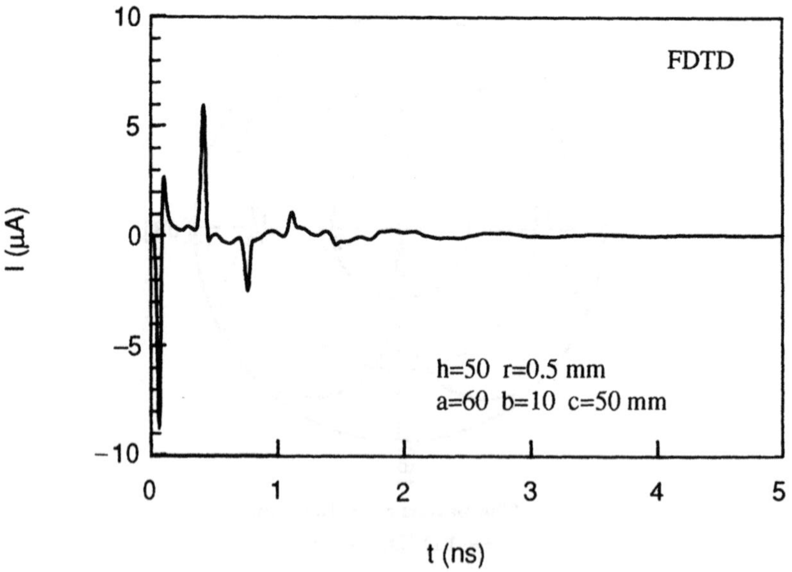

FIGURE 14-26. Transient current flowing in base of monopole centered (w = 30 mm) on smallest (c = 50 mm) conducting box due to 0.13 ns Gaussian pulse voltage source.

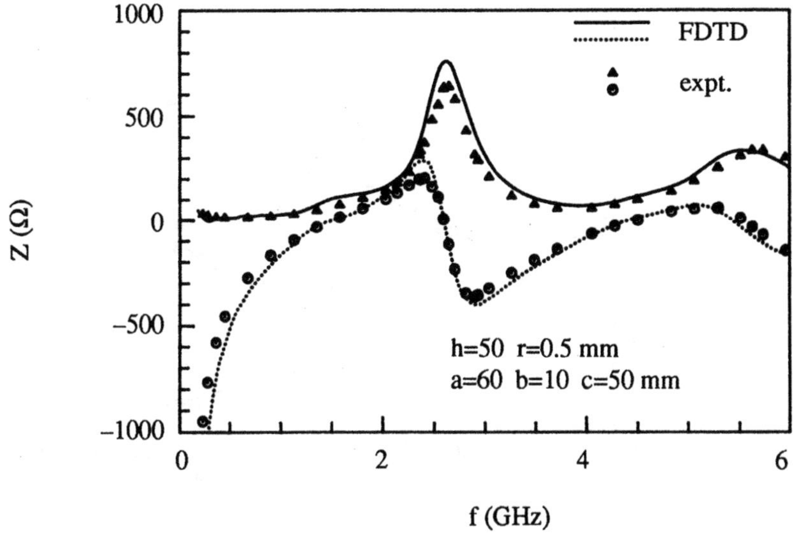

FIGURE 14-27. Input impedance for monopole antenna centered (w = 30 mm) on smallest (c = 50 mm) conducting box computed using transient FDTD results and compared to measurements.

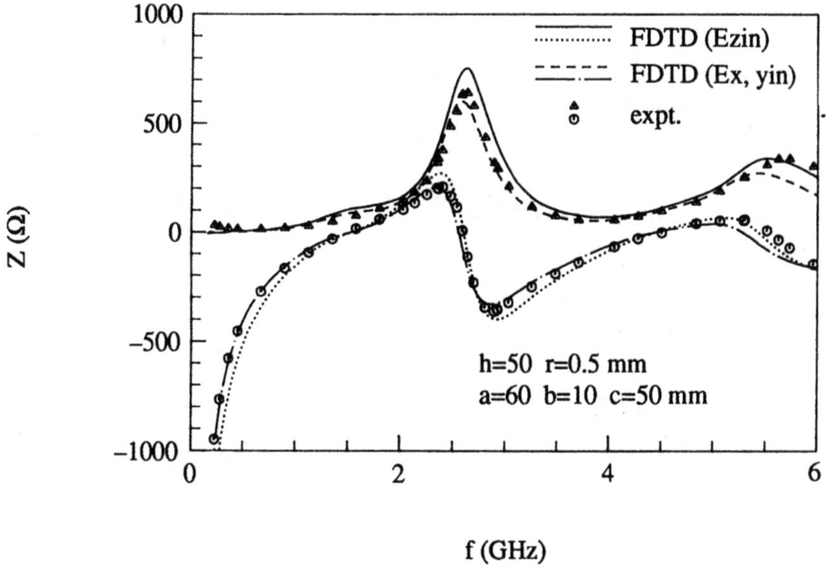

FIGURE 14-28. Input impedance for monopole antenna centered (w = 30 mm) on smallest (c = 50 mm) conducting box computed using transient FDTD results with the monopole fed in two different ways and compared to measurements.

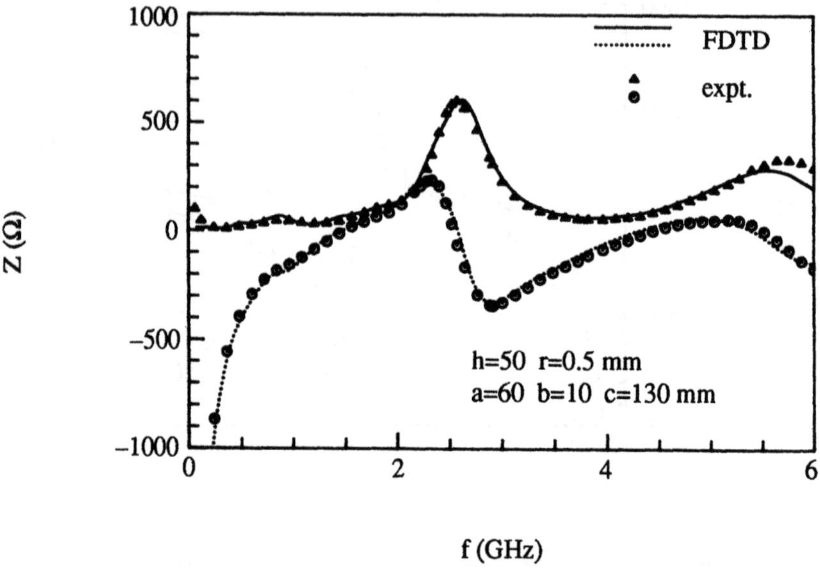

FIGURE 14-29. Input impedance for monopole antenna centered (w = 30 mm) on intermediate size (c = 130 mm) conducting box computed using transient FDTD results and compared to measurements.

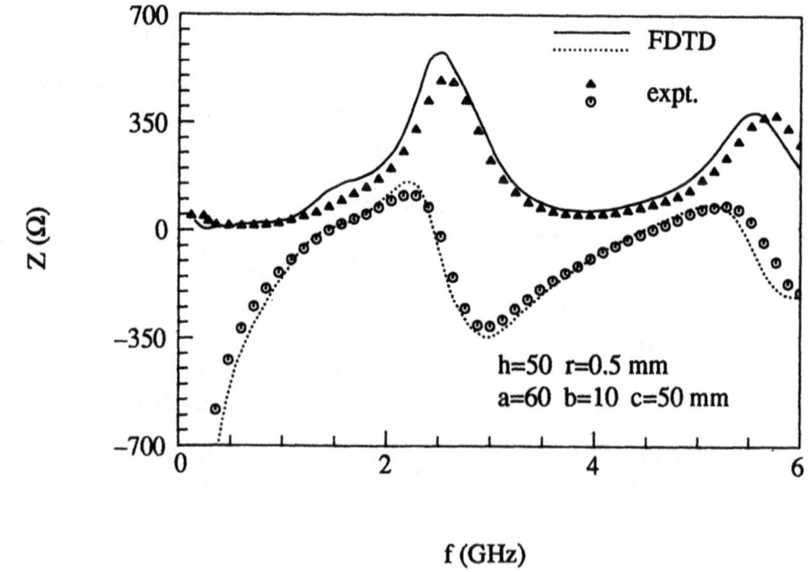

FIGURE 14-30. Input impedance for monopole antenna offset from center (w = 10 mm) on smallest (c = 50 mm) conducting box computed using transient FDTD results and compared to measurements.

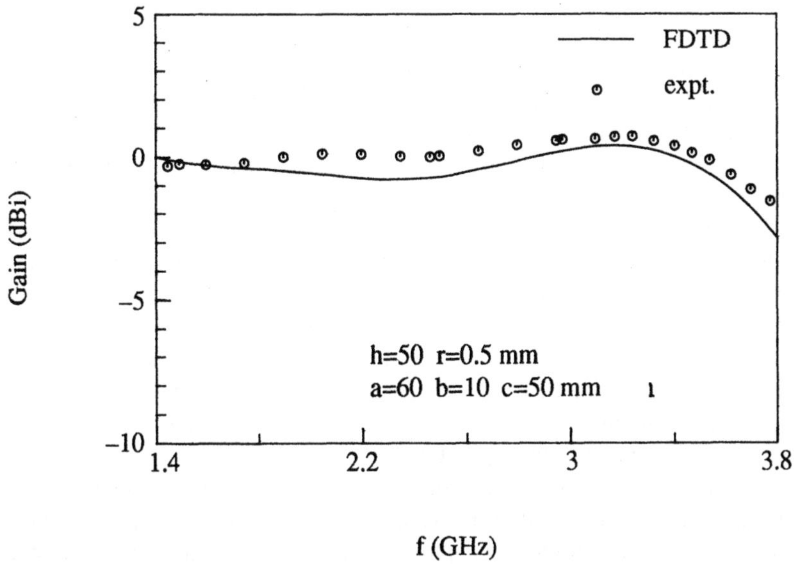

FIGURE 14-31. Absolute gain with respect to an isotropic antenna in the $\theta = 90, \phi = 0$ direction, for the monopole centered ($w = 30$ mm) on the smallest box ($c = 50$ mm) calculated using transient FDTD results and compared with measurements.

Finally consider the results for antenna gain in the $\theta = 90$, $\phi = 0°$ direction. The cell size, time step, and problem space size are the same as used in the impedance calculations. In Figure 14-31 the FDTD absolute gain (with respect to isotropic), obtained from one calculation using pulse excitation, is compared to measured results. The agreement is excellent over the frequency range shown.

In this section the ability of FDTD to calculate radiation patterns, input impedance, and absolute gain for a monopole antenna on a conducting box has been demonstrated. The feed region was modeled using only the usual rectangular Yee cell FDTD electric field components. The sub-cell approach given in Chapter 10 was used to calculate the magnetic field components adjacent to the wire to compensate for the thin wire being smaller than the FDTD cells. Two feed methods, corresponding to MoM wire gap and magnetic frill feeds, were investigated and found to give similar results. The FDTD results were compared both with measurements and with the MoM with good agreement. The computer time required for a single frequency pattern calculation was similar to that for the MoM. However, for wide bandwidth impedance and gain calculations the FDTD method is much faster for the antenna and box sizes considered. Also, the FDTD approach could be extended quite easily to consider antennas including dielectric materials, such as a monopole on a conducting box coated with a layer of lossy dielectric.

14.4 SHAPED-END WAVEGUIDE ANTENNA

In the previous examples the antenna geometries fit conveniently into the FDTD grid. In this section the antenna geometry being considered is based on a circular tube, and does not fit well into the rectangular FDTD grid used. Nevertheless, as will be demonstrated, accurate results for the gain patterns can be obtained.

The antennas structures considered are circular waveguides and shaped-end radiators. A shaped-end radiator (SER or Vlasov radiator) is simply a hollow circular tube with a beveled cut on one end. The beveled cut is designed to provide some directionality to the radiation. References are given[6-8] which provide more detailed information on these antenna structures. Figure 14-32 shows a side view of a SER with the cut angle ξ and the monopole probe used to excite the antenna. Thus, the circular waveguide is a special case of the general SER geometry with $\xi = 90°$.

The antennas are oriented with the cylinder axis parallel to the z axis. The antennas are modeled in the usual stepped-edge (or staircased) fashion (discussed in the previous chapter) with the cells being approximately cubical. To excite the antennas, a monopole probe with one cell air gap was placed above a perfectly conducting endcap (see Figure 14-32). The monopole probe is designed to excite the TM_{01} mode (for sinusoidal excitation), and the probes are $\lambda_g/2$ long, where λ_g is the guide wavelength of the TM_{01} mode. The monopole probe is excited at the air gap by a single electric field component that is an analytically defined excitation. The excitation used is a time harmonic signal with a linear ramping function r(t) that transitions from 0 to 1 over several cycles of the sinusoid. The functional form of this excitation is

$$e_z^i(t) = r(t)\sin(\omega_0 t) \tag{14.12}$$

with ω_0 being the radian frequency of interest.

The first antenna structure analyzed is a simple circular waveguide with cut angle $\xi = 90°$. Figure 14-33 illustrates how the circular waveguide is modeled in FDTD cells. The waveguide is 4.76 cm in diameter, 21 cm in length, and the sinusoidal excitation frequency is 8.6 GHz. The cutoff frequency, f_c, for the TM_{01} mode is defined by[9]

$$f_c = \frac{\chi_{01}}{2\pi a\sqrt{\mu - \varepsilon}} \tag{14.13}$$

where χ_{01} is the first zero of the Bessel function $J_0(x)$, a is the radius of the waveguide and μ and ε are the constitutive parameters of the fill material of the waveguide. The guide wavelength for the TM_{01}, λ_g, is defined as[9]

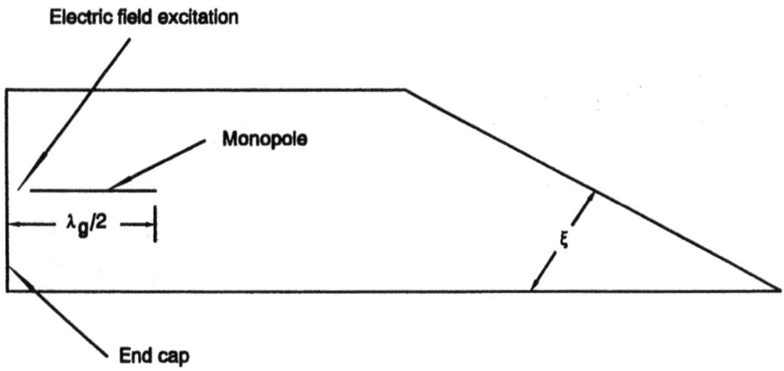

FIGURE 14-32. Side view of Vlasov SER geometry showing cut angle ξ and the monopole probe used for exciting the antenna.

$$\lambda_g = \frac{\lambda_0}{\sqrt{1-\left(f_c/f\right)^2}} \tag{14.14}$$

where λ_0 is the free space wavelength. With the cylinder filled with free space the guide wavelength is calculated to be λ_g = 4.2098 cm. At ten cells per wavelength the cell sizes are chosen to be δz = 4.2098 mm with δx and δy at 4.76 mm so that an integral number spans the cylinder diameter. The time step is 8.76 ps and the total number of time steps is 2048, which represents about 40 cycles of the sinusoidal waveform. The total problem space size was 60 × 60 × 95 cells, which provides a 25-cell border between the cylinder and the outer absorbing boundaries (second order Mur) and a 30-cell border at the radiating end of the cylinder.

In addition to the 90° cut waveguide, antennas with 30° and 60° cut angles ξ are also analyzed. Figures 14-34 and 14-35 illustrate how the 30° and 60° Vlasov radiators are modeled using FDTD cells. The tubes are 4.76 cm in diameter and have a constant reference distance of 17.78 cm from the endcap to the top of the beveled aperture. For these geometries, the sinusoidal frequency, the problem space size, cell sizes, time step, and total number of time steps are the same as for the 90° waveguide radiator.

We desire to calculate the gain patterns from these antennas at the excitation frequency. To obtain the input power, the electric field and the current are recorded vs. time at the air gap of the monopole probe. The input voltage is simply given by

$$v_{in}(t) = e_z^i(t)\Delta z \tag{14.15}$$

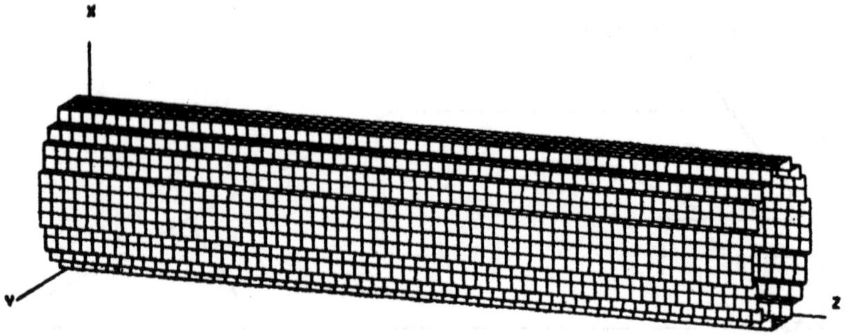

FIGURE 14-33. Perspective view illustrating how a circular waveguide is modeled using FDTD cells.

The input voltage and current and far zone fields in each pattern direction are transformed to the frequency domain, and the frequency component at the excitation frequency is selected. The input power is given by (14.15), and the gain by (14.8).

Figure 14-36 shows the copolarized absolute gain FDTD computations and measurements vs. polar angle θ for the straight-cut circular waveguide. Note the agreement is excellent over the entire angular range. Figure 14-37 shows the electric field distribution across the cylinder cross section. This field distribution is a reasonable approximation to the TM_{01} mode distribution given by $J_0(\beta\rho)$, where β is the propagation constant for the waveguide mode and ρ is the distance from the center of the waveguide tube $\left(\rho = \sqrt{x^2 + y^2}\right)$. Figure 14-38 shows the electric field distribution on the cylinder axis vs. z position along the cylinder. We can see from this figure the wavelength of the TM_{01} mode is exactly 10 cells/λ corresponding to the Δz selected for this problem.

Figures 14-39 and 14-40 show the copolarized absolute gain FDTD computations and measurements vs. azimuth angle ϕ for the 30° and 60° Vlasov radiators, respectively. The agreement between the FDTD computations and measurements is quite good and the curves differ only by 1 to 2 dB, except at the pattern nulls. The larger values for the measured data at these nulls is most likely due to the presence of the cross-polarization component. These data were plotted at $\theta = 32°$, which corresponds to the peak in the polar gain pattern. Figures 14-41 and 14-42 show the copolarized absolute gain FDTD computations and measurements vs. polar angle θ for the 30° and 60° Vlasov SER radiators, respectively. Note in these figures that the FDTD method follows the main beam very well, but has more difficulty tracking the measurements away from the main beam, and especially in the rearward direction ($\theta = \pm 180°$). The larger values of gain for the FDTD computations in the rear direction is most likely due to diffraction effects from the staircased beveled cut on the radiating end of the cylinder.

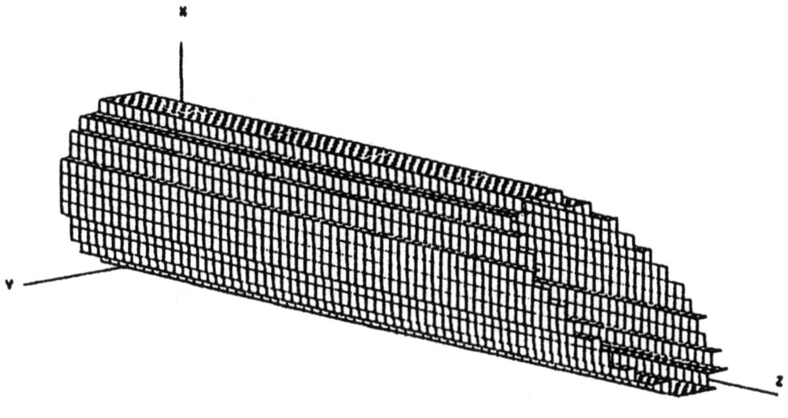

FIGURE 14-34. Perspective view illustrating how a 30° Vlasov SER is modeled using FDTD cells.

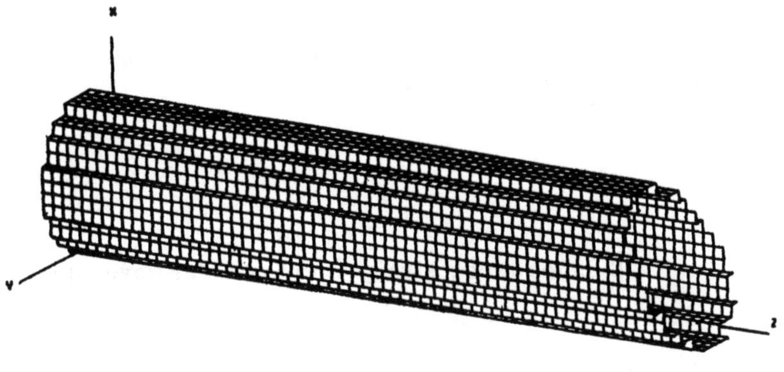

FIGURE 14-35. Perspective view illustrating how a 60° Vlasov SER is modeled using FDTD cells.

This section has demonstrated the application of the FDTD method to a different class of antenna structures involving curved surfaces. The basic FDTD method including staircasing was used and produced good results in comparison to measured data.

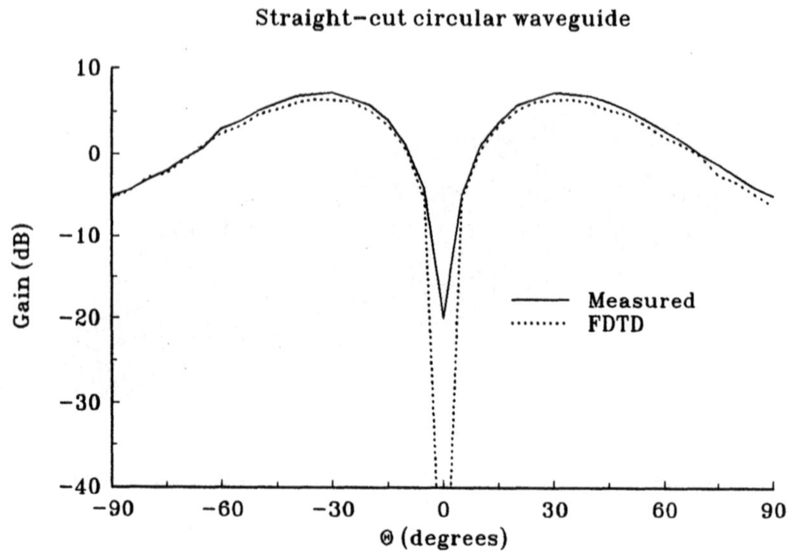

FIGURE 14-36. Copolarized absolute gain measurements and FDTD computation vs. polar angle θ for the straight-cut circular waveguide.

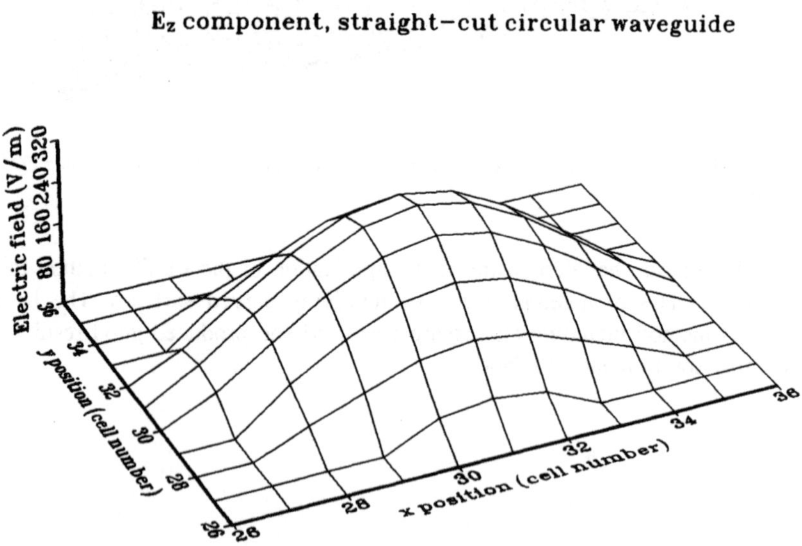

FIGURE 14-37. Copolarized electric field distribution across the cylinder cross section for the straight-cut circular waveguide.

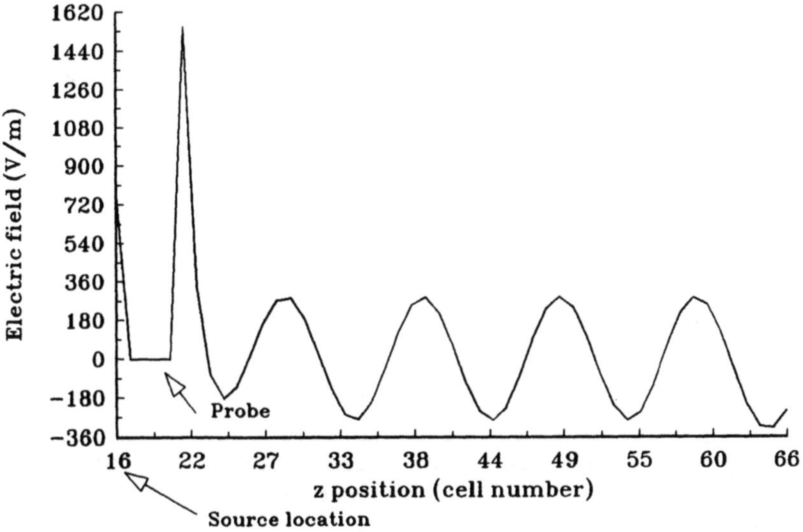

FIGURE 14-38. Copolarized electric field distribution on the cylinder axis vs. z position along the cylinder for the straight-cut circular waveguide.

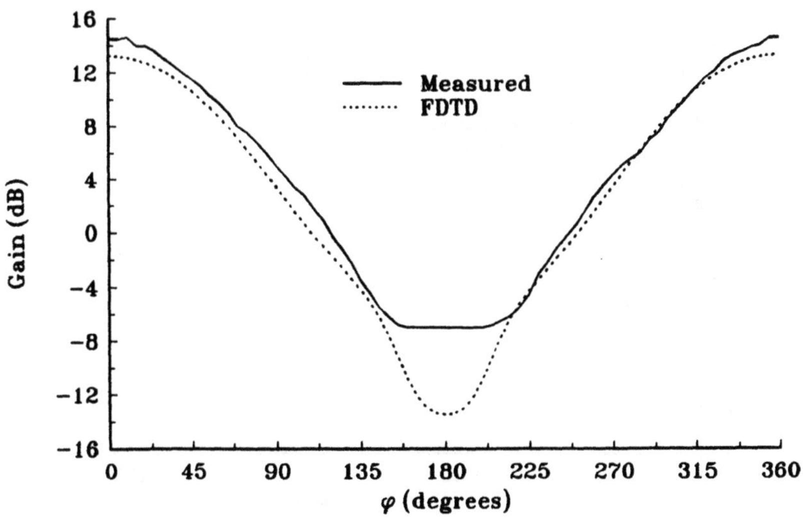

FIGURE 14-39. Copolarized absolute gain measurements and FDTD computations vs. azimuth angle ϕ for the 30° Vlasov SER with $\theta = 32°$.

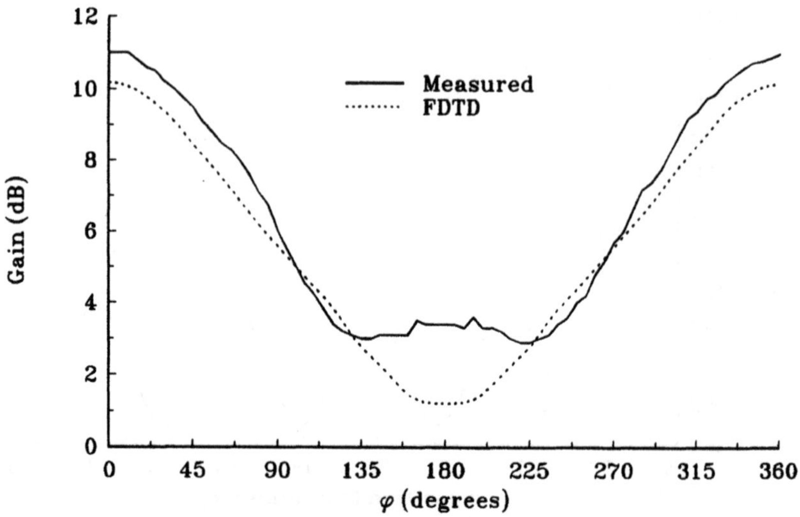

FIGURE 14-40. Copolorized absolute gain measurements and FDTD computations vs. azimuth angle ϕ for the 60° Vlasov SER with $\theta = 32°$.

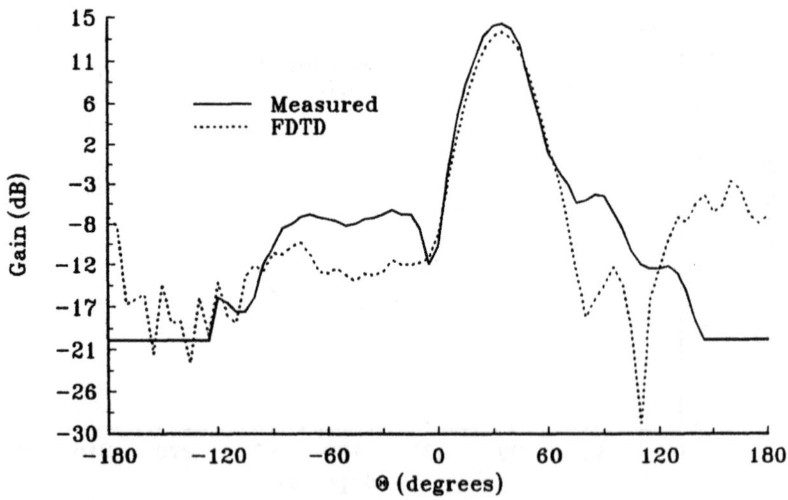

FIGURE 14-41. Copolorized absolute gain measurements and FDTD computations vs. polar angle θ for the 30° Vlasov SER.

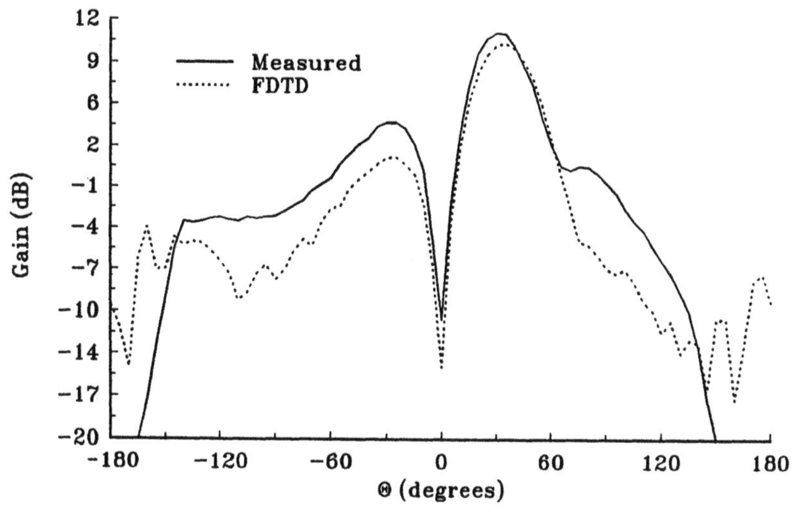

FIGURE 14-42. Copolarized absolute gain measurements and FDTD computations vs. polar angle θ for the 60° Vlasov SER.

REFERENCES

1. **Maloney, G., Smith, G. S., and Scott, W. R., Jr.,** Accurate computation of the radiation from simple antennas using the finite-difference time domain method, *IEEE Trans. Ant. Prop.*, 38, 1059, 1990.
2. **Kunz, K., Luebbers, R., and Hunsberger, F.,** A thin dipole antenna demonstration of the antenna modeling capabilities of the finite difference time domain technique, *Appl. Comput. Electromag. Soc. J.*, 5(1), 2, 1990.
3. **Newman, E.,** A User's Manual for the Electromagnetic Surface Patch Code: ESP Version IV, The Ohio State University Research Foundation, ElectroScience Laboratory, Department of Electrical Engineering, Columbus, OH, 1988.
4. **Chen, L., Uno, T., Adachi, S., Luebbers, R., and Kunz, K.,** FDTD Method Analysis of a Monopole Antenna Mounted on a Conducting Rectangular Box, *IEEE AP-S Int. Symp. Dig.*, 3, 1670, 1992.
5. **Yamaguchi, R., Sawaya, K., Fujino, Y., and Adachi, S.,** *IEE Jpn. Tech. Rep.* 91-80, 41, 1991.
6. **Vlasov S. N. and Orlova, I. M.,** Quasioptical transformer which transforms the waves in a waveguide having circular cross-section into a highly directional wave beam, *Radiophys. Quant. Electr.*, 17(1), 115, 1975.
7. **Sealy P. J. and Vernon, R. J.,** Vlasov launchers for TE_{0n} and high-azimuthal-index rotating TE modes, *14th Annu. Int. Conf. Infrared and Millimeter Waves Dig.*, 1, 158, 1989.
8. **Ruth, B. G., Dalstrom, R. K., Schlesiger, C. D., and Libelo, L. F.,** Design and low-power testing of a microwave Vlasov mode converter, *1989 IEEE MTT-s Int. Microwave Symp. Dig.*, 3, 1277, 1989.
9. **Balanis, C.,** *Advanced Engineering Electromagnetics,* John Wiley & Sons, New York, 1989, 479.

Chapter 15

GYROTROPIC MEDIA

15.1 INTRODUCTION

Gyrotropic is used to indicate materials which are anisotropic and have a strong frequency dependence in one of their constitutive parameters. When subjected to a constant magnetic bias field, both plasmas and ferrites exhibit anisotropic constitutive parameters. For electronic plasmas this anisotropy must be described by using a permittivity tensor in place of the usual scalar permittivity. Each member of this tensor is also very frequency dependent. Like plasmas, ferrites are a class of materials which become anisotropic when subjected to a static magnetic field. Unlike plasmas, which are dispersive even without the biasing field, dispersion in ferrites is coupled with the presence of the biasing field. When this field disappears the permeability is no longer frequency dependent and the only resonances that exist are comparatively weak.

While the ferrite permeability tensor is very similar in form to the magnetized plasma permittivity tensor, the two tensors are not identical due to the different asymptotic frequency behavior of the respective materials. In addition, because of the difference in boundary conditions for the two materials, substitution of one material for the other will not always yield completely dual behavior.

This chapter describes an FDTD formulation which incorporates both anisotropy and frequency dispersion at the same time, enabling the wideband transient analysis of magnetoactive plasma and magnetized ferrites. The frequency dispersive nature of these materials is included using the methods described in Chapter 8. However, the susceptibility functions which must be considered here are more complicated than any considered in that chapter. In addition, the anisotropy of the materials forces another extension to the FDTD method, since orthogonal components of electric or magnetic fields are directly coupled by the tensor permittivity or permeability. While in this chapter results are presented only for a 1-D geometry with field components in 2-D, 3-D results could be obtained by extension of the approach described here. If the Yee cell arrangement of field components is used, this would involve spatial averaging in order to obtain the necessary field components. The material in this chapter is based closely on References 1 and 2.

15.2 MAGNETIZED PLASMA

Cold plasmas, metals at optical frequencies, and semiconductors are often described as a free electron (or ion) gas. The equation of motion for the charged carriers (electrons) in response to an electric field yields the usual Drude frequency domain permittivity expression as discussed in Chapter 8.

There are several consequences of this expression. Among them are the possibility of relative permittivities <1 and that for frequencies below the plasma frequency ω_p, electromagnetic waves do not propagate very far into a plasma.

It should be noted that FDTD is generally a macroscopic approach, and the primary interest here is how electromagnetic waves interact with an isotropic magnetized plasma. Modeling of microscopic phenomena is not attempted here and is left to more mature particle-in-cell (PIC) simulation codes which are better suited for modeling plasma dynamics.[3]

The well-known permittivity expression for unmagnetized plasmas is given in Chapter 8 and reproduced here for convenience as

$$\hat{\varepsilon}(\omega) = 1 + \frac{\omega_p^2}{\omega(j\nu_c - \omega)} \tag{15.1}$$

Notice that for lossless plasmas the permittivity approaches zero at the plasma frequency (ω_p), and below the plasma frequency the real part of the permittivity of the plasma is negative. When a plasma is subjected to a steady magnetic field, such as in the earth's ionosphere, the electrons will rotate about the magnetic field vector. The plasma will now become nonreciprocal and the scalar relationship between electric polarization and electric field must be replaced by the tensor relationship

$$D_i(\omega) = \varepsilon_0\, \bar{\varepsilon}_{ij}(\omega) \cdot E_j(\omega) \tag{15.2}$$

with the tensor permittivity

$$\bar{\varepsilon}_{ij}(\omega) = \begin{vmatrix} \varepsilon_{xx}(\omega) & j\varepsilon_{xy}(\omega) & 0 \\ -j\varepsilon_{yx}(\omega) & \varepsilon_{yy}(\omega) & 0 \\ 0 & 0 & \varepsilon_{zz}(\omega) \end{vmatrix} \tag{15.3}$$

where it has been assumed the biasing field is parallel to the z axis. The components of this tensor are

$$\varepsilon_{xx}(\omega) = \varepsilon_{yy}(\omega) = 1 - \frac{(\omega_p/\omega)^2 [1-(j\nu_c/\omega)]}{[1-(j\nu_c/\omega)]^2 - (\omega_b/\omega)^2} \tag{15.4}$$

$$\varepsilon_{xy}(\omega) = \varepsilon_{yx}(\omega) = \frac{(\omega_p/\omega)^2 (\omega_b/\omega)}{[1-(j\nu_c/\omega)]^2 - (\omega_b/\omega)^2} \tag{15.5}$$

while ε_{zz} is identical to (15.1) since the biasing magnetic field will not affect wave behavior in that direction. In these expressions ω_p is the plasma frequency, ω_b is the cyclotron frequency (proportional to the static field B_z), and ν_c is the electron collision frequency representing the loss mechanism. The real and imaginary parts of (15.4) and (15.5) are plotted in Figure 15-1.

The consequences of (15.2) are many and are best left to other sources. Extensive texts on the subject include that of Ginzburg.[4] Among the electromagnetic phenomena made possible by the circulating electrons is Faraday rotation of waves propagating along the biasing field direction. Because of the circulating electrons, circularly polarized (CP) waves will interact with the medium depending on the handedness of the wave with respect to the circulation direction. If the CP wave electric field vector rotates with the direction of electron motion, there will be a net transfer of energy between the wave and the electrons and possibly a resonance mechanism. If the CP electric field vector rotates in the direction opposite that of the electrons, then there will be minimal interaction between the wave and electrons.

We now proceed to develop the FDTD treatment of this dispersive and anisotropic media. To simplify the process we deal with total fields with propagation in the $\pm z$ direction, and with only x and y components of the electric and magnetic fields.

The first step is to determine the time domain susceptibility functions, which as described in Chapter 8 can be obtained from the frequency domain permittivity functions by inverse Fourier transformation. The time domain susceptibility function corresponding to (15.1) (except at zero frequency) is given in (8.27) and reproduced below as

$$\chi_e(\tau) = \frac{\omega_p^2}{\nu_c}\left[1 - \exp\left(-\nu_c \tau\right)\right]U(\tau) \tag{15.6}$$

where $U(\tau)$ is the unit step function. Using the Fourier transform pairs given in (8.38) and (8.49) the time domain susceptibility functions corresponding to the permittivity functions in (15.3) and (15.4) are found to be

$$\chi_{xx}(\tau) = \chi_{yy}(\tau)$$
$$= \frac{\omega_p^2}{\nu_c^2 + \omega_b^2}\left(\nu_c - e^{-\nu_c \tau}\left[\nu_c \cos\left(\omega_b \tau\right) - \omega_b \sin\left(\omega_b \tau\right)\right]\right)U(\tau) \tag{15.7}$$

$$\chi_{xy}(\tau) = \chi_{yx}(\tau)$$
$$= \frac{\omega_p^2}{\nu_c^2 + \omega_b^2}\left(\omega_b - e^{-\nu_c \tau}\left[\omega_b \cos\left(\omega_b \tau\right) + \nu_c \sin\left(\omega_b \tau\right)\right]\right)U(\tau) \tag{15.8}$$

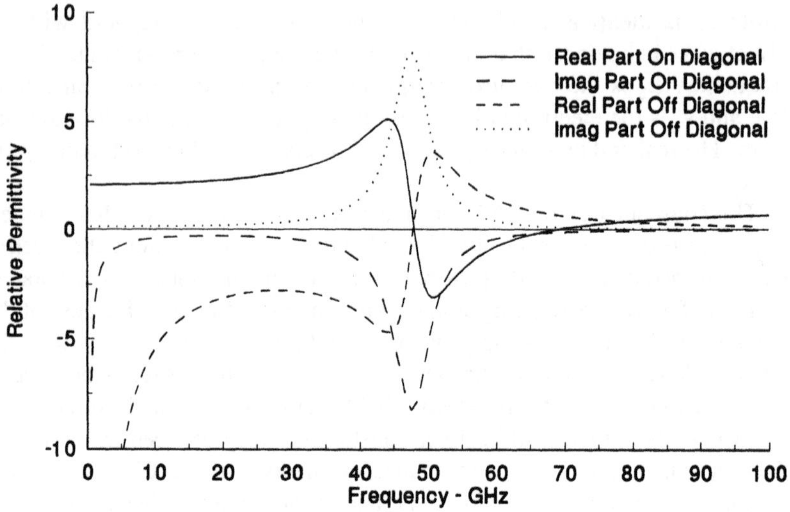

FIGURE 15-1. Real and imaginary parts of the on- and off-diagonal frequency domain permittivities. The plasma parameters are $\omega_p = 2\pi \cdot 50 \times 10^9$ rad/s, $\omega_b = 6 \times 10^{11}$ rad/s, and $\nu_c = 2 \times 10^{10}$ rad/s.

where the time domain susceptibility functions are transforms of the frequency domain susceptibilities appearing in

$$\varepsilon_{xx}(\omega) = 1 + \chi_{xx}(\omega) \leftrightarrow \delta(\tau) + \chi_{xx}(\tau) \tag{15.9}$$

$$j\varepsilon_{xy}(\omega) = j\chi_{xy}(\omega) \leftrightarrow \delta(\tau) + \chi_{xy}(\tau) \tag{15.10}$$

where $\leftrightarrow$ denotes a Fourier transform.

Notice that when the plasma is not magnetized (ω_b =0) the off-diagonal susceptibilities vanish and the on-diagonal elements reduce to the isotropic plasma permittivities, as expected. A plot of these functions is shown in Figure 15-2.

As discussed in Chapter 8, in the time domain the relationship of (15.2) becomes a convolution involving (15.6), (15.7), or (15.8). These convolutions will be reevaluated recursively by introduction of a complex susceptibility function. Only the real part of this function is used when updating the electric fields. This will require storing a single complex number for each susceptibility function convolution summation. The complex susceptibilities corresponding to (15.7) and (15.8) are

$$\begin{aligned} \hat{\chi}_{xx}(\tau) &= \hat{\chi}_{yy}(\tau) \\ &= \frac{\omega_p^2}{\nu_c^2 + \omega_b^2}(\nu_c + j\omega_b)\left[1 - e^{-(\nu_c - j\omega_b)\tau}\right]U(\tau) \end{aligned} \tag{15.11}$$

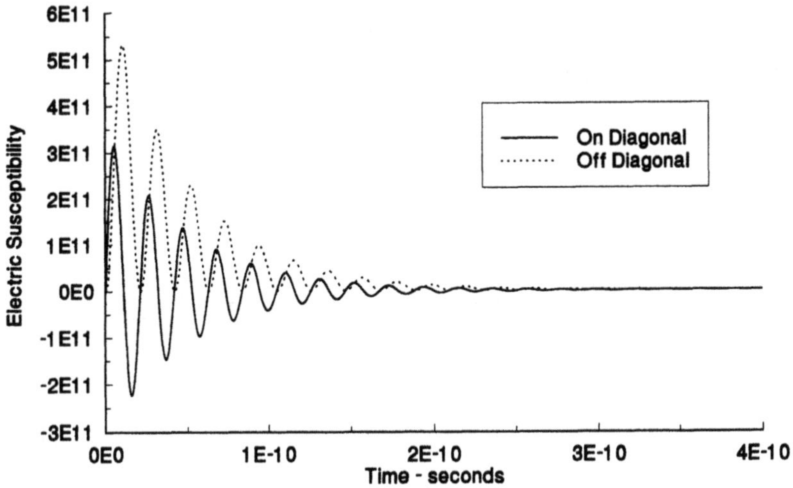

FIGURE 15-2. Time domain electric susceptibilities corresponding to the frequency domain functions in Figure 1.

$$\hat{\chi}_{xy}(\tau) = \hat{\chi}_{yx}(\tau)$$
$$= \frac{\omega_p^2}{\nu_c^2 + \omega_b^2}(\omega_b - j\nu_c)\left[1 - e^{-(\nu_c - j\omega_b)\tau}\right]U(\tau) \quad (15.12)$$

with

$$\chi_{xx}(\tau) = \mathrm{Re}\left[\hat{\chi}_{xx}(\tau)\right] \quad (15.13)$$

$$\chi_{xy}(\tau) = \mathrm{Re}\left[\hat{\chi}_{xy}(\tau)\right] \quad (15.14)$$

where Re is the real operation, and the "^" symbol designates complex quantities.

Efficient implementation of the time domain complex susceptibility convolution is not the lone difficulty which magnetoactive plasmas present to FDTD. The other problem is that the FDTD equations for two field components must be solved simultaneously since they are coupled by the off-diagonal permittivity terms. The constitutive relations in this case are

$$D_x(\omega) = \varepsilon_0(1 + \chi_{xx}(\omega))E_x(\omega) - j\chi_{xy}(\omega)\varepsilon_0 E_y(\omega) \quad (15.15)$$

$$D_y(\omega) = \varepsilon_0(1 + \chi_{yy}(\omega))\,E_y(\omega) + j\chi_{yx}(\omega)\varepsilon_0 E_x(\omega) \quad (15.16)$$

$$D_z(\omega) = \varepsilon_0\left(1 + \chi_{zz}(\omega)\right) E_z(\omega) \tag{15.17}$$

or in the time domain

$$\frac{D_x(t)}{\varepsilon_0} = E_x(t) + \chi_{xx}(\tau) * E_x(t) - \chi_{xy}(\tau) * E_y(t) \tag{15.18}$$

$$\frac{D_y(t)}{\varepsilon_0} = E_y(t) + \chi_{yy}(\tau) * E_y(t) - \chi_{yx}(\tau) * E_x(t) \tag{15.19}$$

$$\frac{D_z(t)}{\varepsilon_0} = E_z(t) + \chi_{zz}(\tau) * E_z(t) \tag{15.20}$$

where (*) denotes the convolution operation, and $\chi_{zz}(\tau)$ is the isotropic plasma susceptibility function of (15.6).

By use of the convolution integral, (15.18) and (15.19) imply

$$\frac{D_x(t)}{\varepsilon_0} = E_x(t) + \int_0^t E_x(t-\tau)\chi_{xx}(\tau)d\tau - \int_0^t E_y(t-\tau)\chi_{xy}(\tau)d\tau \tag{15.21}$$

$$\frac{D_y(t)}{\varepsilon_0} = E_y(t) + \int_0^t E_y(t-\tau)\chi_{yy}(\tau)d\tau + \int_0^t E_x(t-\tau)\chi_{yx}(\tau)d\tau \tag{15.22}$$

or, in terms of discrete time steps and complex susceptibilities

$$\frac{D_x^n}{\varepsilon_0} = E_x^n + \mathrm{Re}\left[\int_0^{n\Delta t} E_x(n\Delta t - \tau)\hat{\chi}_{xx}(\tau)d\tau\right] - \mathrm{Re}\left[\int_0^{n\Delta t} E_y(n\Delta t - \tau)\hat{\chi}_{xy}(\tau)d\tau\right] \tag{15.23}$$

$$\frac{D_y^n}{\varepsilon_0} = E_y^n + \mathrm{Re}\left[\int_0^{n\Delta t} E_y(n\Delta t - \tau)\hat{\chi}_{yy}(\tau)d\tau\right] + \mathrm{Re}\left[\int_0^{n\Delta t} E_x(n\Delta t - \tau)\hat{\chi}_{yx}(\tau)d\tau\right] \tag{15.24}$$

If we assume E_x and E_y are constant over each time step, the integrals become summations

$$\frac{D_x^{n+1}}{\varepsilon_0} = E_x^{n+1} + \mathrm{Re}\left[\sum_{m=0}^{n} E_x^{n+1-m} \int_{m\Delta t}^{(m+1)\Delta t} \hat{\chi}_{xx}(\tau)d\tau - \sum_{m=0}^{n} E_y^{n+1-m} \int_{m\Delta t}^{(m+1)\Delta t} \hat{\chi}_{xy}(\tau)d\tau\right] \tag{15.25}$$

and it follows that

$$\begin{aligned}\frac{D_x^{n+1} - D_x^{n}}{\varepsilon_0} &= E_x^{n+1} - E_x^{n} + \chi_{xx}^{o} E_x^{n+1} - \chi_{xy}^{o} E_y^{n+1} \\ &+ \mathrm{Re}\left[\sum_{m=1}^{n} E_x^{n+1-m} \int_{m\Delta t}^{(m+1)\Delta t} \hat{\chi}_{xx}(\tau)d\tau\right] - \mathrm{Re}\left[\sum_{m=1}^{n-1} E_x^{n-m} \int_{m\Delta t}^{(m+1)\Delta t} \hat{\chi}_{xx}(\tau)d\tau\right] \\ &- \mathrm{Re}\left[\sum_{m=1}^{n} E_y^{n+1-m} \int_{m\Delta t}^{(m+1)\Delta t} \hat{\chi}_{xy}(\tau)d\tau\right] + \mathrm{Re}\left[\sum_{m=1}^{n-1} E_y^{n-m} \int_{m\Delta t}^{(m+1)\Delta t} \hat{\chi}_{xy}(\tau)d\tau\right]\end{aligned} \tag{15.26}$$

which becomes

$$\begin{aligned}\frac{D_x^{n+1} - D_x^{n}}{\varepsilon_0} &= \left(1 + \chi_{xx}^{o}\right) E_x^{n+1} - E_x^{n} - \chi_{xy}^{o} E_y^{n+1} \\ &+ \mathrm{Re}\left[\sum_{m=0}^{n-1} E_x^{n-m}\left[\int_{(m+1)\Delta t}^{(m+2)\Delta t} \hat{\chi}_{xx}(\tau)d\tau - \int_{m\Delta t}^{(m+1)\Delta t} \hat{\chi}_{xx}(\tau)d\tau\right]\right] \\ &- \mathrm{Re}\left[\sum_{m=0}^{n-1} E_y^{n-m}\left[\int_{(m+1)\Delta t}^{(m+2)\Delta t} \hat{\chi}_{xy}(\tau)d\tau - \int_{m\Delta t}^{(m+1)\Delta t} \hat{\chi}_{xy}(\tau)d\tau\right]\right]\end{aligned} \tag{15.27}$$

where

$$\hat{\chi}_{xx}^{m} = \int_{m\Delta t}^{(m+1)\Delta t} \hat{\chi}_{xx}(\tau)d\tau \tag{15.28}$$

$$\hat{\chi}_{xy}^{m} = \int_{m\Delta t}^{(m+1)\Delta t} \hat{\chi}_{xy}(\tau)d\tau \tag{15. 29}$$

and where it is understood that $\chi_{xy}^{m} = \mathrm{Re}\left[\hat{\chi}_{xy}^{m}\right]$, etc. Defining

$$\Delta\hat{\chi}_{xx}^{m} = \hat{\chi}_{xx}^{m} - \hat{\chi}_{xx}^{m+1} \tag{15.30}$$

$$\Delta\hat{\chi}_{xy}^{m} = \hat{\chi}_{xy}^{m} - \hat{\chi}_{xy}^{m+1} \tag{15.31}$$

(15.27) becomes

$$\begin{aligned} \frac{D_x^{n+1} - D_x^n}{\varepsilon_0} &= \left(1 + \chi_{xx}^{o}\right)E_x^{n+1} - E_x^n - \chi_{xy}^{o}E_x^{n+1} \\ &- \mathrm{Re}\left[\sum_{m=0}^{n-1} E_x^{n-m}\Delta\hat{\chi}_{xx}^{m} + \sum_{m=0}^{n-1} E_y^{n-m}\Delta\hat{\chi}_{xy}^{m}\right] \end{aligned} \tag{15.32}$$

Before solving for the updated field values let us consider the evaluation of the convolution summations, and introduce some simplifying notation. As in Chapter 8, but generalized to off-diagonal terms, we define the values of the complex discrete convolutions at time $n\Delta t$ as

$$\hat{\Psi}_{xy}^{n} = \sum_{m=0}^{n-1} E_y^{n-m}\Delta\hat{\chi}_{xy}^{m} \tag{15.33}$$

Because $\Delta\hat{\chi}_{xy}^{m}$ has an exponential time dependence we can evaluate the convolutions recursively as

$$\hat{\Psi}_{xy}^{n+1} = \sum_{m=0}^{n} E_y^{n+1-m} \cdot \Delta\hat{\chi}_{xy}^{m} = E_y^{n}\Delta\hat{\chi}_{xy}^{0} + e^{-(\nu_c - j\omega_b)\Delta t} \cdot \hat{\Psi}_{xy}^{n} \tag{15.34}$$

Therefore, each discrete convolution requires the storage of one complex number per cell, with two complex multiplications and one addition per update operation.

Solving (15.32) for the new field value we obtain

$$E_x^{n+1} = \frac{E_x^n + \chi_{xy}^{o}E_y^{n+1} + \Psi_{xx}^{n} - \Psi_{xy}^{n} + \frac{\Delta t}{\varepsilon_o}\left(\nabla \times \vec{H}\right)_x}{1 + \chi_{xx}^{o}} \tag{15.35}$$

where the subscript x on the curl operation denotes the x component, and it is understood that Ψ_{xy}^{n} is the real part of the complex recursive convolution summation $\hat{\Psi}_{xy}^{n}$. By the same procedure we obtain

$$E_y^{n+1} = \frac{E_y^n - \chi_{yx}^o E_y^{n+1} + \Psi_{yy}^n - \Psi_{yx}^n + \frac{\Delta t}{\varepsilon_o}\left(\nabla \times \vec{H}\right)_y}{1 + \chi_{yy}^o} \tag{15.36}$$

These two equations can be solved simultaneously to yield

$$\begin{aligned} E_x^{n+1} &= \frac{1+\chi_{xx}^o}{\left(1+\chi_{xx}^o\right)^2 + \left(\chi_{xy}^o\right)^2} E_x^n + \frac{\chi_{xy}^o}{\left(1+\chi_{xx}^o\right)^2 + \left(\chi_{xy}^o\right)^2} E_y^n \\ &+ \frac{1+\chi_{xx}^o}{\left(1+\chi_{xx}^o\right)^2 + \left(\chi_{xy}^o\right)^2}\left[\Psi_{xx}^n - \Psi_{xy}^n + \frac{\Delta t}{\varepsilon_o}\left(\nabla \times \vec{H}\right)_x\right] \\ &+ \frac{\chi_{xy}^o}{\left(1+\chi_{xx}^o\right)^2 + \left(\chi_{xy}^o\right)^2}\left[\Psi_{yy}^n + \Psi_{yx}^n + \frac{\Delta t}{\varepsilon_o}\left(\nabla \times \vec{H}\right)_y\right] \end{aligned} \tag{15.37}$$

$$\begin{aligned} E_y^{n+1} &= \frac{1+\chi_{yy}^o}{\left(1+\chi_{yy}^o\right)^2 + \left(\chi_{yx}^o\right)^2} E_y^n - \frac{\chi_{yx}^o}{\left(1+\chi_{yy}^o\right)^2 + \left(\chi_{yx}^o\right)^2} E_x^n \\ &+ \frac{1+\chi_{yy}^o}{\left(1+\chi_{xx}^o\right)^2 + \left(\chi_{xy}^o\right)^2}\left[\Psi_{yy}^n + \Psi_{yx}^n + \frac{\Delta t}{\varepsilon_o}\left(\nabla \times \vec{H}\right)_y\right] \\ &+ \frac{\chi_{yx}^o}{\left(1+\chi_{yy}^o\right)^2 + \left(\chi_{yx}^o\right)^2}\left[\Psi_{xx}^n - \Psi_{xy}^n + \frac{\Delta t}{\varepsilon_o}\left(\nabla \times \vec{H}\right)_x\right] \end{aligned} \tag{15.38}$$

The FDTD update equations for the magnetic fields are the same as for free space because no magnetic materials are present.

To demonstrate this FDTD formulation, we present the 1-D problem of reflection and transmission of a Gaussian-derivative pulsed plane wave normally incident on a magnetized plasma layer. The Gaussian derivative pulse is described in (8.35) and (8.36). The direction of propagation is parallel to the biasing field direction. The FDTD problem space consists of 350 cells, the magnetoactive plasma occupying cells 200 through 320, with free space in the other cells. Each cell is 75 μm long so that the plasma layer is 9.0 mm thick. Both ends of the cell space are terminated in a first order Mur absorbing boundary condition. For these simulations the plasma parameters were

$$\begin{aligned} \omega_p &= (2\pi)\cdot 50.0\times 10^9 \ \text{rad/s} \\ \omega_b &= 3.0\times 10^{11} \ \text{rad/s} \\ \nu_c &= 2.0\times 10^{10} \ \text{rad/s} \end{aligned}$$

The simulations were allowed to run for 6500 time steps ($\Delta t = 0.125$ ps) and the electric field time histories (both co- and cross-polarized) were recorded just in front of and behind the slab. These time histories were then transformed to the frequency domain and combined to yield right and left circularly polarized (RCP and LCP) reflection/transmission coefficients at each frequency:

$$T_{RCP}(\omega) = \hat{E}_{xt}(\omega) + j \cdot \hat{E}_{yt}(\omega) \tag{15.39}$$

$$T_{LCP}(\omega) = \hat{E}_{xt}(\omega) - j \cdot \hat{E}_{yt}(\omega) \tag{15.40}$$

with T (and t) replaced by R (and r) for the reflection coefficients, and the t or r subscript on the complex electric field values denoting locations at either the transmission or reflection side of the plasma slab. For comparison analytically computed reflection and transmission coefficients for a magnetized plasma layer following the method found in Ginzburg[4] are provided. Figures 15-3 through 15-10 show the agreement between the FDTD results and the analytic values for both RCP and LCP polarizations. The notation $(FD)^2TD$ indicates that the frequency-dependent FDTD formulation presented here is used for computing the FDTD results. Both magnitudes and phases exhibit excellent agreement.

This section has described how a material possessing both dispersive and anisotropic material properties can be efficiently implemented by the FDTD method. For a plasma subjected to a constant magnetic field, the resultant complex permittivity tensor will yield causal time domain susceptibility expressions which can be implemented with FDTD. For 1-D wave propagation, as presented here, the off-diagonal susceptibilities will create coupling between orthogonal electric field components, resulting in polarization rotation.

Excellent agreement between pulsed wave simulations and analytic frequency domain results was demonstrated over a wide frequency range where the material properties of the plasma were greatly varying. In the next section a similar derivation will be applied to magnetized ferrite materials.

15.3 MAGNETIZED FERRITES

The derivation of the FDTD equations for updating the magnetic fields in a magnetized ferrite will now be derived. This derivation is quite similar to the above for the magnetized plasmas, but the two materials are not duals and the FDTD formulation derived in the last section cannot be directly applied. In terms of the dispersive FDTD treatment, the dissimilarity between the constitutive parameter tensors necessitates finding new time-domain susceptibility functions and incorporating them into a new FDTD algorithm. The procedure for developing this algorithm parallels that presented in the previous section.

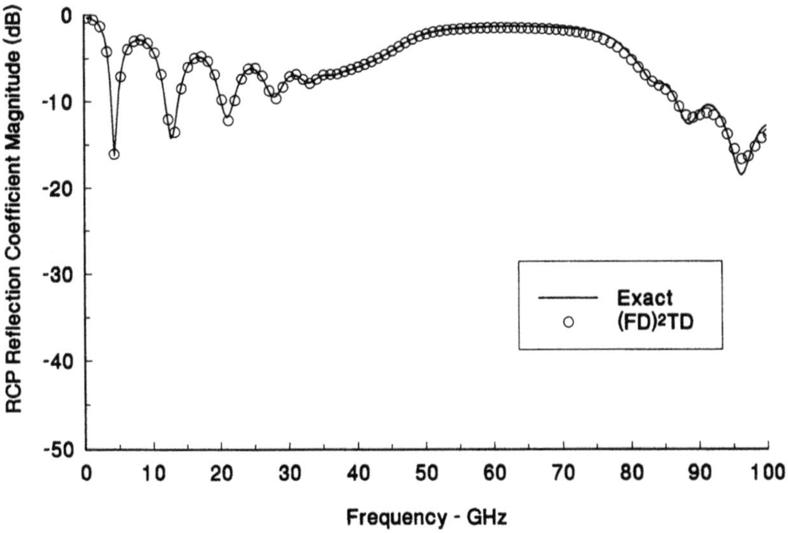

FIGURE 15-3. FDTD and exact reflection coefficient magnitude vs. frequency for a plane wave incident on a plasma slab for RCP.

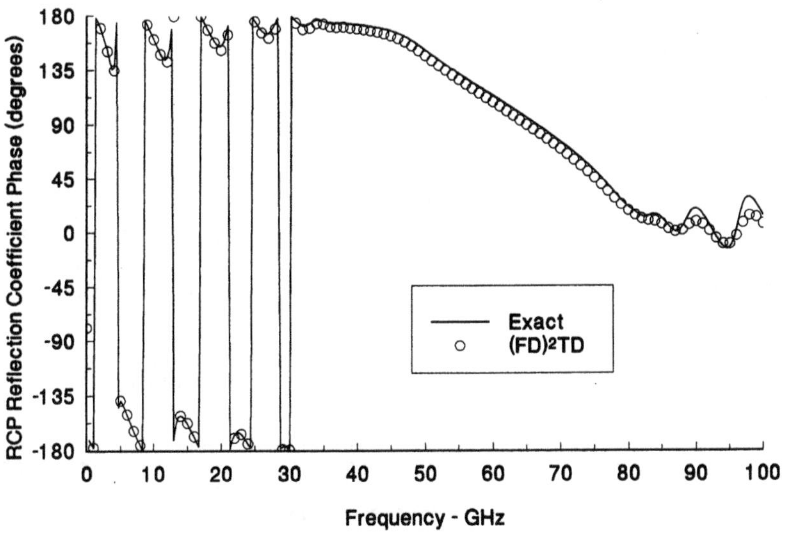

FIGURE 15-4. FDTD and exact reflection coefficient phase vs. frequency for a plane wave incident on a plasma slab for RCP.

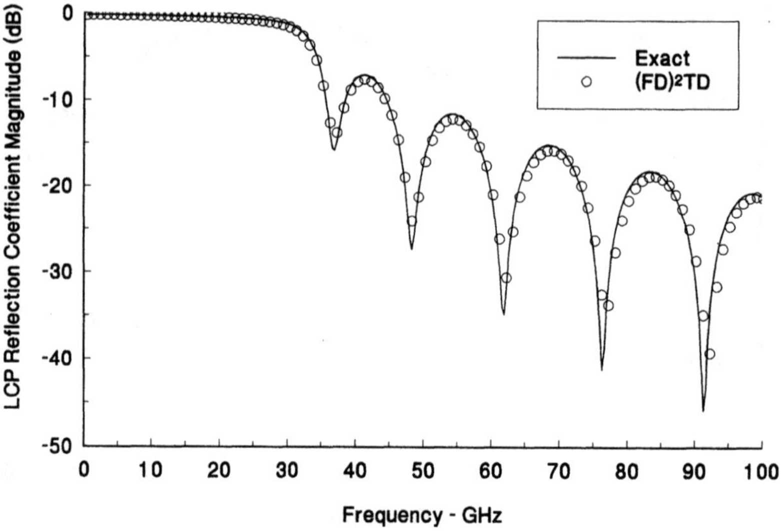

FIGURE 15-5. FDTD and exact reflection coefficient magnitude vs. frequency for a plane wave incident on a plasma slab for LCP.

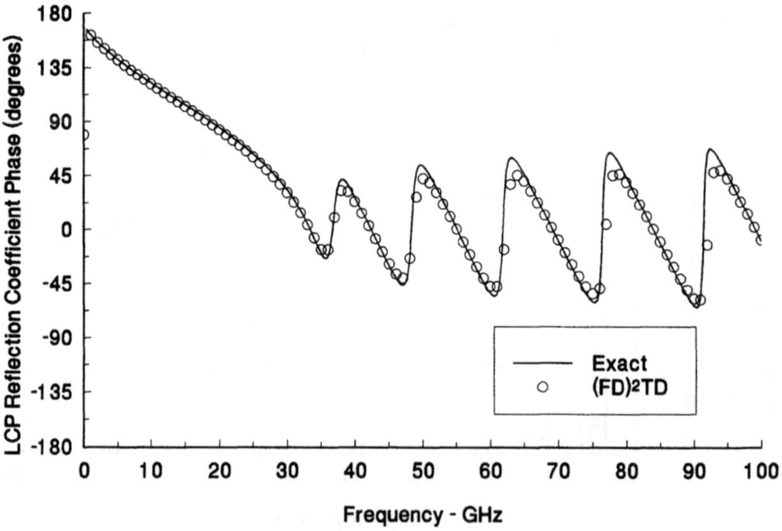

FIGURE 15-6. FDTD and exact reflection coefficient phase vs. frequency for a plane wave incident on a plasma slab for LCP.

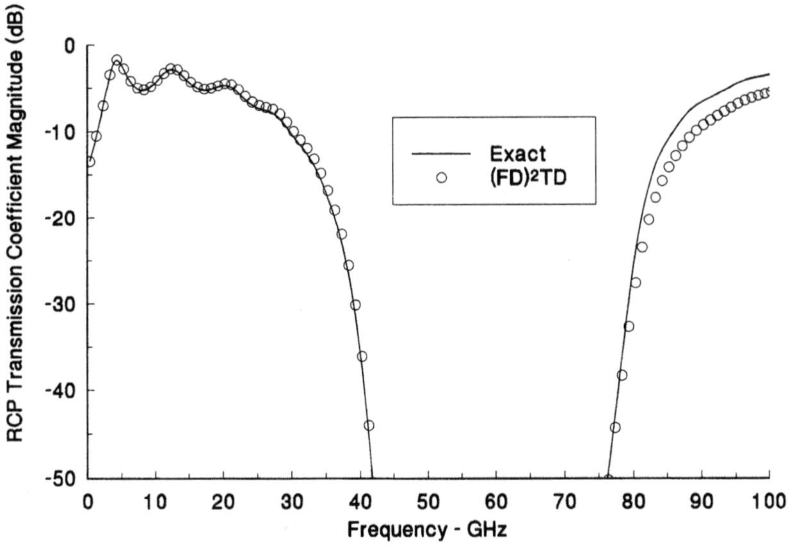

FIGURE 15-7. FDTD and exact transmission coefficient magnitude vs. frequency for a plane wave incident on a plasma slab for RCP.

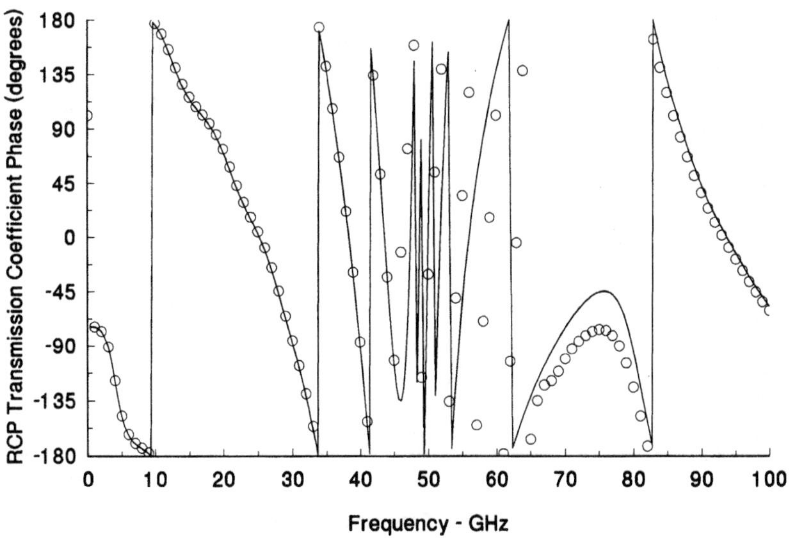

FIGURE 15-8. FDTD and exact transmission coefficient phase vs. frequency for a plane wave incident on a plasma slab for RCP.

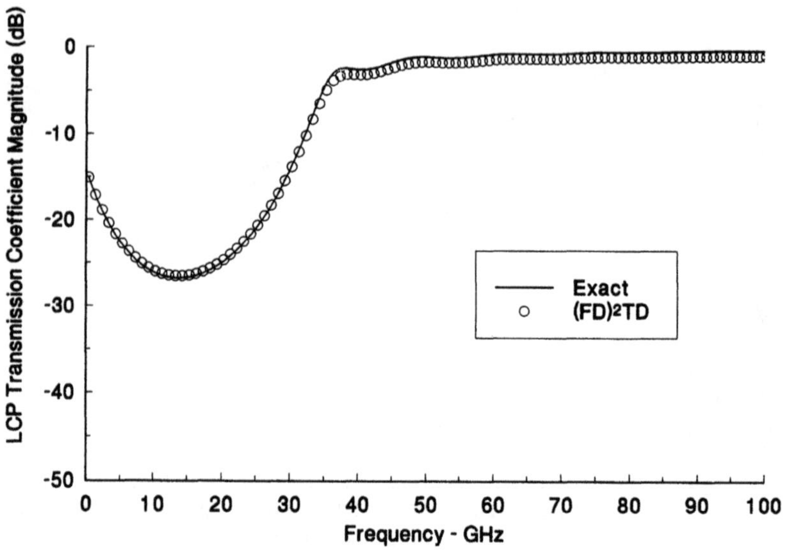

FIGURE 15-9. FDTD and exact transmission coefficient magnitude vs. frequency for a plane wave incident on a plasma slab for LCP.

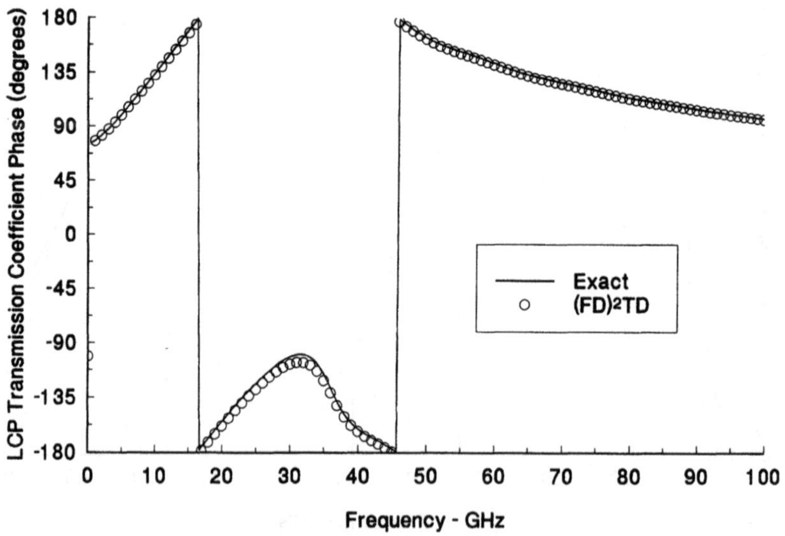

FIGURE 15-10. FDTD and exact transmission coefficient phase vs. frequency for a plane wave incident on a plasma slab for LCP.

In order to simplify the derivation the same situation will be assumed as in the previous section, propagation in the z direction with only x and y components of the electric and magnetic fields present.

Chemically, ferrites are (most often) described by the empirical equation $MeO{\cdot}Fe_2O_3$, where Me is some divalent metal ion (nickel, manganese, nickel-zinc and many others).[5] Variations of this basic formula exist, many times with the Fe ion replaced by some other element. The selection of one type of ferrite over another depends on the intended application of the required electrical qualities.

Ferrites without a biasing field normally have one or more resonances (called "natural" resonances) due to microscopic structure and loss. However, it is when the ferrite is subjected to a strong static magnetic biasing field that the material becomes both strongly dispersive and strongly anisotropic.

In determining the interactions between the electromagnetic fields and the ferrite material an important parameter is the orientation between the transient fields and the static biasing field. There are two basic situations. The first is when the magnetic field vector of the wave is oriented parallel to the static field vector. In this case the result is an additional increase in ferrite magnetization along the biasing field (again, assuming the transient field strength to be small compared to biasing field). In terms of ferrite device operation, this situation is uninteresting as it yields no unusual propagation phenomena. However, if the magnetic field of the transient wave is located in a direction other than that of the static field, the permeabilities will depend on the strength of the biasing field, and the ferrite will be strongly anisotropic. The exact form of these permeabilities is addressed next.

Assuming that the geometry is such that the applied static magnetic field, H_0, is $\hat{z}$-directed, then the permeability of infinite, homogeneous ferrite can be expressed by

$$\mu*(\omega) = \mu_0\hat{\mu}_{ij}(\omega) = \mu_0\begin{bmatrix} 1+\chi_m(\omega) & -j\kappa(\omega) & 0 \\ j\kappa(\omega) & 1+\chi_m(\omega) & 0 \\ 0 & 0 & 1 \end{bmatrix} \tag{15.41}$$

where the susceptibilities are

$$\chi_m(\omega) = \frac{(\omega_0 + j\omega\alpha)\omega_m}{(\omega_0 + j\omega\alpha)^2 - \omega^2} \tag{15.42}$$

$$\kappa(\omega) = \frac{-\omega\omega_m}{(\omega_0 + j\omega\alpha)^2 - \omega^2} \tag{15.43}$$

with α the damping constant, $\omega_0 = \gamma_m H_0$ the precessional frequency, where γ_m (= 28 GHz/kO) is the gyromagnetic ratio and H_0 is the static magnetic biasing field, and $\omega_m = \gamma_m 4\pi M_0$ where M_0 is the static magnetization. The real and imaginary parts of (15.42) and (15.43) are shown in Figure 15-11 for the ferrite used in the demonstration at the end of this section.

The form of the above tensor is a result of the static biasing field being in the z direction. If the biasing field is along some other direction, then the tensor must be altered. Comparing this tensor and the permittivity tensor for plasmas shows that there is much similarity, most notably the equal but unsymmetric off-diagonal elements. Notice that unlike magnetoactive plasmas, at low frequencies the imaginary parts remain finite. This is a consequence of the lack of free magnetic carriers, so that there can be no zero frequency magnetic "conduction" current.

A number of propagation phenomena result from this permeability tensor. The first is the interaction between ferrite media and CP waves. For example, let the static magnetic field be parallel to the direction of wave propagation. Since the magnetic field components of the wave are transverse to the magnetic field, they both experience coupled frequency-dependent permeabilities. In this situation the wavenumbers for RCP and LCP will be different, and unequal attenuations and propagation velocities result.

Using the above tensor, the relationship between $\vec{B}$ and $\vec{H}$ is

$$B_x(\omega) = \mu_0\left(1 + \chi_m(\omega)\right)H_x(\omega) - j\kappa(\omega)\mu_0 H_y(\omega) \tag{15.44}$$

$$B_y(\omega) = \mu_0\left(1 + \chi_m(\omega)\right)H_y(\omega) + j\kappa(\omega)\mu_0 H_x(\omega) \tag{15.45}$$

$$B_z(\omega) = \mu_0 H_z(\omega) \tag{15.46}$$

Time-domain versions of these relationships are

$$\frac{B_x(t)}{\mu_0} = H_x(t) + \chi_m(\tau) * H_x(t) - \kappa(\tau) * H_y(t) \tag{15.47}$$

$$\frac{B_y(t)}{\mu_0} = H_y(t) + \chi_m(\tau) * H_y(t) + \kappa(\tau) * H_x(t) \tag{15.48}$$

$$B_z(t) = \mu_0 H_z(t) \tag{15.49}$$

where "*" denotes the convolution operation and where $\chi_m(\tau)$and $\kappa(\tau)$ are the Fourier transforms of $\chi_m(\omega)$ and, $\kappa(\omega)$, respectively.

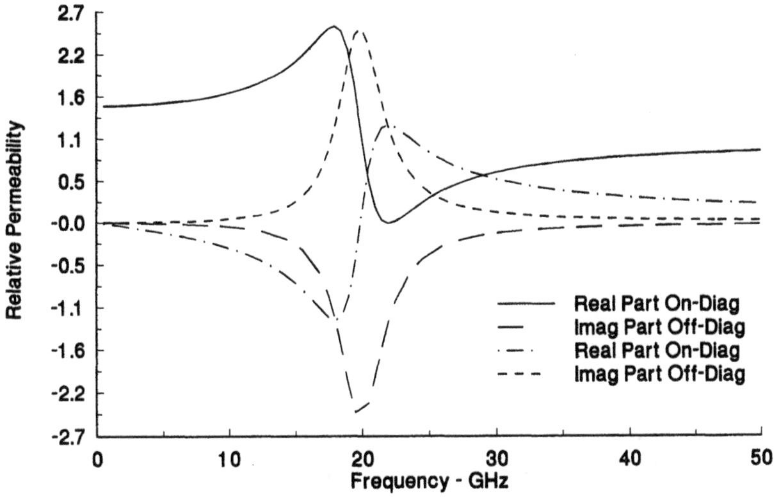

FIGURE 15-11. Real and imaginary parts of the on- and off-diagonal frequency domain permeabilities. The magnetized ferrite parameters are $\omega_0 = 2\pi \cdot 20 \times 10^9$ rad/s, $\omega_m = 2\pi \cdot 10 \times 10^9$ rad/s, and $\alpha = 0.10$.

Before these equations can be put into an FDTD algorithm, two difficulties must be addressed. The first is that (15.47) and (15.48) are not independent equations; they are coupled, and therefore any solution for one variable will require the simultaneous solution of the other. The second impediment is that $\kappa(\omega)$ does not transform into a real, causal time-domain function. However, as with the off-diagonal permittivities of the previous section, the quantity $j\kappa(\omega)$ does yield a causal, real-valued time function.

The time-domain susceptibilities that correspond to the above frequency-domain expressions are obtained from the application of (8.38) and (8.49) as

$$\chi_m(t) = \frac{\omega_m}{1+\alpha^2} \exp\left(\frac{-\omega_0 \alpha}{1+\alpha^2} t\right) \left[\alpha \cos\left(\frac{\omega_0}{1+\alpha^2} t\right) + \sin\left(\frac{\omega_0}{1+\alpha^2} t\right)\right] U(t) \tag{15.50}$$

$$\kappa(t) = \frac{\omega_m}{1+\alpha^2} \exp\left(\frac{-\omega_0 \alpha}{1+\alpha^2} t\right) \left[\alpha \sin\left(\frac{\omega_0}{1+\alpha^2} t\right) - \cos\left(\frac{\omega_0}{1+\alpha^2} t\right)\right] U(t) \tag{15.51}$$

The progression from (15.47) and (15.48) to an FDTD algorithm is fairly straightforward, and quite similar to the derivation given in the previous section. Expressing the convolutions explicitly in discrete time we have

$$\begin{aligned} B_x^n = \mu_0 H_x^n + \mu_0 \int_0^{n\Delta t} \chi_m(\tau) H_x(n\Delta t - \tau) d\tau \\ - \mu_0 \int_0^{n\Delta t} \kappa(\tau) H_y(n\Delta t - \tau) d\tau \end{aligned} \tag{15.52}$$

$$\begin{aligned} B_y^n = \mu_0 H_y^n + \mu_0 \int_0^{n\Delta t} \chi_m(\tau) H_y(n\Delta t - \tau) d\tau \\ + \mu_0 \int_0^{n\Delta t} \kappa(\tau) H_x(n\Delta t - \tau) d\tau \end{aligned} \tag{15.53}$$

Assuming that each magnetic field component is constant over each time increment, the convolution integrals in these equations can be simplified to

$$\begin{aligned} \frac{B_x^n}{\mu_0} = H_x^n + \sum_{m=0}^{n-1} H_x^{n-m} \int_{m\Delta t}^{(m+1)\Delta t} \chi_m(\tau) d\tau \\ - \sum_{m=0}^{n-1} H_y^{n-m} \int_{m\Delta t}^{(m+1)\Delta t} \kappa(\tau) d\tau \end{aligned} \tag{15.54}$$

$$\begin{aligned} \frac{B_y^n}{\mu_0} = H_y^n + \sum_{m=0}^{n-1} H_y^{n-m} \int_{m\Delta t}^{(m+1)\Delta t} \chi_m(\tau) d\tau \\ + \sum_{m=0}^{n-1} H_x^{n-m} \int_{m\Delta t}^{(m+1)\Delta t} \kappa(\tau) d\tau \end{aligned} \tag{15.55}$$

For the next time step

$$\begin{aligned} \frac{B_x^{n+1}}{\mu_0} H_x^{n+1} + \sum_{m=0}^{n} H_x^{n+1+m} \int_{m\Delta t}^{(m+1)\Delta t} \chi_m(\tau) d\tau \\ - \sum_{m=0}^{n} H_x^{n+1-m} \int_{m\Delta t}^{(m+1)\Delta t} \kappa(\tau) d\tau \end{aligned} \tag{15.56}$$

$$\begin{aligned} \frac{B_y^{n+1}}{\mu_0} = H_y^{n+1} + \sum_{m=0}^{n} H_y^{n+1-m} \int_{m\Delta t}^{(m+1)\Delta t} \chi_m(\tau) d\tau \\ + \sum_{m=0}^{n} H_x^{n+1-m} \int_{m\Delta t}^{(m+1)\Delta t} \kappa(\tau) d\tau \end{aligned} \tag{15.57}$$

In preparation for the finite difference approximation to the time derivative of the magnetic field we combine (15.54) and (15.56) to obtain

$$\frac{B_x^{n+1} - B_x^n}{\mu_0} = H_x^{n+1} - H_x^n$$

$$+ H_x^{n+1} \int_0^{\Delta t} \chi_m(\tau)d\tau - H_y^{n+1} \int_0^{\Delta t} \kappa(\tau)d\tau \tag{15.58}$$

$$+ \sum_{m=1}^{n} H_x^{n+1-m} \int_{m\Delta t}^{(m+1)\Delta t} \chi_m(\tau)d\tau - \sum_{m=0}^{n-1} H_x^{n-m} \int_{m\Delta t}^{(m+1)\Delta t} \chi_m(\tau)d\tau$$

$$- \sum_{m=1}^{n} H_y^{n+1-m} \int_{m\Delta t}^{(m+1)\Delta t} \kappa(\tau)d\tau + \sum_{m=0}^{n-1} H_y^{n-m} \int_{m\Delta t}^{(m+1)\Delta t} \kappa(\tau)d\tau$$

Introducing the usual simplifying notation

$$\chi^m = \int_{m\Delta t}^{(m+1)\Delta t} \chi_m(\tau)d\tau$$

$$\kappa^m = \int_{m\Delta t}^{(m+1)\Delta t} \kappa(\tau)d\tau$$

$$\Delta\chi^m = \chi^m - \chi^{m+1}$$

$$\Delta\kappa^m = \kappa^m - \kappa^{m+1}$$

the above equation can be compactly written as

$$\frac{B_x^{n+1} - B_x^n}{\mu_0} = \left(1 + \chi^0\right)H_x^{n+1} - H_x^n - \kappa^0 H_y^{n+1}$$

$$- \sum_{m=0}^{n-1} H_x^{n-m} \Delta\chi^m + \sum_{m=0}^{n-1} H_y^{n-m} \Delta\kappa^m \tag{15.59}$$

Using the finite difference approximation

$$\frac{B_x^{n+1} - B_x^n}{\Delta t} \approx -\left(\nabla \times \vec{E}\right)_x$$

the reduced update equation becomes

$$H_x^{n+1} = \frac{H_x^n - \kappa^0 H_y^{n+1} + \sum_{m=0}^{n-1} H_x^{n-m} \Delta\chi^m - \sum_{m=0}^{n-1} H_y^{n-m} \Delta\kappa^m - \frac{\Delta t}{\mu_0}\left(\nabla \times \vec{E}\right)_x}{1 + \Delta\chi^0} \quad (15.60)$$

The same derivation applied to (15.55) and (15.57) will yield

$$H_y^{n+1} = \frac{H_y^n - \kappa^0 H_x^{n+1} + \sum_{m=0}^{n-1} H_y^{n-m} \Delta\chi^m + \sum_{m=0}^{n-1} H_x^{n-m} \Delta\kappa^m - \frac{\Delta t}{\mu_0}\left(\nabla \times \vec{E}\right)_y}{1 + \Delta\chi^0} \quad (15.61)$$

These equations are now ready to be solved simultaneously. Doing so yields:

$$\begin{aligned} H_x^{n+1} = {} & \frac{1+\chi^0}{\left(1+\chi^0\right) + \left(\kappa^0\right)^2} H_x^n + \frac{\kappa^0}{\left(1+\chi^0\right)^2 + \left(\kappa^0\right)^2} H_y^n \\ & + \frac{\kappa^0}{\left(1+\chi^0\right)^2 + \left(\kappa^0\right)^2} \left[\sum_{m=0}^{n-1} H_y^{n-m} \Delta\chi^m + \sum_{m=0}^{n-1} H_x^{n-m} \Delta\kappa^m \right] \\ & + \frac{1+\chi^0}{\left(1+\chi^0\right)^2 + \left(\kappa^0\right)^2} \left[\sum_{m=0}^{n-1} H_x^{n-m} \Delta\chi^m - \sum_{m=0}^{n-1} H_y^{n-m} \Delta\kappa^m \right] \\ & - \frac{\Delta t}{\mu_0} \frac{\kappa^0}{\left(1+\chi^0\right)^2 + \left(\kappa^0\right)^2} \left(\nabla \times \vec{E}\right)_y - \frac{\Delta t}{\mu_0} \frac{1+\chi^0}{\left(1+\chi^0\right)^2 + \left(\kappa^0\right)^2} \left(\nabla \times \vec{E}\right)_x \end{aligned} \quad (15.62)$$

$$\begin{aligned} H_y^{n+1} = {} & \frac{1+\chi^0}{\left(1+\chi^0\right)^2 + \left(\kappa^0\right)^2} H_y^n - \frac{\kappa^0}{\left(1+\chi^0\right)^2 + \left(\kappa^0\right)^2} H_x^n \\ & - \frac{\kappa^0}{\left(1+\chi^0\right)^2 + \left(\kappa^0\right)^2} \left[\sum_{m=0}^{n-1} H_x^{n-m} \Delta\chi^m - \sum_{m=0}^{n-1} H_y^{n-m} \Delta\kappa^m \right] \\ & + \frac{1+\chi^0}{\left(1+\chi^0\right)^2 + \left(\kappa^0\right)^2} \left[\sum_{m=0}^{n-1} H_y^{n-m} \Delta\chi^m + \sum_{m=0}^{n-1} H_x^{n-m} \Delta\kappa^m \right] \end{aligned}$$

$$+\frac{\Delta t}{\mu_0}\frac{\kappa^0}{\left(1+\chi^0\right)^2+\left(\kappa^0\right)^2}\left(\nabla\times\vec{E}\right)_x-\frac{\Delta t}{\mu_0}\frac{1+\chi^0}{\left(1+\chi^0\right)^2+\left(\kappa^0\right)^2}\left(\nabla\times\vec{E}\right)_y \quad (15.63)$$

These equations show how the FDTD update equations are modified for the magnetic field components transverse the biasing field direction. It is evident that these equations reduce to the usual FDTD equations when $\omega_m = 0$ (i.e., the biasing field is removed), as expected.

The above equations are almost ready for implementation into FDTD. What remains is the transformation of the summation terms in (15.62) and (15.63) into recursive accumulations. As with the other dispersive materials presented so far, the damped exponential forms of the susceptibility functions allow the summations to be efficiently compacted into one variable, rather than requiring inordinate amounts of computer storage and execution time. The procedure is the same as that performed for the magnetoactive plasmas, except for the functional differences in the susceptibility functions.

Using complex variables and mathematical operations, the updates can be performed as presented in Chapter 8 and in the previous section. The real-valued expressions in (15.62) and (15.63) are expressed as the real parts of the complex-valued functions

$$\hat{\chi}(\tau)=\frac{\omega_m}{1+\alpha^2}(\alpha-j)\exp\left[-\frac{\omega_0}{1+\alpha^2}(\alpha-j)t\right] \quad (15.64)$$

$$\hat{\kappa}(\tau)=\frac{\omega_m}{1+\alpha^2}(1+j\alpha)\exp\left[-\frac{\omega_0}{1+\alpha^2}(\alpha-j)t\right] \quad (15.65)$$

where again the superscript "^" denotes a complex variable, and $j=\sqrt{-1}$. The values for χ^0 and κ^0 are given by

$$\chi^0=\frac{\omega_m}{\omega_0}\mathrm{Re}\left\{1-\exp\left[-\frac{\omega_0}{1+\alpha^2}(\alpha-j)t\right]\right\} \quad (15.66)$$

$$\kappa^0=\frac{\omega_m}{\omega_0}\mathrm{Re}\left\{\frac{1+j\alpha}{\alpha-j}\left[1-\exp\left[-\frac{\omega_0}{1+\alpha^2}(\alpha-j)t\right]\right]\right\} \quad (15.67)$$

and the values for $\Delta\chi^0$ and $\Delta\kappa^0$ are

$$\Delta\chi^0=\frac{\omega_m}{\omega_0}\left[1-\exp\left[-\frac{\omega_0}{1+\alpha^2}(\alpha-j)t\right]\right]^2 \quad (15.68)$$

$$\Delta\kappa^0 = \frac{\omega_m}{\omega_0}\frac{1+j\alpha}{\alpha-j}\left[1-\exp\left[-\frac{\omega_0}{1+\alpha^2}(\alpha-j)\Delta t\right]\right]^2 \tag{15.69}$$

with the recursive relation

$$\Delta\begin{Bmatrix}\chi^{m+1}\\ \kappa^{m+1}\end{Bmatrix} = \exp\left(-\frac{\omega_0}{1+\alpha^2}(\alpha-j)\Delta t\right)\Delta\begin{Bmatrix}\chi^{m}\\ \kappa^{m}\end{Bmatrix} \tag{15.70}$$

being the basis for replacing the summations in (15.62) and (15.63) with recursive accumulations as was done in the previous section in (15.33) and (15.34). The procedure is identical, and is explained in Chapter 8, so it will not be repeated here. The electric fields are updated in the usual way, depending on the permittivity of the materials involved.

Let us now demonstrate the application of this FDTD formulation. As was stated at the beginning of this chapter, magnetized ferrites will have different effects on waves depending on the orientation of the static field and the propagation direction of the wave. For demonstration purposes, the most interesting situation will be used as an example. In this orientation, called longitudinal propagation, the propagation direction and biasing field vector are identical (parallel to the $\hat{z}$ axis). Like magnetoplasmas, the ferrite has two eigenmodes (allowed propagation modes) with different propagation constants. Solution of the wave equation shows that the two modes are RCP and LCP waves. At each frequency there will be a unique propagation constant for each of the two CP waves, signifying unequal propagation velocities and unequal attenuations. As a result, a linearly polarized wave will undergo a net rotation as it passes through the material.

Using the ferrite permeability tensor and the two propagation modes, it can be shown that the two propagation constants are[6]

$$\kappa_{RCP}(\omega) = \frac{\omega}{c}\sqrt{1+\chi_m(\omega)-\kappa(\omega)}$$

$$\kappa_{LCP}(\omega) = \frac{\omega}{c}\sqrt{1+\chi_m(\omega)+\kappa(\omega)}$$

For frequencies much less or greater than the resonance frequency ω_0, the two wavenumbers will be nearly identical. Near the resonance frequency, however, the two constants will be significantly different, leading to very different behavior between the two waves.

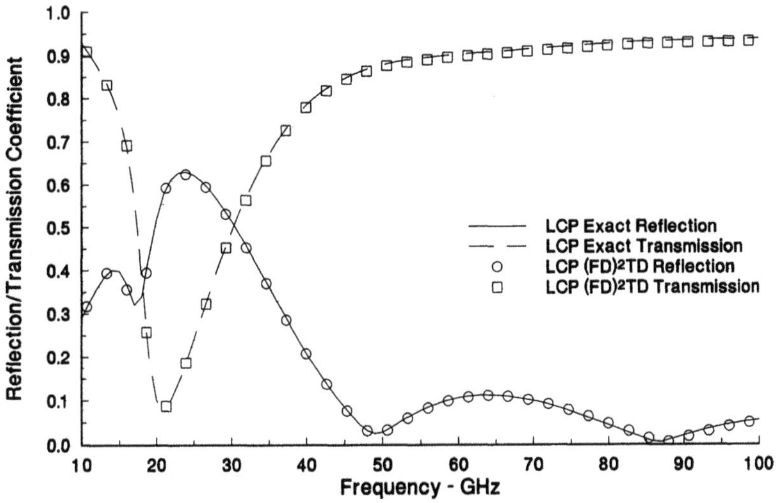

FIGURE 15-12. FDTD and exact transmission and reflection coefficient magnitudes vs. frequency for a plane wave incident on a magnetized ferrite slab for LCP.

The FDTD problem space for this demonstration consisted of 800 1-D cells 75μm in width with $\Delta t = 0.125$ ps. The center 200 cells were filled with ferrite, and the magnetic fields calculated using the above FDTD formulation. The ferrite material parameters used are

$$\omega_0 = (2\pi) \cdot 20 \times 10^9$$

$$\omega_m = (2\pi) \cdot 10 \times 10^9$$

and

$$\alpha = 0.10$$

The excitation for this problem was the Gaussian-derivative pulse used in the previous section. The procedure for determining the transmission coefficients is also the same as that presented in the last section. Figures 15-12 and 15-13 show the FDTD calculated reflection and transmission coefficient magnitudes for both polarizations vs. the expected analytic results found in Reference 6. Again, the notation $(FD)^2TD$ indicates that the frequency-dependent FDTD formulation presented here is used for computing the FDTD results. The agreement is excellent.

Although magnetized ferrites and magnetized plasmas produce gyrotropic effects by different microscopic mechanisms, there are many similarities between the two materials. The most important similarity is in the symmetry between constitutive parameter matrices.

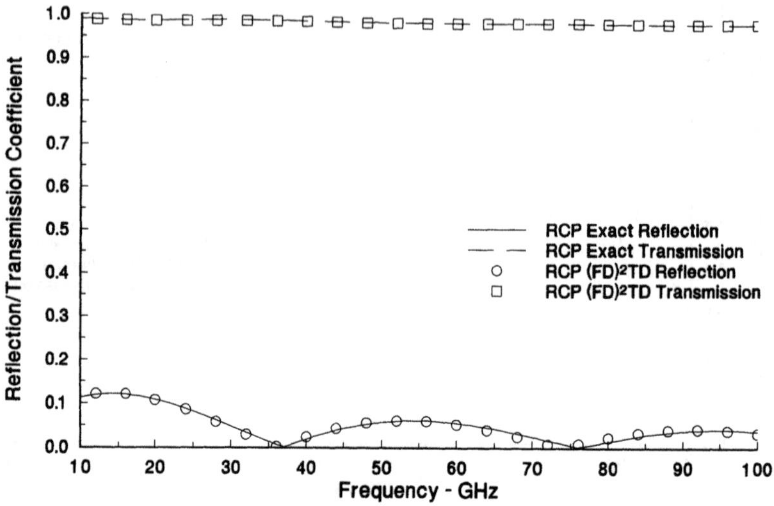

FIGURE 15-13. FDTD and exact transmission and reflection coefficient magnitudes vs. frequency for a plane wave incident on a magnetized ferrite slab for RCP .

The magnetized ferrite formulation presented here is also very similar to that developed for magnetized plasmas. However, it was shown that because ferrites do not have the equivalent of free conducting electrons, the frequency- and time-domain constitutive parameters behave differently. A more basic difference is that ferrites are a dispersive magnetic material while the plasma is a dispersive dielectric material.

In this chapter the time domain susceptibility functions for these two materials were presented, and the recursive convolution method was applied to efficiently compute the convolution as a running sum. In addition, the anisotropy of the material was incorporated into the FDTD algorithm for 1-D propagation with both transverse field components present. Extension to 2- and 3-D will involve some spatial averaging of the fields in order to include the anisotropic effects if a Yee type of spatial field distribution is used.

REFERENCES

1. **Hunsberger, F. P., Jr.,** Extensions of the Finite Difference Time-Domain Method to Gyrotropic Media, Ph.D. dissertation, The Pennsylvania State University, University Park, May 1991.
2. **Hunsberger, F., Luebbers R., and Kunz, K.,** Finite-difference time-domain analysis of gyrotropic media. I. Magnetized plasma, *IEEE Trans. Ant. Prop.*
3. **Drobot, A., Ed.,** *Computer Applications in Plasma Science and Engineering,* Springer-Verlag, New York, 1991.
4. **Ginzburg, V. L.,** *The Propagation of Electromagnetic Waves in Plasmas,* 2nd ed., Pergamon Press, Oxford, 1970.

5. **Lax B. and Button, K.,** *Microwave Ferrites and Ferrimagnetics,* McGraw-Hill, New York, 1962.
6. **Sodha M. and Srivastava, N.,** *Microwave Propagation in Ferrites,* Plenum Press, New York, 1981.

PART 5: MATHEMATICAL BASIS OF FDTD AND ALTERNATE METHODS

Chapter 16

DIFFERENCE EQUATIONS IN GENERAL

16.1 INTRODUCTION

The differencing scheme employed in the preceding chapters is a central difference scheme in space and time, typically referred to as the leapfrog method. It is stable and accurate to second order. A number of other differencing approaches have been formulated. Some are unstable; others, while stable, provide a lower order of accuracy; and finally there are those schemes that are stable and of a higher order of accuracy, but at the cost of a much more complex and computationally expensive expression.

In the differencing schemes discussed above only hyperbolic equations such as the Maxwell equations and the wave equation, which can be directly derived from the Maxwell equations, are considered. More generally we can consider the broad class of second order partial differential equations that describe most physical phenomena. In 2-D, where we typically assume one is spatial and the other temporal, a general second order spatial differential equation may be expressed as

$$a\frac{\partial^2 A}{\partial x^2}+b\frac{\partial^2 A}{\partial x\partial y}+c\frac{\partial^2 A}{\partial y^2}+d\frac{\partial A}{\partial x}+e\frac{\partial A}{\partial y}+fA+g=0 \qquad (16.1)$$

where a, b, c, d, e, f, and g may depend on x and y. There are three categories, one of which the above equation must fall into:

$$\text{Hyperbolic - } \left(b^2-4ac\right)>0 \qquad (16.2a)$$

$$\text{Parabolic - } \left(b^2-4ac\right)=0 \qquad (16.2b)$$

$$\text{Elliptic - } \left(b^2-4ac\right)<0 \qquad (16.2c)$$

The 1-D wave equation

$$\frac{\partial^2\psi}{\partial x^2}-\frac{1}{c_0^2}\frac{\partial^2\psi}{\partial t^2}=0 \qquad (16.3)$$

has $a = 1$ and $c = -1/c_0^2$ so that $b^2 - 4ac = -4ac = 4/c_0^2 > 0$ and the equation is hyperbolic. The diffusion equation which can in its simplest form be expressed as

$$\frac{\partial \psi}{\partial t} - k\frac{\partial^2 \psi}{\partial x^2} = 0 \tag{16.4}$$

has a = –k and c = 1 so that $b^2 - 4ac = 0$ and the equation is parabolic. The Schrödinger equation

$$i\hbar\frac{\partial \psi}{\partial t} = \frac{\partial^2 \psi}{\partial x^2} \tag{16.5}$$

is also a parabolic equation. Poisson's equation in two spatial dimensions

$$\nabla^2 \psi = \left(\frac{\partial^2}{\partial x^2} + \frac{\partial^2}{\partial y^2}\right)\psi = -\rho \tag{16.6}$$

has a = 1 and c = 1 and $b^2 - 4ac = -4ac = -4 < 0$ so that the equation is elliptical. Elliptical equations such as Poisson's and Laplace's equations are not of interest here because there is no time dependence and the behavior is static.

There are two broad approaches to formulating difference schemes for any of these equations: explicit and implicit differencing.

In an explicit scheme an updated component at time $t = (n+1)\Delta t$ at location $x = I\Delta x$ (one dimension is used here for simplicity) labeled ψ^{n+1} (I) is found from an explicit operation on prior time components, i.e., ψ^n, ψ^{n-1}, ..., $\phi^{n+1/2}$, $\phi^{n-1/2}$,.... Note the use of more than one component. This is a generalization of the commonly expressed definitions[1,2] where the Maxwell equations are not the primary focus. The offset in times between ψ and ϕ is only meant to be suggestive of the possible prior time schemes. In some explicit schemes the offset would not be present. By explicit we mean simple mathematical operations, addition, subtraction, multiplication, division, etc. that can be immediately evaluated to yield the updated field component. Note too that the prior time values need not be at the same spatial location.

An implicit scheme expresses the updated component in terms of prior time components at the same and adjoining locations just as for the explicit scheme, but in addition in terms of adjoining spatial components at the same updated time. This results in what is typically a large set of coupled first order equations which may be cast into matrix form and solved simultaneously for the updated values at each time step by traditional methods. Because the matrix itself does not explicitly yield the solution the formulation is referred to as an implicit differencing scheme. A prominent example, often used in heat flow or parabolic diffusion equations, is the Crank-Nicholson equation:

$$\psi^{n+1}(I) = \psi^{n}(I) + \frac{k\Delta t}{2(\Delta x)^2}\left[\psi^{n+1}(I+1) - 2\psi^{n+1}(I) + \psi^{n+1}(I-1)\right]$$
$$+ \frac{k\Delta t}{2(\Delta x)^2}\left[\psi^{n}(I+1) - 2\psi^{n}(I) - \psi^{n}(I-1)\right] \tag{16.7}$$

We are only interested in explicit schemes because of their use in FDTD and because of their simplicity and efficiency. Interestingly, some difference schemes can be formulated in either implicit or explicit form, for example,[2] the Dufort-Frankel scheme, also used for parabolic diffusion equations:

Implicit:

$$\psi^{n+1}(I) = \psi^{n-1}(I) + \frac{2k\Delta t}{(\Delta x)^2}\left\{\psi^{n}(I+1) - \left[\psi^{n+1}(I) - \psi^{n-1}(I)\right] + \psi^{n}(I-1)\right\} \tag{16.8}$$

Explicit:

$$\psi^{n+1}(I) = \left(\frac{1-\alpha}{1+\alpha}\right)\psi^{n-1}(I) + \frac{\alpha}{1+\alpha}\left[\psi^{n}(I+1) + \psi^{n}(I-1)\right]$$
$$\text{where } \alpha = \frac{2k\Delta t}{(\Delta x)^2} \tag{16.9}$$

The explicit form is of interest for solving the Maxwell Equations when $\partial D/\partial t$ is small compared to σE and can be ignored, leading to a diffusion equation. The Dufort-Frankel method allows any size time steps so that when diffusion conditions pertain it may be employed to economically solve for the field behavior. Unfortunately, in the case of large time steps an oscillation due to errors in the scheme is present and limits the accuracy of the predictions.

Turning now to explicit schemes exclusively we can finish our survey of parabolic difference equations by examining some common schemes, namely explicit first order and the leapfrog method, that are readily formulated for parabolic equations as:

Explicit first order:

$$\psi^{n+1}(I) = \psi^{n}(I) + \frac{k\Delta t}{(\Delta x)^2}\left[\psi^{n}(I+1) - 2\psi^{n}(I) + \psi^{n}(I+1)\right] \tag{16.10}$$

Leapfrog:

$$\psi^{n+1}(I) = \psi^{n-1}(I) + \frac{2k\Delta t}{(\Delta x)^2}\left[\psi^{n}(I+1) - 2\psi^{n}(I) + \psi^{n}(I+1)\right] \tag{16.11}$$

The explicit first order method is stable for $\Delta t \leq \frac{1}{2}(\Delta x)^2/(k)$ while the leapfrog method is always unstable for parabolic equations.

Hyperbolic equations (this equation type governs electromagnetics when oscillatory behavior is present as opposed to when the system is static with $\nabla^2 \phi = -\rho$ or is a slowly varying system that behaves diffusively as when $\partial D/\partial t \sim 0$) behave oppositely for the two schemes just discussed. The explicit first order method is unstable for hyperbolic equations, while the leapfrog method is stable for sufficiently small time steps. In summary, similar techniques can be formulated for hyperbolic and parabolic equations, but they may well work differently for each equation type. Having dispensed with elliptical equations as uninteresting because they are static and parabolic equations because they do not describe harmonically varying electromagnetic systems we now focus on hyperbolic equations.

16.2 DIFFERENCING SCHEMES FOR HYPERBOLIC EQUATIONS

The wave equation

$$\nabla^2 \psi - \frac{1}{c_0^2} \frac{\partial^2 \psi}{\partial t^2} = 0 \tag{16.12}$$

in one dimension

$$\frac{\partial^2 \psi}{\partial x^2} - \frac{1}{c_0^2} \frac{\partial^2 \psi}{\partial t^2} = 0 \tag{16.13}$$

can be factored into

$$L^+ L^- \psi = 0 \tag{16.14}$$

where

$$L^+ = \frac{\partial}{\partial x} + \frac{1}{c_0} \frac{\partial}{\partial t} \tag{16.15}$$

$$L^- = \frac{\partial}{\partial x} - \frac{1}{c_0} \frac{\partial}{\partial t} \tag{16.16}$$

While the wave equation admits waves propagating in either direction, L^+ and L^- above allow for waves traveling only in the positive or negative x direction, i.e.,

$$L^+ \psi^+ = 0 \tag{16.17}$$

$$L^{-}\psi^{-} = 0 \tag{16.18}$$

where ψ^+ is positive going and ψ^- is negative going. These equations are called the one-way wave equations. Because L^+ and L^- commute, any wave of the form $\psi = A\psi^+ + B\psi^-$ satisfies $L^+ L^- \psi = 0$.

Starting with the one-way wave equation in the form

$$\frac{\partial \psi}{\partial t} + a\frac{\partial \psi}{\partial x} = 0 \tag{16.19}$$

and two alternate formulations of the derivative in difference form given here for the time derivative

$$\frac{\partial \psi}{\partial t} \approx \frac{\psi^{n+1}(I) - \psi^{n}(I)}{\Delta t} \quad \text{(forward difference)} \tag{16.20}$$

or

$$\frac{\partial \psi}{\partial t} \approx \frac{\psi^{n+1}(I) - \psi^{n-1}(I)}{2\Delta t} \quad \text{(centered difference)} \tag{16.21}$$

and spatial derivative

$$\frac{\partial \psi}{\partial x} \approx \frac{\psi^{n}(I) - \psi^{n}(I-1)}{\Delta x} \quad \text{(forward difference)} \tag{16.22}$$

or

$$\frac{\partial \psi}{\partial x} \approx \frac{\psi^{n}(I+1) - \psi^{n}(I-1)}{2\Delta x} \quad \text{(center difference)} \tag{16.23}$$

the following five finite difference schemes can be obtained:[1]

$$\frac{\psi^{n+1}(I) - \psi^{n}(I)}{\Delta t} + a\frac{\psi^{n}(I+1) - \psi^{n}(I)}{\Delta x} = 0$$
$$\text{(forward time and space scheme)} \tag{16.24a}$$

$$\frac{\psi^{n+1}(I) - \psi^{n}(I)}{\Delta t} + a\frac{\psi^{n}(I) - \psi^{n}(I-1)}{\Delta x} = 0 \tag{16.24b}$$
$$\text{(forward time, backward space scheme)}$$

$$\frac{\psi^{n+1}(I)-\psi^{n}(I)}{\Delta t}+a\frac{\psi^{n}(I+1)-\psi^{n}(I-1)}{2\Delta x}=0 \tag{16.24c}$$

(forward time, centered space scheme)

$$\frac{\psi^{n+1}(I)-\psi^{n-1}(I)}{2\Delta t}+a\frac{\psi^{n}(I+1)-\psi^{n}(I-1)}{2\Delta x}=0 \tag{16.24d}$$

leapfrog (centered space and time scheme)

$$\frac{\psi^{n+1}(I)-\frac{1}{2}\left[\psi^{n}(I+1)-\psi^{n}(I-1)\right]}{\Delta t}+a\frac{\psi^{n}(I+1)-\psi^{n}(I-1)}{2\Delta x}=0 \tag{16.24e}$$

(Lax - Friedrichs scheme)

This categorization makes clear what has been meant in earlier chapters by centered differencing and also clarifies the term leapfrog scheme. However, while the one-way wave equation will be used in the derivation of the Mur boundary condition in Chapter 18, the Maxwell curl equations in component form appear as

$$\frac{\partial\psi}{\partial t}+\frac{\partial\phi}{\partial x}=0 \tag{16.25}$$

Using this form a similar set of difference equations directly related to the Maxwell curl equations can be formulated:[2]

$$\psi^{n+1}(I)=\psi^{n}(I)-\left[\phi^{n}(I+1)-\phi^{n}(I-1)\right]\frac{\Delta t}{2\Delta x} \tag{16.26}$$

first order explicit scheme

$$\psi^{n+1}(I)=\frac{1}{2}\left[\psi^{n}(I+1)+\psi^{n}(I-1)\right]-\left[\phi^{n}(I+1)-\phi^{n}(I-1)\right]\frac{\Delta t}{2\Delta x} \tag{16.27}$$

Lax scheme

$$\psi^{n+1}(I)=\psi^{n-1}(I)-\frac{\Delta t}{\Delta x}\left[\phi^{n}(I+1)-\phi^{n}(I-1)\right] \tag{16.28}$$

Leapfrog scheme

The earlier forward time and space scheme, forward time and backward space scheme, and the forward time and centered space scheme are variants of the first order scheme presented here. It is always unstable for hyperbolic equations. The leapfrog schemes are much the same as are the Lax schemes; of the Lax and leapfrog schemes the leapfrog scheme provides the more accurate results.[1]

Having narrowed the field to the leapfrog method for hyperbolic equations we can now look at permutations on this technique.

16.3 PERMUTATIONS ON THE LEAPFROG TECHNIQUE

Following Strikwerda[1] we can obtain fourth order accurate approximations to the first derivative operator using a Taylor series expansion. We start with the center difference operator δ_0 defined by

$$\delta_0 V_m = \frac{1}{2}\left(\delta_+ V_m + \delta_- V_m\right) = \frac{V_{m+1} - V_{m-1}}{2h} \tag{16.29}$$

$$\text{where} \quad \delta + V_m = \frac{V_{m+1} - V_m}{h} \quad \text{(forward difference)}$$

$$\delta_- V_m = \frac{V_m - V_{m-1}}{h} \quad \text{(backward difference)}$$

The Taylor series expansion gives

$$\begin{aligned} \delta_0 u &= \frac{du}{dx} + \frac{h^2}{6}\frac{d^3u}{dx^3} + 0(h^4) \\ &= \left(1 + \frac{h^2}{6}\frac{d^2}{dx^2}\right)\frac{du}{dx} + 0(h^4) \\ &= \left(1 + \frac{h^2}{6}\delta^2\right)\frac{du}{dx} + 0(h^4) \end{aligned} \tag{16.30}$$

where $\delta^2 = (\delta_+ - \delta_-)/h$ or

$$\delta^2 V_m = \frac{V_{m+1} - 2V_m + V_{m-1}}{h^2}$$

and $d^2f/dx^2 = \delta^2 f + 0(h^2)$ is used to obtain the last expression of (16.30).

Symbolically we can rewrite (16.30) as

$$\frac{du}{dx} = \left(1 + \frac{h^2}{6}\delta^2\right)^{-1} \delta_0 u + 0\left(h^4\right) \tag{16.31}$$

and for

$$\frac{du}{dx} = f \tag{16.32}$$

we obtain to fourth order

$$\left(1 + \frac{h^2}{6}\delta^2\right)^{-1} \delta_0 u(X_m) = f(X_m) \tag{16.33}$$

Thus,

$$\delta_0{}^u(X_m) = \left(1 + \frac{h^2}{6}\delta^2\right) f(X_m) \tag{16.34}$$

or finally

$$\begin{aligned} \frac{V_{m+1} - V_{m-1}}{2h} &= f_m + \frac{1}{6}\left(f_{m+1} - 2f_m + f_{m-1}\right) \\ &= \frac{1}{6}\left(f_{m+1} + 4f_m + f_{m-1}\right) \end{aligned} \tag{16.35}$$

This is in contrast to the second order accurate expression

$$\frac{V_{m+1} - V_{m-1}}{2h} = f_m \tag{16.36}$$

One can use these higher order accuracy expressions to obtain a higher order accuracy expression of the Maxwell curl equations but at the cost of a higher computational expense. Generally we have not found the tradeoff advantageous.

REFERENCES

1. **Strikwerda, J.,** *Finite Difference Schemes and Partial Differential Equations,* Wadsworth and Brooles/Cole Mathematics Series, Belmont, CA, 1989.
2. **Potter, D.,** *Computational Physics,* John Wiley & Sons, New York, 1973.

Chapter 17

STABILITY, DISPERSION, ACCURACY

17.1 INTRODUCTION

Chapter 17 formally establishes the stability condition for the leapfrog method as applied to hyperbolic partial differential equations. This stability condition, known as the Courant stability condition, is the key to this chapter. It sets a condition on the Δx and Δt difference sizes that result in a stable scheme. The Courant stability condition was presented in Chapter 3, but here we derive and justify it based on a complete mathematical treatment. One warning is that satisfying Courant stability does not guarantee unconditional stability (which is possible with some implicit schemes applied to parabolic partial differential equations, such as the Crank-Nicholson scheme); rather it is a conditional stability. For hyperbolic partial differential equations modeled with any of the various explicit differencing schemes, at best only conditional stability is possible, and the stability condition is typically of a form similar to the Courant stability condition for the leapfrog method.

In the electromagnetic applications of FDTD this conditional stability can be illustrated easily. Suppose we have an FDTD code that has a time step size set at or below the Courant limit. We still must not excite the FDTD problem space at a frequency high enough so that the additional constraint of many FDTD cells per wavelength is violated. Let us nonetheless violate this constraint. Since the spatial sampling will then be insufficient the code will be unstable, even though the Courant limit is satisfied.

Lack of stability for a hyperbolic set of equations, as occurs when using an explicit first order differencing scheme or the leapfrog method with too large a Δt, results in dramatically increasing high frequency (wavelengths on the order of the grid size) oscillation. In short, the predicted response "blows up" and is quite useless. For usable predictions the correct difference scheme must be employed under conditions that ensure stability.

While an unstable scheme is definitely inaccurate, different stable schemes will provide different orders of accuracy. The order of accuracy can be determined straightforwardly. Related to accuracy is the question of dispersion — how the signal disperses because of the use of a finite grid and its numerical effects. Note this is numerical dispersion due to finite spatial grid steps and not physical dispersion as arises from propagation in a frequency-dependent material media. The effects of grid dispersion are examined in more detail here.

Formally we want the leapfrog scheme to yield predictions that approximate the exact solutions of the hyperbolic partial differential equations. Further, we want this approximation to approach the exact solution as the spatial grid, Δx, ..., and temporal grid, Δt, approach zero. If this is indeed the case we say that the predictions are converging and that the difference scheme is convergent.

The Lax-Richtmyer equivalence theorem[1] states that a consistent finite difference scheme for a partial differential equation is convergent if and only if it is stable.

Consistency is the requirement that

$$L\psi - L_{\Delta t,\Delta x}\psi \to 0 \quad \text{for } \Delta x, \Delta t \to 0$$

where L is the differential operator in the hyperbolic equation $L\psi = 0$ and $L_{\Delta t,\Delta x}$ is the corresponding difference scheme operator. The selection of the leapfrog method assures consistency. Therefore, the desired convergence results when the scheme is stable.

It is for this reason we concentrate on stability, then look at dispersion and accuracy in terms of the order of the difference approximation.

17.2 STABILITY

Consider a system satisfying $(d\vec{u})/(dt) = L\vec{u}$, where L is a spatial differential operator and $\vec{u}(r,t)$ is the state vector known at $t = 0$, i.e., $\vec{u}\,(r,0) = \vec{u}_0$ so the equation is an initial value equation. In difference form this can be written

$$\vec{u}^{n+1} = T(\Delta t, \Delta x)\vec{u}^n \tag{17.1}$$

or

$$\vec{u}^{n+1} = \vec{u}^n + \textstyle\int_{t^n}^{t^{n+1}} L\vec{u}\, dt'$$

or using a Taylor series expansion

$$= \vec{u}^n + \int_{t^n}^{t^{n+1}} \left\{ \sum_{r=0}^{p-1} \left[\frac{d^r}{dt^r}(L\vec{u}) \right]_{t^n} \frac{t'^r}{r!} + 0\left(t'^p\right) \right\} dr'$$

$$\cong \vec{u}^n + L\vec{u}\Delta t = (I + \Delta t L)\vec{u}^n \equiv T(\Delta t,\ \Delta x)\vec{u}^n$$

where I is the unitary matrix. The final result is the expression for an explicit difference scheme.

Now assume a system where linear superposition holds. Solutions can be expressed in terms of Fourier modes of the form for compactness position is given by x_j rather than x(J)

$$\vec{u}_j^n = \hat{u}^n e^{ikxj}$$

where by substitution into (17.1) there results

$$\hat{u}^{n+1}e^{ikxj} = T(\Delta t, \Delta x)\hat{u}^n e^{ikxj'}$$

or

$$\hat{u}^{n+1} = e^{-ikxj}T(\Delta t, \Delta x)e^{ikxj'}\hat{u}^n = G(\Delta t, \Delta x)\hat{u}^n \tag{17.2}$$

G is the amplification matrix of the difference scheme for the Fourier mode of wavenumber k.

We can express the Fourier modes as an expansion in the eigenvectors of the amplification matrix G. In particular let $\hat{u}_\mu^o$ be the amplitude of the vector Fourier mode along the eigenvector $\vec{S}^{(\mu)}$ at time step zero, then

$$\hat{u}^o = \sum_\mu \hat{u}_\mu^o \vec{S}^{(\mu)}$$

From $\hat{u}^o$ and (17.2) we can find $\hat{u}^n$ as

$$\hat{u}^n = (G)^n \hat{u}^o = (G)^n \sum_\mu \hat{u}_\mu^o \vec{S}^{(\mu)} \tag{17.3}$$

Now $\vec{S}^{(\mu)}$ as an eigenvector of G satisfies

$$\left(G - g_\mu\right)S^{(\mu)} = 0 \text{ or } G\, S^{(\mu)} = g_\mu S^{(\mu)} \tag{17.4}$$

Thus, from (17.3)

$$\hat{u}^n = \sum \hat{u}_\mu^o \left(g_\mu\right)^n S^{(\mu)}$$

For $|g_\mu| \leq 1$ this behavior for $\hat{u}^n$ is stable. Equivalently we could require

$$g_\mu^* g_\mu = \left|g_\mu\right|^2 \leq 1$$

Our task is now to formulate the time centered leapfrog method into a form with G, find the g_μ and see under what conditions $|g_\mu| < 1$.

The difference equations in 1-D are of the form

$$u_j^{n+1/2} = u_j^{n-1/2} - \frac{\Delta t}{\Delta x}\left(f_{j+1}^n - f_{j-1}^n\right)$$

or

$$E_j^{n+1/2} = E_j^{n-1/2} - \frac{\Delta t}{\varepsilon \Delta x}\left(H_{j+1/2}^{n} - H_{j-1/2}^{n}\right) \text{ from } \left(\frac{\partial D}{\partial t} + \frac{\partial H}{\partial x} = 0\right) \quad (17.5)$$

$$H_{j+1/2}^{n} = H_{j+1/2}^{n-1} - \frac{\Delta t}{\mu \Delta x}\left(E_{j+1}^{n-1/2} - E_j^{n-1/2}\right) \text{ from } \left(\frac{\partial B}{\partial t} + \frac{\partial E}{\partial x} = 0\right) \quad (17.6)$$

Now (17.5) may be rewritten by substituting (17.6) into (17.5) as

$$E_j^{n+1/2} = E_j^{n-1/2} - \frac{\Delta t}{\varepsilon \Delta x}$$

$$\cdot\left[H_{j+1/2}^{n-1} - \frac{\Delta t}{\mu \Delta x}\left(E_{j+1}^{n-1/2} - E_j^{n-1/2}\right) - H_{j-1/2}^{n-2} + \frac{\Delta t}{\mu \Delta x}\left(E_j^{n-1/2} - E_{j-1}^{n-1/2}\right)\right]$$

Thus,

$$E_j^{n+1/2} = E_j^{n-1/2} + \left(\frac{c\Delta t}{\Delta x}\right)^2\left(E_{j+1}^{n-1/2} - 2E_j^{n-1/2} + E_{j-1}^{n-1/2}\right)$$
$$- \frac{\Delta t}{\varepsilon \Delta x}\left(H_{j+1/2}^{n-1} - H_{j-1/2j}^{n-1}\right) \quad (17.6a)$$

$$H_{j+1/2}^{n} = H_{j+1/2}^{n-1} - \frac{\Delta t}{\mu \Delta x}\left(E_{j+1}^{n-1/2} - E_j^{n-1/2}\right)$$

Now let

$$\vec{u}^{n} = \begin{pmatrix} \hat{E}^{n-1/2} e^{ikx} \\ \hat{H}^{n-1} e^{ik(x+1/2\Delta x)} \end{pmatrix}$$

so that $\hat{u}^{n+1} = G\hat{u}^{n}$ can be written as

$$\vec{u}^{n+1} = \begin{bmatrix} 1+\left(\frac{c\Delta t}{\Delta x}\right)\left[e^{ik\Delta x} - 2 + e^{-ik\Delta x}\right] & \frac{-\Delta t}{\varepsilon \Delta x}\left(1 - e^{-ik\Delta x}\right) \\ -\frac{\Delta t}{\mu \Delta x}\left(e^{-ik\Delta x} - 1\right) & 1 \end{bmatrix} \hat{u}_n \quad (17.7)$$

where we have used

$$E_{j+1}^{n-½} - E_j^{n-½} = \left(\hat{E}^{n-½} e^{ikxj}\right)\left[e^{ik\Delta x} - 1\right]$$

$$E_{j+1}^{n-½} - 2E_j^{n-½} + E_{j-1}^{n-½} = \left(E^{n-½} e^{ikxj}\right)\left[e^{ik\Delta x} - 2 + e^{ik\Delta x}\right]$$

Thus, the eigen values of (17.4) may be found from the determinant of $|G - g|$, by setting the determinant to zero so that

$$\left\{\left[1 + \left(\frac{c\Delta t}{\Delta x}\right)\right]^2 \left(e^{ik\Delta x} - 2 + e^{-ik\Delta x}\right) - g\right\}[1 - g] - \left(\frac{c\Delta t}{\Delta x}\right)^2 \left(e^{ik\Delta x} - 2 + e^{-ik\Delta x}\right) = 0 \tag{17.8}$$

this simplifies to

$$q^2 - \left[2t\left(\frac{c\Delta t}{\Delta x}\right)^2 \left(e^{ik\Delta x} - z + e^{-ik\Delta x}\right)\right] g + 1 = 0 \tag{17.9}$$

For $(c\Delta t)/(\Delta x) = 1$ or equivalently $\Delta t = \Delta x/c$ the expression further simplifies to

$$g^2 - 2(\cos k\Delta x)g + 1 = 0 \tag{17.10}$$

and

$$\begin{aligned} g &= \frac{2(\cos k\Delta x) \pm \sqrt{4\cos^2 k\Delta x - 4}}{2} \\ &= \cos k\Delta \pm \sqrt{\cos^2 k\Delta x - 1} \\ &= \cos k\Delta x \pm i \sin k\Delta x \end{aligned} \tag{17.11}$$

Therefore $|g| = 1$ and the scheme is stable for $\Delta t = \Delta x/c$ in 1-D. In 3-D and for cubical cells the stability condition generalizes to $\Delta t = \Delta x/\sqrt{3}c$.

17.3 NUMERICAL DISPERSION

Start with the six free space finite difference equations:

$$\begin{aligned} E_x^{s,\, n+½}(I,J,K) &= E_x^{s,\, n-½}(I,J,K) \\ &+ \frac{\Delta t}{\varepsilon_0}\left[\frac{H_z^{s,\, n}(I,J,K) - H_z^{s,\, n}(I,J-1,K)}{\Delta y} - \frac{H_y^{s,\, n}(I,J,K) - H_y^{s,\, n}(I,J,K-1)}{\Delta z}\right] \end{aligned} \tag{17.12a}$$

$$E_y^{s,\,n+\frac{1}{2}}(I,J,K) = E_y^{s,\,n-\frac{1}{2}}(I,J,K) + \frac{\Delta t}{\varepsilon_0}\left[\frac{H_x^{s,\,n}(I,J,K) - H_x^{s,\,n}(I,J,K-1)}{\Delta z} - \frac{H_z^{s,\,n}(I,J,K) - H_z^{s,\,n}(I-1,J,K)}{\Delta x}\right] \tag{17.12b}$$

$$E_y^{s,\,n+\frac{1}{2}}(I,J,K) = E_y^{s,\,n-\frac{1}{2}}(I,J,K) + \frac{\Delta t}{\varepsilon_0}\left[\frac{H_x^{s,\,n}(I,J,K) - H_x^{s,\,n}(I,J,K-1)}{\Delta z} - \frac{H_x^{s,\,n}(I,J,K) - H_x^{s,\,n}(I,J-1,K)}{\Delta y}\right] \tag{17.12c}$$

$$H_x^{s,\,n}(I,J,K) = H_x^{s,\,n-1}(I,J,K) + \frac{\Delta t}{\mu_0}\left[\frac{E_y^{s,\,n-\frac{1}{2}}(I,J,K+1) - E_y^{s,\,n-\frac{1}{2}}(I,J,K)}{\Delta z} - \frac{E_z^{s,\,n-\frac{1}{2}}(I,J+1,K) - E_x^{s,\,n-\frac{1}{2}}(I,J,K)}{\Delta y}\right] \tag{17.12d}$$

$$H_y^{s,\,n}(I,J,K) = H_y^{s,\,n-1}(I,J,K) + \frac{\Delta t}{\mu_0}\left[\frac{E_z^{s,\,n-\frac{1}{2}}(I+1,J,K) - E_z^{s,\,n-\frac{1}{2}}(I,J,K)}{\Delta x} - \frac{E_x^{s,\,n-\frac{1}{2}}(I,J,K+1) - E_x^{s,\,n-\frac{1}{2}}(I,J,K)}{\Delta z}\right] \tag{17.12e}$$

$$H_z^{s,\,n}(I,J,K) = H_z^{s,\,n-1}(I,J,K) + \frac{\Delta t}{\mu_0}\left[\frac{E_x^{s,\,n-\frac{1}{2}}(I,J+1,K) - E_x(I,J,K)}{\Delta y} - \frac{E_y^{s,\,n-\frac{1}{2}}(I+1,J,K) - E_x^{s,\,n-\frac{1}{2}}(I,J,K)}{\Delta x}\right] \tag{17.12f}$$

and the Fourier modes for each component:

$$E^{s,\,n+\frac{1}{2}}_{x;\,a,b,c}(I,J,K) = e_{x;\,a,b,c}e^{j\left\{\omega(n+\frac{1}{2})\Delta t-\left[a(I+\frac{1}{2})\Delta x+bJ\Delta y+cK\Delta z\right]\right\}} \quad (17.13a)$$

$$E^{s,\,n+\frac{1}{2}}_{y;\,a,b,c}(I,J,K) = e_{y;\,a,b,c}e^{j\left\{\omega(n+\frac{1}{2})\Delta t-\left[a\,I\Delta x+b(J+\frac{1}{2})\Delta y+cK\Delta z\right]\right\}} \quad (17.13b)$$

$$E^{s,\,n+\frac{1}{2}}_{z;\,a,b,c}(I,J,K) = e_{z;\,a,b,c}e^{j\left\{\omega(n+\frac{1}{2})\Delta t-\left[a\,I\Delta x+bJ\Delta y+c(K+\frac{1}{2})\Delta z\right]\right\}} \quad (17.13c)$$

$$H^{s,\,n}_{z;\,a,b,c}(I,J,K) = h_{z;\,a,b,c}e^{j\left\{\omega n\Delta t-\left[a\,I\Delta x+b(J+\frac{1}{2})\Delta y+c(K+\frac{1}{2})\Delta z\right]\right\}} \quad (17.13d)$$

$$H^{s,\,n}_{y;\,a,b,c}(I,J,K) = h_{y;\,a,b,c}e^{j\left\{\omega\, n\Delta t-\left[a(I+\frac{1}{2})\Delta x+bJ\Delta y+c(K+\frac{1}{2})\Delta z\right]\right\}} \quad (17.13e)$$

$$H^{s,\,n}_{z;\,a,b,c}(I,J,K) = h_{z;\,a,b,c}e^{j\left\{\omega\, n\Delta t-\left[a(I+\frac{1}{2})\Delta x+b(J+\frac{1}{2})\Delta y+cK\Delta z\right]\right\}} \quad (17.13f)$$

By direct substitution of the appropriate Fourier mode into the six free space finite difference equations we obtain the requisite six equations and six unknowns. We will assume $e_{x;a,b,c}$ is given along with a,b,c; then the five remaining Fourier mode amplitudes and ω are unknown. What is wanted is ω in terms of a,b,c.

The six equations can now be found, but first a few simplifications are in order. All Fourier modes share the term $e^{j[\omega n\Delta t - (aI\Delta x + bJ\Delta y + cK\Delta z)]}$ which can be divided out at the start. Also, the a,b,c subscript can be suppressed for brevity. Then:

$$e_x e^{j(\omega 1/2\Delta t - a1/2\Delta x)} = e_x e^{j(-\omega 1/2\Delta t - a1/2\Delta x)} +$$

$$\frac{\Delta t}{\varepsilon_0}\left[\frac{h_z e^{j(-a1/2\Delta x - b1/2\Delta y)} - h_z e^{j(-a1/2\Delta x + b1/2\Delta y)}}{\Delta y}\right.$$
$$\left.\frac{h_y e^{j(-a1/2\Delta x - c1/2\Delta z)} - h_y e^{j(-a1/2\Delta x + c1/2\Delta z)}}{\Delta z}\right] \quad (17.14a)$$

$$e_y e^{j(\omega\, 1/2\Delta t - b1/2\Delta y)} = e_y e^{j(-\omega\, 1/2\Delta t - b1/2\Delta y)}$$
$$+\frac{\Delta t}{\varepsilon_0}\left[\frac{h_x e^{j(-b1/2\Delta y - c1/2\Delta z)} - h_x e^{j(-b1/2\Delta y + c1/2\Delta z)}}{\Delta z}\right.$$
$$\left.-\frac{h_z e^{j(-a1/2\Delta x - b1/2\Delta y)} - h_z e^{j(-a1/2\Delta x + b1/2\Delta y)}}{\Delta x}\right] \quad (17.14b)$$

$$
\begin{aligned}
& e_z e^{j(\omega\, 1/2\Delta t - c1/2\Delta z)} = e_z e^{j(-\omega\, 1/2\Delta t - c1/2\Delta z)} \\
& + \frac{\Delta t}{\varepsilon_0}\left[\frac{h_y e^{j(-a1/2\Delta x - c1/2\Delta z)} - h_y e^{j(-a1/2\Delta x + c1/2\Delta z)}}{\Delta x}\right. \\
& \left. - \frac{h_x e^{j(-b1/2\Delta y - c1/2\Delta z)} - h_x e^{j(-b1/2\Delta y + c1/2\Delta z)}}{\Delta y}\right]
\end{aligned}
\tag{17.14c}
$$

$$
\begin{aligned}
& h_x e^{j(-b\, 1/2\Delta y - c1/2\Delta z)} = h_x e^{j(-\omega\, \Delta t - b1/2\Delta y - c1/2\Delta z)} \\
& + \frac{\Delta t}{\mu_0}\left[\frac{e_y e^{j(-\omega\, 1/2\Delta t - b1/2\Delta y - c\Delta z)} - e_y e^{j(-\omega\, 1/2\Delta t - b1/2\Delta y)}}{\Delta z}\right. \\
& \left. - \frac{e_z e^{j(-\omega\, 1/2\Delta t - b\Delta y - c1/2\Delta z)} - e_z e^{j(-\omega\, 1/2\Delta t - c1/2\Delta z)}}{\Delta y}\right]
\end{aligned}
\tag{17.14d}
$$

$$
\begin{aligned}
& h_y e^{j(-a\, 1/2\Delta x - c1/2\Delta z)} = h_y e^{j(-\omega\, \Delta t - a1/2\Delta x - c1/2\Delta z)} \\
& + \frac{\Delta t}{\mu_0}\left[\frac{e_z e^{j(-\omega\, 1/2\Delta t - a\Delta x - c1/2\Delta z)} - e_z e^{j\left(-\omega\, 1/2\Delta t - {}^{b}\, 1/2\Delta z\right)}}{\Delta x}\right. \\
& \left. - \frac{e_x e^{j(-\omega\, 1/2\Delta t - a1/2\Delta x - c\Delta z)} - e_x e^{j\left(-\omega\, 1/2\Delta t - {}^{a}\, 1/2\Delta x\right)}}{\Delta z}\right]
\end{aligned}
\tag{17.14e}
$$

$$
\begin{aligned}
& h_z e^{j(-a1/2\Delta x - b1/2\Delta y)} = h_x e^{j(-\omega\, \Delta t - a1/2\Delta x - b1/2\Delta y)} \\
& + \frac{\Delta t}{\mu_0}\left[\frac{e_x e^{j\left(-\omega\, 1/2\Delta t - a1/2\Delta x - b\Delta^{y}\right)} - e_x e^{j(-\omega\, 1/2\Delta t - a1/2\Delta x)}}{\Delta y}\right. \\
& \left. - \frac{e_y e^{j(-\omega\, 1/2\Delta t - a\Delta x - b1/2\Delta y)} - e_y e^{j(-\omega\, 1/2\Delta t - b1/2\Delta y)}}{\Delta x}\right]
\end{aligned}
\tag{17.14f}
$$

The result of dividing out common terms and a small amount of manipulation is

$$
e_x\left[2\sin(\omega\Delta t/2)\right] = \frac{\Delta t}{\varepsilon_0}\left\{h_y\left[\frac{2\sin(c\Delta z/2)}{\Delta z}\right] - h_z\left[\frac{2\sin(b\Delta y/2)}{\Delta y}\right]\right\} \tag{17.15a}
$$

$$e_y\left[2\sin(\omega\Delta t/2)\right] = \frac{\Delta t}{\varepsilon_0}\left\{h_z\left[\frac{2\sin(a\Delta x/2)}{\Delta x}\right] - h_y\left[\frac{2\sin(c\Delta z/2)}{\Delta z}\right]\right\} \tag{17.15b}$$

$$e_z\left[2\sin(\omega\Delta t/2)\right] = \frac{\Delta t}{\varepsilon_0}\left\{h_x\left[\frac{2\sin(b\Delta x/2)}{\Delta y}\right] - h_y\left[\frac{2\sin(a\Delta x/2)}{\Delta x}\right]\right\} \tag{17.15c}$$

$$h_x\left[2\sin(\omega\Delta t/2)\right] = \frac{\Delta t}{\mu_0}\left\{e_z\left[\frac{2\sin(b\Delta y/2)}{\Delta y}\right] - e_y\left[\frac{2\sin(c\Delta z/2)}{\Delta z}\right]\right\} \tag{17.15d}$$

$$h_y\left[2\sin(\omega\Delta t/2)\right] = \frac{\Delta t}{\mu_0}\left\{e_x\left[\frac{2\sin(c\Delta z/2)}{\Delta z}\right] - e_y\left[\frac{2\sin(a\Delta x/2)}{\Delta x}\right]\right\} \tag{17.15e}$$

$$h_z\left[2\sin(\omega\Delta t/2)\right] = \frac{\Delta t}{\mu_0}\left\{e_y\left[\frac{2\sin(a\Delta x/2)}{\Delta x}\right] - e_y\left[\frac{2\sin(b\Delta y/2)}{\Delta y}\right]\right\} \tag{17.15f}$$

In the limit $\Delta t, \Delta x, \Delta y, \Delta z \rightarrow 0$ these expressions yield exactly the same results for a Fourier mode as the original partial differential equations expressing the Maxwell equations in component form, i.e., $2\sin(\omega\Delta t/2)/\Delta t \rightarrow \omega$, $2\sin(a\Delta x/2)/\Delta x \rightarrow a$, etc.

This set of coupled equations can be solved using conventional linear algebra. A solution requires the determinant to be zero and this condition allows w to be found in terms of a, b, and c.

Following this approach:

$$\begin{bmatrix} W & 0 & 0 & 0 & -\left[\frac{C}{\varepsilon_0}\right] & \left[\frac{B}{\varepsilon_0}\right] \\ 0 & W & 0 & \left[\frac{C}{\varepsilon_0}\right] & 0 & -\left[\frac{A}{\varepsilon_0}\right] \\ 0 & 0 & W & -\left[\frac{B}{\varepsilon_0}\right] & \left[\frac{A}{\varepsilon_0}\right] & 0 \\ 0 & \left[\frac{C}{\mu_0}\right] & -\left[\frac{B}{\mu_0}\right] & W & 0 & 0 \\ -\left[\frac{C}{\mu_0}\right] & 0 & \left[\frac{A}{\mu_0}\right] & 0 & W & 0 \\ \left[\frac{B}{\mu_0}\right] & -\left[\frac{A}{\mu_0}\right] & 0 & 0 & 0 & W \end{bmatrix} \begin{bmatrix} e_x \\ e_y \\ e_z \\ h_x \\ h_y \\ h_z \end{bmatrix} = 0$$

$$\begin{vmatrix} \\ \\ \end{vmatrix} \quad \text{M} \tag{17.16}$$

where

$$W = \frac{2\sin(\omega\Delta t/2)}{\Delta t}$$

$$A = \frac{2\sin(a\Delta x/2)}{\Delta x}$$

$$B = \frac{2\sin(b\Delta y/2)}{\Delta y}$$

$$C = \frac{2\sin(c\Delta z/2)}{\Delta z}$$

Setting $|M| = 0$ yields

$$W^6 - W^4\left(A^2 + B^2 + C^2\right)/\varepsilon_0\mu_0 = 0 \tag{17.17}$$

or

$$W^2 - \left(A^2 + B^2 + C^2\right)/\varepsilon_0\mu_0 = 0 \tag{17.18}$$

In the limit as the cell size and time step go to zero, $W = \omega$, $A = k_x$, $B = k_y$, and $C = k_z$ and the expression becomes

$$\omega^2 - k^2c^2 = 0 \tag{17.19}$$

the exact result as expected for a continuum. If we are not in the continuum limit and our cell size is finite, then

$$\left(\frac{2\sin(W\Delta t/2)}{\Delta t}\right)^2 = \frac{1}{\varepsilon_0\mu_0}\left[\left(\frac{2\sin(a\Delta x/2)}{\Delta x}\right)^2 + \left(\frac{2\sin(b\Delta y/2)}{\Delta y}\right)^2 + \left(\frac{2\sin(c\Delta z/2)}{\Delta z}\right)^2\right] \tag{17.20}$$

For 1-D

$$\frac{\sin(\omega\Delta t/2)}{\Delta t} = \frac{1}{\sqrt{\varepsilon_0\mu_0}}\frac{2\sin(k\Delta x/2)}{\Delta x} \tag{17.21}$$

If $\Delta t \equiv (\Delta x)/(\alpha c)$ and $\alpha \leq 1$ then the 1-D dispersion relation (17.21) becomes

$$\frac{\sin\left(\frac{\omega\Delta x/2}{c\alpha}\right)}{\frac{\Delta x}{\alpha c}} = \frac{1}{\sqrt{\varepsilon_0\mu_0}}\sin\frac{(k\Delta x/2)}{\Delta x} \tag{17.22}$$

Only for $\alpha = 1$ is the above relationship satisfied by $\omega/c = k$ so that there is no dispersion. This is a very special case. If $\alpha \neq 1$ we have a different dispersion relationship between ω and k, although 1 is still approximated by $\omega/c = k$. If an individual is working in a higher dimensionality then they can at best achieve dispersionless behavior along a particular direction.

For example, in 3-D let $\Delta t = \Delta x/\sqrt{3}\,c$, $k_x = k_y = k_z$, and $\Delta x, = \Delta y = \Delta z$, then

$$\left(\frac{2\sin\left(\omega\Delta x\sqrt{3}c\right)/2}{\Delta x/\sqrt{3}c}\right)^2 = \frac{1}{\sqrt{\varepsilon_0\mu_0}}\left[3\left(\frac{2\sin\left(k_x\Delta x/2\right)}{\Delta x}\right)^2\right] \tag{17.23a}$$

or

$$3c^2\left(\frac{2\sin\left(\omega\sqrt{3}c\right)/2}{\Delta x}\right)^2 = \frac{3}{\sqrt{\varepsilon_0\mu_0}}\left(\frac{2\sin\left(k_x\Delta x/2\right)}{\Delta x}\right) \tag{17.23b}$$

which holds so long as

$$c^2 = \frac{1}{\varepsilon_0\mu_0},\ k_x = \omega/\sqrt{3}c \text{ or } k^2 = k_x^2 + k_y^2 + k_z^2 = 3\left(\omega/\sqrt{3}c\right)^2 = \frac{\omega^2}{c^2}$$

Thus, along the direction of $k_x = k_y = k_z$, or more generally, $|k_x| = |k_y| = |k_z|$ and with $\Delta t = \Delta x/\sqrt{3}\,c$ there is no numerical dispersion. In any other direction there will, even under these special circumstances, be numerical dispersion.

17.4 ACCURACY

For a Fourier mode given by

$$u = he^{ikx}$$

the exact first derivative in space yields

$$\frac{du}{dx} = ikhe^{ikx}$$
$$= iku$$

The center difference approximation may be written

$$\frac{\Delta u}{\Delta x} = \frac{h\,e^{ikx(J+1)} - h\,e^{eikx(J-1)}}{2\Delta x} \tag{17.24}$$

or with x(J+1) = x(J) + Δx and x(J–1) = x(J) – Δx this becomes

$$\frac{\Delta u}{\Delta x} = \frac{h}{\Delta x} e^{ikx(J)} \frac{1}{2}\left(e^{ik\Delta x} - e^{-ik\Delta x}\right)$$
$$= \frac{iu}{\Delta x}\sin(k\Delta x)\,.\quad i\,k\,u \text{ for small } k\Delta x \tag{17.25}$$

For larger kΔx we can expand sin (kΔx) in a Taylor series to obtain

$$\frac{\Delta u}{\Delta x} = \frac{iu}{\Delta x}\left(k\Delta x - \frac{(k\Delta)^3}{6} + 0\left[(k\Delta x)^5\right]\right)$$
$$= i\,u\,k\left(1 - \frac{(k\Delta x)^2}{6} + 0\left[(k\Delta x)^4\right]\right) \tag{17.26}$$

Thus, the center differencing employed above is referred to as being second order accurate in kΔx. The same results holds for the centered time difference.

REFERENCE

1. **Potter, D.,** *Computational Physics,* John Wiley & Sons, New York, 1973.

Chapter 18

OUTER RADIATION BOUNDARY CONDITIONS

18.1 INTRODUCTION

In Chapter 3 the most commonly used outer radiation boundary conditions (ORBC), known as first and second order Mur, were presented in a form suitable for application but without explanation or justification. In this chapter we present a short history of absorbing boundary conditions that have been used, explain the basis of the Mur absorbing boundary, and present some generalizations of it.

As discussed briefly in Chapter 3, an ORBC is needed because the FDTD technique uses a finite problem space. This space may be relatively large, say $100 \times 100 \times 100$ cells or 1 million cells in all, yet it must of necessity have components on the surface of the problem space. These components, unlike the remaining interior components, are not completely surrounded by their neighbors. At least one neighboring component is missing because the problem space is terminated at this outermost or surface component. As a result when calculating the update of a surface component according to the FDTD difference equations given in Chapter 2 there is not enough information to correctly calculate the updated value.

An ORBC must be employed to approximate the missing field components, or more precisely the missing space surrounding the problem space. If the approximation were perfect then the outermost component would be updated as if the scattered field passed through the surface component's location unaffected by the truncation of the problem space. Thus, a perfect ORBC lets the scattered field appear to radiate outward to infinity unimpeded. Any real ORBC is an approximation; it will only approach the ideal and will generate some reflected radiation, typically small, from the outermost or outer boundary components.

The use of an ORBC cannot be readily avoided. If the outermost components are E field components, as is often the case, and no ORBC is employed, then all the energy reaching the outermost components will be reflected back into the problem space, severely corrupting the predicted responses. The mechanism behind the reflection is easily envisioned using a 1-D transmission line analog based on the 1-D rectangular coordinate system set of equations of Appendix A. The outermost E field component acts like the end of an open transmission line in this 1-D analog. As such the radiating fields reaching this outermost location are reflected in their entirety without a sign change.

We assume in this discussion that the FDTD calculations involve scatterers in free space. If, for example, the FDTD space is contained within a conducting cavity then no ORBC will be required. Even for some limited applications where the scatterer is surrounded by free space it is possible to make FDTD predictions without an ORBC. The approach is to use a small enough scatterer in a big enough problem space to obtain sufficient "clear time" before the reflected signal returns to the scatterer to obtain meaningful responses in or on the scatterer. This approach is quite wasteful and now that ORBCs have been developed that work well their use is virtually universal.

It is well worth at least a brief examination of how ORBCs have evolved, and so a short review of some of the techniques employed is given in the following section. The latest commonly employed approach, called the Mur radiating boundary condition, is discussed at greater length. It can be implemented at any order, with first order the simplest and least demanding of resources and second order the most favored among FDTD practitioners. This is the ORBC presented briefly in Chapter 3. Higher order implementations have also been tried and are part of the ongoing quest for the "perfect" ORBC. This chapter concludes with a look at these efforts to find a better ORBC.

18.2 EVOLUTION OF ORBCS

A number of ORBCs were developed prior to the now-widely used Mur radiation condition. These include a conducting outer layer of the problem space, Legendre polynomial analytic continuations, impedance matching conditions at the outer boundary, and simple look-back schemes. The Mur condition is itself a look-back scheme but, at least in its higher order formulations, is anything but simple.

The conducting layer approach was used by Taflove and Brodwin[1] in some early work with the FDTD technique. Typically, a total field formulation was employed and the field oscillated at a single frequency. By adjusting the conductivity of a several-cell-thick layer at the extremities of the problem space the field could be mostly absorbed. The disadvantage to this method was that it entailed using a single frequency and was very dependent on the skill of the user in setting the conductivity levels and rate of spatial variation to optimize absorption. This approach is little used.

Legendre polynomial analytic continuation was an approach espoused by C. Longmire and examined by Gilbert[2] along with several other of the approaches discussed here, including impedance matching. Classical in nature and quite capable of yielding accurate results, this approach nonetheless was much too formidable to implement easily and required inordinately large resources to do its job. It is not employed in any operational code known to the authors.

The impedance matching technique is particularly appealing in its simplicity. In effect all that is required is to assume a wave propagating outward and relate E and H via the impedance of free space, Z_0 = 377 Ω. In spherical coordinates this can be expressed as

$$E_\theta(r,t) = Z_0 H_\phi(r,t) \tag{18.1a}$$

$$E_\phi(r,t) = Z_0 H_\theta(r,t) \tag{18.1b}$$

The unfortunate reality is that with spatially interleaved Es and Hs, required for stability, there are spatial offsets between components. Attempts to correct for these offsets lead to complications equivalent in complexity to look-back schemes and this approach has been dropped in favor of the look-back schemes themselves.

One of the earliest look-back schemes was developed first by Merewether[3] in 2-D and then extended by Holland[4] to 3-D. It was this latter implementation that was used in the F-111 modeling detailed in Chapter 4. It assumes that a far field radiation behavior describes the scattered fields, i.e.,

$$E_{\binom{\theta}{\phi}}(r,t) = \frac{1}{r} f_E(r - ct) \tag{18.2}$$

or

$$H_{\binom{\theta}{\phi}}(r,t) = \frac{1}{r} f_H(r - ct) \tag{18.3}$$

Thus, either E or H are found at the surface of the problem space from the immediately interior field points whose behavior is extrapolated to the outermost components according to the above prescription that attributes a 1/r field dependence and a wave propagation velocity of c.

Note that there is no calculation of field components beyond the problem space. Rather, the surface field components are updated according to the above prescription and correspond, at least approximately, to an outgoing radiated wave. The "overwriting" of the outermost components as opposed to the introduction of "missing" components is a common feature of all the schemes presented here.

The terminology "look-back scheme" comes from the fact that the outermost components are overwritten using field components within the problem space, components that are found by looking back toward the scatterer. The simple look-back scheme outlined here gives results roughly comparable to first order Mur and is in fact one of a myriad of look-back schemes that can be formulated using a general approach[5] that encompasses the simple scheme, Mur of any order, and other variations.

18.3 MUR OUTER RADIATION BOUNDARY CONDITION

Solution of the coupled curl equations via the FDTD technique is equivalent to the solution of the wave equation for either field component.[6] Mur's[7] derivation of an ORBC as well as more generalized treatments[5] start with the wave equation. They treat either E or H, working with the individual field components. The wave equation is approximated according to an approach first developed by Engquist and Majda[8] for the scalar wave equation. The approximation is referred to as a one-way wave equation or a look-back scheme.

Following Mur's treatment the wave equation for a single field component W may be written as

$$\left(\partial_x^2 + \partial_y^2 + \partial_z^2 - c_0^{-2}\partial_t^2\right)W = 0 \tag{18.4}$$

where $\partial_x^2 = \partial^2 / \partial x^2$, etc., and c_0 is the velocity of propagation. Assuming the FDTD problem space or mesh is located at $0 \le x$ there is a boundary at $x = 0$ that a scattered wave will reach and be reflected unless an ORBC is imposed.

This ORBC is found from the above wave equation. The scattered wave has velocity components v_x, v_y, v_z such that $v_x^2 + v_y^2 + v_y^2 = c_0^2$. It is also possible to define inverse velocity components S_x, S_y, S_z such that

$$S_x = \frac{\partial_x}{\partial_t} = \frac{\partial / \partial x}{\partial / \partial t} = \frac{I}{\partial x / \partial t}, \quad S_y = \frac{1}{(\partial y / \partial t)}, \quad S_x = \frac{1}{(\partial z / \partial t)}$$

$$S_x^2 + S_y^2 + S_z^2 = \frac{1}{c_0^2} \tag{18.5}$$

The scattered wave can be approximated locally by a plane wave where

$$W = \mathrm{Re}\left[\psi\left(t + S_x x + S_y y + S_z z\right)\right] \tag{18.6}$$

and, by expressing S_x as $(1/c_0^2 - S_y^2 - S_z^2)^{\frac{1}{2}}$, this becomes

$$W = \mathrm{Re}\left\{\psi\left[t + \left(c_0^2 - S_y^2 - S_z^2\right)^{\frac{1}{2}} x + S_y y + S_z z\right]\right\} \tag{18.7}$$

When $\mathrm{Re}\left(c_0^{-2} - S_y^2 - S_z^2\right) \ge 0$, i.e., $S_x \ge 0$ the wave is traveling in the $-x$ direction at some unspecified angle; i.e., S_y and S_z are not specified. This wave satisfies, as can be seen by direct substitution, the first order boundary condition at the boundary $x = 0$

$$\left\{\frac{\partial}{\partial x} - c_0^{-1}\left[1 - (c_0 S_y)^2 - (c_0 S_z)^2\right]^{\frac{1}{2}} \frac{\partial}{\partial t}\right\} W\Big|_{x=0} = 0 \tag{18.8}$$

The W satisfying the above equation is therefore a wave traveling in the –x direction and is outgoing and may therefore be characterized as absorbed. Because S_y and S_z are not specified, a solution is not determined for the above first order boundary condition. Approximations of first and second order will be examined following Mur that eliminate this ambiguity and allow W to be found at the boundary x = 0.

The first order approximation is that

$$\left[1 - (c_0 S_y)^2 - (c_0 S_z)^2\right]^{\frac{1}{2}} = 1 + 0\left[(c_0 S_y)^2 - (c_0 S_z)^2\right] \approx 1 \tag{18.9}$$

so that the first order boundary condition is approximated to first order by

$$\left(\frac{\partial}{\partial x} - \frac{1}{c_0}\frac{\partial}{\partial t}\right) W\Big|_{x=0} = 0 \tag{18.10}$$

The second order approximation is

$$\begin{aligned}\left[1 - (c_0 S_y)^2 - (c_0 S_z)^2\right]^{\frac{1}{2}} &\approx 1 - \frac{1}{2}\left[(c_0 S_y)^2 - (c_0 S_z)^2\right] \\ + 0\left[(c_0 S_y)^2 + (c_0 S_z)^2\right]^2 &\approx 1 - \frac{1}{2}\left[(c_0 S_y)^2 - (c_0 S_z)^2\right]\end{aligned} \tag{18.11}$$

so that the first order boundary condition is approximated to second order by

$$\left[\frac{\partial}{\partial x} - \frac{1}{c_0}\left\{1 - \frac{1}{2}\left[(c_0 S_y)^2 + (c_0 S_z)^2\right]\right\}\frac{\partial}{\partial t}\right] W\Big|_{x=0} = 0 \tag{18.12}$$

or by taking the time derivative of the equation

$$\left\{\frac{\partial^2}{\partial x \partial t} - \frac{1}{c_0}\frac{\partial^2}{\partial t^2} + \frac{1}{2} c_0 \left[\frac{\partial}{\partial t}(S_y)^2 + \frac{\partial}{\partial t}(S_z)^2\right]\frac{\partial}{\partial t}\right\} W\Big|_{x=0} = 0 \tag{18.13}$$

which becomes upon multiplication by $1/c_0$ Mur's expression (12)

$$\left(c_0^{-1}\,\partial_{xt}^2 - c_0^{-2}\,\partial_t^2 + \frac{1}{2}\left(\partial_y^2 + \partial_z^2\right)\right)W\Big|_{x=0} = 0 \tag{18.14}$$

The first order approximation for E_z is discretized as

$$\begin{aligned} &E_z^{n+1}(0,j,k+1/2) = E_z^n(0,j,k+1/2) \\ &+\frac{c_0\Delta t - \Delta x}{c_0\Delta t + \Delta x}\left[E_z^{n+1}(1,j,k+1/2) - E_z^n(0,j,k+1/2)\right] \end{aligned} \tag{18.15}$$

The second order approximation is discretized for E_z at the boundary $x = 0$ as

$$\begin{aligned} &E_z^{n+1}(0,j,k+1/2) = -\,E_z^{n-1}(1,j,k+1/2) \\ &+\frac{c_0\Delta t - \Delta x}{c_0\Delta t + \Delta x}\left[E_z^{n+1}(1,j,k+1/2) + E_z^{n-1}(0,j,k+1/2)\right] \\ &+\frac{2\Delta x}{c_0\Delta t + \Delta x}\left[E_z^n(0,j,k+1/2) + E_z^n(1,j,k+1/2)\right] \\ &+\frac{\Delta x(c_0\Delta t)^2}{2(\Delta y)^2(c_0\Delta t + \Delta x)}\cdot\left[E_z^n(0,j+1,k+1/2) - 2E_z^n(0,j,k+1/2)\right. \\ &+E_z^n(0,j-1,k+1/2) + E_z^n(1,j+1,k+1/2) \\ &\left.-\,2E_z^n(1,j,k+1/2) + E_z^n(1,j-1,k+1/2)\right] \\ &+\frac{\Delta x(c_0\Delta t)^2}{2(\Delta z)^2(c_0\Delta t + \Delta x)}\cdot\left[E_z^n(0,j,k+3/2) - 2E_z^n(0,j,k+1/2)\right. \\ &+E_z^n(0,j,k-1/2) - E_z^n(1,j,k+3/2) \\ &\left.-\,2E_z^n(1,j,k+1/2) + E_z^n(1,j,k-1/2)\right] \end{aligned} \tag{18.16}$$

where we have extended Mur's original result to allow for noncubical cells.

Mur examined the efficiency of these two approximations in a 2-D space with a monochromatic point source. He examined how well the circular radiation pattern was maintained (Figures 18-1 and 18-2) for different source locations. Mur concluded that the second order approximation worked best and recommended its use. The disadvantage of the second order Mur is that prior time values of the outer E field components must be stored. Additionally, first

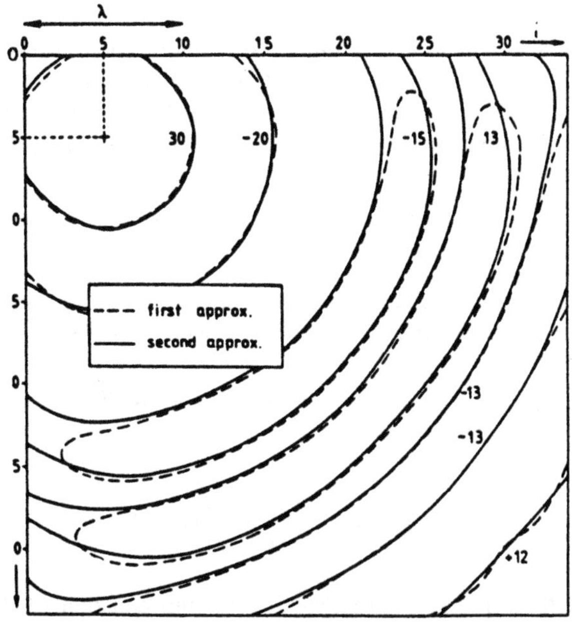

FIGURE 18-1. Contour plot of the radiation pattern, after 141 time steps, of an isotropic source located at node (5,5) of a 35*35-node mesh (arbitrary units).

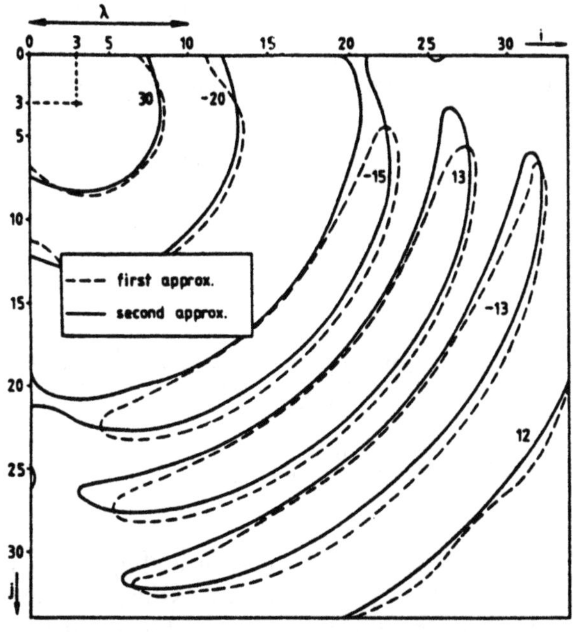

FIGURE 18-2. Contour plot of the radiation pattern, after 141 time steps, of an isotropic source located at node (3,3) of a 35*35-node mesh (arbitrary units).

order Mur has been shown in the interior coupling study of Chapter 4 to have excellent stability, running for 1 million time steps in that investigation.

Another approach to evaluating the efficiency of an ORBC is to examine the reflection as a function of angle[9] in a 2-D formulation. It starts by treating a single frequency component of a single field component in a 2-D space

$$u_{lm}^{n} = e^{j\omega n\Delta t}\left(e^{jk_x l\Delta x} + R_e^{-jk_x l\Delta x}\right)e^{-jk_y m\Delta y} \tag{18.17}$$

i.e., the sum of the incident and reflected scattered field at the outer boundary. Here, k_x and k_y are positive and R is the reflection coefficient that is to be found.

The process continues with the assumption of an ORBC, starting with Mur's first order approximation expressed by

$$u_{0m}^{n+1} = u_{1m}^{n} - \frac{\Delta x - c\Delta t}{\Delta x + c\Delta t}\left(u_{1m}^{n+1}\right) \tag{18.18}$$

Substituting the expression for $u_{lm}{}^{n}$ above into this equation yields

$$R = -e^{jk_x\Delta x}\frac{\Delta x\sin(\omega\Delta t/2)\cos\left(k_x\Delta x/2\right) - c\Delta t\cos(\omega\Delta t/2)\sin\left(k_x\Delta x/2\right)}{\Delta x\sin(\omega\Delta t/2)\cos\left(k_x\Delta x/2\right) + c\Delta t\cos(\omega\Delta t/2)\sin\left(k_x\Delta x/2\right)} \tag{18.19}$$

Following the same procedure for Mur's second order approximation yields

$$R = -e^{jk_x\Delta x}\left(\frac{a-b}{a+b}\right) \tag{18.20}$$

where

$$a = \Delta x\left[\sin^2\frac{\omega\Delta t}{2} - \frac{1}{2}\left(\frac{c\Delta t}{\Delta y}\right)^2\sin^2\left(\frac{k_y\Delta t}{2}\right)\right]\cos\frac{k_x\Delta x}{2}$$

$$b = c\Delta t\cos\frac{\omega\Delta t}{2}\sin\frac{\omega\Delta t}{2}\sin\frac{k_x\Delta x}{2}$$

The reflection coefficient can be evaluated (Figure 18-3) for typical conditions: $\Delta x = \Delta y$, $c\Delta t = \Delta x/2$ (Δt slightly less than the Courant stability condition) and $\omega/c = 2\pi/(10\Delta x)$. Because of discretization the reflection coefficient is not exactly zero at normal incidence ($\theta = 0$). By varying the

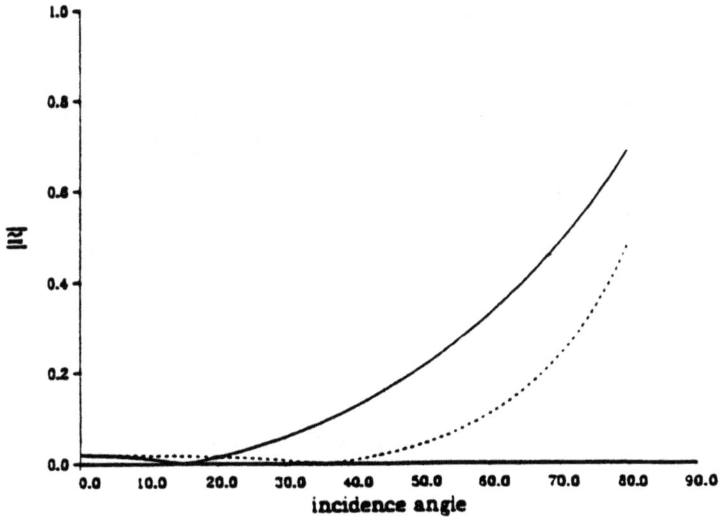

FIGURE 18-3. Magnitude of the radiation coefficient vs. incident angle θ for various discretized RBCs. Solid line is the first order of impedance RBC and the dashed line is the second order RBC (commonly called the Mur condition). Numerical parameters are $\Delta z = \Delta y$, $c\Delta t = \Delta x/2$, and $\omega/c = 2\pi/(10\Delta x)$.

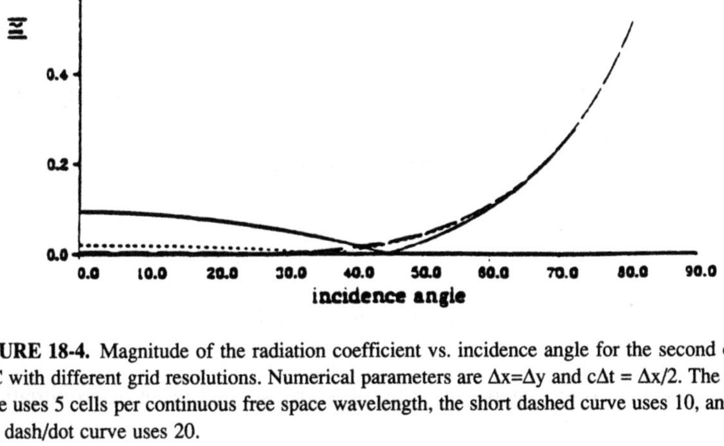

FIGURE 18-4. Magnitude of the radiation coefficient vs. incidence angle for the second order RBC with different grid resolutions. Numerical parameters are $\Delta x = \Delta y$ and $c\Delta t = \Delta x/2$. The solid curve uses 5 cells per continuous free space wavelength, the short dashed curve uses 10, and the long dash/dot curve uses 20.

number of cells per wavelength (Figure 18-4) the dependence on this parameter can also be examined.

18.4 ALTERNATE FORMULATIONS OF THE ONE-WAY WAVE EQUATIONS

Mur's approach was to "motivate" a first order boundary condition or one-way wave equation. The key to his approximation is how he expands the term $\sqrt{1-(c_0 S_y)^2-(c_0 S_z)^2}$ in what is in effect a Taylor series expansion. The presence of the square root term in the first order boundary condition classifies the operator

$$L^{-} \equiv \frac{\partial}{\partial x} - c_0^{-1}\left[1-\left(c_0 S_y\right)^2-\left(c_0 S_z\right)^2\right]^{1/2} \frac{\partial}{\partial t} \tag{18.21}$$

in the equation

$$L^{-}W\big|_{x=0} = 0 \tag{18.22}$$

as a pseudo-differential operator in that it is not local in time and space.

Alternate formulations using expansions other than the Taylor series expansion of Mur have been investigated[5] in 2-D problem spaces. The one-way wave equation is expressed as

$$L^{-} \equiv \partial_x - \frac{\partial_t}{c}\sqrt{1-S^2} \tag{18.23}$$

$$L^{-1}W\big|_{x=0} = 0 \tag{18.24}$$

The expansion of the term $\sqrt{1-S^2}$ can take on many forms and with each form the one-way wave equation is expressed differently. This behavior has already been seen with the first and second order approximations used by Mur (18.9 and 18.10, 18.11 and 18.14).

In general one can write

$$\sqrt{1-S^2} = p_m(s)/q_n(s) \tag{18.25}$$

One approximation, a generalization of Mur's second order approximation is

$$\sqrt{1-S^2} \,.\, p_0 + p_2 S^2 \tag{18.26}$$

TABLE 18-1
Coefficients for Third Order Conditions

Type of approximation	p_0	p_2	q_2	Angles of exact absorption (degrees)
Padé	1.00000	–0.75000	–0.25000	0.00
L^{∞}_{α} ($\alpha = 45°$)	0.99973	–0.80864	–0.31657	11.7, 31.9, 43.5
Chebyshev points	0.99650	–0.91296	–0.47258	15.0, 45.0, 75.0
L^2	0.99250	–0.92233	–0.51084	18.4, 51.3, 76.6
C-P	0.99030	–0.94314	–0.55556	18.4, 53.1, 81.2
Newman points	1.00000	–1.00000	–0.66976	0.0, 60.5, 90.0
L^-	0.95651	-0.94354	-0.07385	26.9, 66.6, 87.0

Note: q_0 = 1.00000 for each approximant.

which results in a second order one-way wave equation

$$W_{xt} - \frac{p_0}{c} W_{tt} - cp_2 W_{yy} = 0 \quad \text{at} \quad x = 0 \tag{18.27}$$

Alternatively the approximation

$$\sqrt{1 - S^2} \cong \frac{p_0 + p_2 S^2}{q_0 + q_2 S^2} \tag{18.28}$$

yields the third order one-way wave equation or absorbing boundary condition

$$q_0 W_{xtt} + c^2 q_2 W_{yy} - \frac{p_0}{c} W_{tt} - cp_2 W_{tyy} = 0 \quad \text{at} \quad x = 0 \tag{18.29}$$

These approximations can be categorized into types (2,0) and (2,2) based on the order of the polynomials $p_m(S)$ and $q_n(S)$ used in the approximation. Studies[10] have been made of approximations of type (0,0), (4,2), and (4,4) as well as (2,0) and (2,2). Formally, Mur's second order approximation is a Padé (2,0) type approximation with $p_0 = 1$ and $p_2 = 1/2$. Choosing a Padé (2,2) type with $p_0 = q_0 = 1$, $p_2 = -3/4$, $q_2 = -1/4$ results in the third order absorbing boundary condition of Engquist and Majda.[8]

For type (0,0) where $\sqrt{1 - S^2} \simeq 1$ we have Mur's first order boundary condition and the wave is exactly absorbed at normal incidence. The other two types we are examining here also have certain angles where absorption is complete. This can be seen Reference 5 in Tables 18.1 and 18.2 for third order boundary conditions, type (2,2), and second order boundary conditions, type (2,0).

It is evident that it is possible to obtain perfect results at more than one angle. However, when the overall absorption is evaluated for all angles the

TABLE 18-2
Coefficients for Second Order Conditions

Type of approximation	p_o	p_2	Angles of exact absorption (degrees)
Padé	1.00000	–0.5000	0.00
L^{∞}_{x} ($\alpha = 20°$)	1.00023	–0.51555	7.6, 18.7
Chebyshev points	1.03597	–0.76537	22.5, 67.5
L^2	1.03084	–0.73631	22.1, 64.4
C-P	1.06103	–0.84883	25.8, 73.9
Newman points	1.00000	–1.00000	0.0, 90.0
L^{∞}	1.12500	–1.00000	31.4, 81.6

more complex approximations do not best second order Mur. This accounts for the continuing popularity of the Mur absorbing radiation boundary conditions. The search still goes on nonetheless for a significantly improved ORBC.

REFERENCES

1. **Taflove, A. and Brodwin, M.,** Numerical solution of steady state electromagnetic scattering problems using the time-dependent Maxwell's equation, *IEEE Trans. MTT,* 23, 623, 1975.
2. **Gilbert, J.,** Studies in Outer Boundary Conditions for Finite Difference SCEMP Codes, Tech. Rep. MRC-R-296, Mission Research Corporation, November 1976.
3. **Merewether, D. E.,** Transient currents induced on a metallic body of revolution by an electromagnetic pulse, *IEEE Trans. Electromag. Compat.,* 13, 41, 1971.
4. **Holland, R.,** THREDE: a free field EMP coupling and scattering code, *IEEE Trans. Nucl. Sci.,* 24, 2416, 1977.
5. **Blaschak J. and Kriegsmann, G.,** A comparative study of absorbing boundary conditions, *J. Comput. Phys.,* 77, 1988.
6. **Potter, D.,** *Computational Physics,* John Wiley & Sons, New York, 1973.
7. **Mur, G.,** Absorbing boundary conditions for the finite-difference approximation of the time domain electromagnetic field equations, *IEEE Trans. EMC,* 23(4), 377, 1981.
8. **Engquist B. and Majda, A.,** Absorbing boundary conditions for the numerical simulation of waves, *Math. Comput.,* 629, 1977.
9. **Ray, S.,** Characterization of Radiation Boundary Conditions Used in the Finite-Difference Time-Domain Method, *IEEE Antennas and Propagation Society International Symp.,* San Jose, CA, June 26–30, 1989.
10. **Halpern, L. and Trefethen, L.,** Numerical Analysis Report 86-5, Department of Mathematics, Massachusetts Institute of Technology, Cambridge, 1986.

Chapter 19

ALTERNATE FORMULATIONS

19.1 INTRODUCTION

FDTD as developed here solves the Maxwell curl equations for the scattered E and H fields using a broadband analytically described transient incident E and H waveform. The numerical solutions are based on the leapfrog technique. There are many alternate formulations of FDTD to this approach. We feel none are superior in general and most, if not all, have some weakness. It is nonetheless important to survey these alternate formulations so as to provide an appreciation of the possibilities and perhaps to discover a particularly apt approach to a particular problem.

The alternate formulations include

- Total as opposed to scattered field formulation
  - Transient formulation
  - CW formulation
- Potential formulation $(\vec{A}, \phi)$ as opposed to field formulation (E, H)
- Implicit methods
- High frequency imaging approximations
- Acoustic analog/scalar equivalent

19.2 TOTAL FIELD FORMULATION

The total field formulation is strongly associated with Taflove and Umashankar[1] and has a history as long and varied as the scattered field formalism presented in this text. The total field formalism does not separate the field into incident and scattered components, but treats only the total field when interacting with a scattering or coupling geometry. This simplifies the interaction process in that the governing equation now looks just like the Maxwell equations for free space (Chapter 2), except $\varepsilon_0 \rightarrow \varepsilon$ and $\mu_0 \rightarrow \mu$ for the fields in the interaction object where conductivity σ also now may be different than zero.

Along with the simplicity of this formulation is the reputed advantage of better interior coupling predictions;[2] i.e., no noise as is present in the scattered field predictions. This is a somewhat moot point in that when incident and scattered fields are sufficiently out of step due to grid dispersion in the scattered field components, the scattered field components are at high frequencies approaching the Nyquist frequency limit, where the predictions are of doubtful utility in any case. Further, the total field itself must transverse the grid and is subject to dispersion. In a problem with multiple scattering centers the timing of the scattering from each center will be affected, causing subtle phase shift distortions.

The scattered field must be "resurrected" as the total field leaves the region of the scattered or interaction object so as to apply an outer radiation boundary condition (ORBC). For example, the Mur ORBC assumes an outwardly directed wave, satisfying the one-way wave equation. This requires the incident field, which is the same on the total field entering the problem space, to be subtracted from the total field at a subboundary placed just within the problem space. The implementation of this "stripping" process and the "launching" of the total field at some extremity of the problem space is not difficult, but does imply a conceptual overhead of its own. It may be argued that one approach or the other is superior, but overall they are virtually equal.

We have used primarily the separate field formalism in the text, while on occasion generating a limited version of the total field formulation where the incident field is zeroed and the total field appears in a small region such as in the gap of a dipole antenna. This appeals to us as the best mixture of the two techniques, but either can be employed for virtually any problem with success.

It should be mentioned that the earliest version of the total field FDTD formulation employed a continuous wave (CW) or sinusoidal excitation, i.e., a sinusoidal wave was introduced into the problem space from one side. To this was added an absorbing boundary condition in the form of an outer conducting layer in the problem space. This formulation has been superseded by its modern version, as well it should. The latest absorbing boundary conditions such as Mur are superior and fast Fourier transforms (FFTs) return data from a transient at many frequencies, a much more efficient approach than creating an FDTD for a single frequency.

19.3 POTENTIAL FORMULATION

Rather than begin with Maxwell equations an equivalent set of starting equations are the coupled potential equations expressed in c.q.s. units

$$\nabla^2\phi + \frac{1}{c}\frac{\partial}{\partial t}(\nabla \cdot A) = -4\pi\rho \tag{19.1}$$

$$\nabla^2 A - \frac{1}{c}\frac{\partial^2}{\partial t^2}A - \nabla\left(\nabla \cdot A + \frac{1}{c}\frac{\partial\phi}{\partial t}\right) = -4\pi\frac{J}{c} \tag{19.2}$$

where

$$E = -\left(\frac{1}{c}\frac{\partial A}{\partial t} + \nabla\phi\right) \tag{19.3}$$

and

$$B = \nabla \times A \tag{19.4}$$

Note that

$$J = \sigma E \tag{19.5}$$

or

$$J = -\sigma\left(\frac{1}{c}\frac{\partial A}{\partial t} + \nabla\phi\right) \tag{19.6}$$

and that

$$\nabla \cdot J + \frac{\partial \rho}{\partial t} = 0 \tag{19.7}$$

By taking the time derivative of 19.1, one obtains

$$\nabla^2\phi + \frac{1}{c}\frac{\partial^2}{\partial t^2}(\nabla \cdot A) = -4\pi\dot{\rho} = 4\pi\nabla \cdot J = -4\pi\ \nabla \cdot \sigma\left(\frac{1}{c}\frac{\partial A}{\partial t} + \nabla\phi\right) \tag{19.8}$$

while taking the curl of equation 19.2 results in

$$\begin{aligned} &\nabla^2(\nabla \times A) - \frac{1}{c^2}\frac{\partial^2(\nabla \times A)}{\partial t^2} \\ &= -\frac{4\pi}{c}\nabla \times J = \frac{4\pi}{c}\nabla \times \sigma\left(\frac{1}{c}\frac{\partial A}{\partial t} + \nabla\phi\right) \end{aligned} \tag{19.9}$$

The two equations (19.8) and (19.9) may be written as

$$\nabla^2(\nabla \cdot A) - \frac{1}{c^2}\frac{\partial^2}{\partial t^2}(\nabla \cdot A) = \frac{4\pi\sigma}{c^2}\frac{\partial}{\partial t}(\nabla \cdot A) \tag{19.10}$$

$$\nabla^2(\nabla \times A) - \frac{1}{c^2}\partial^2\frac{(\nabla \times A)}{\partial t^2} = \frac{4\pi\sigma}{c^2}\frac{\partial}{\partial t}(\nabla \times A) \tag{19.11}$$

when σ is spatially constant and

$$\nabla^2\frac{\Phi}{c} + \frac{4\pi\sigma}{c}\nabla^2\phi = \nabla^2(\nabla \cdot A) \tag{19.12}$$

or

$$\frac{\Phi}{c} + \frac{4\pi\sigma}{c}\Phi = -\nabla \cdot A \tag{19.13}$$

Equations (19.10) and (19.11) are equivalent to

$$\nabla^2 A - \frac{1}{c^2}\frac{\partial^2}{\partial t^2}A = \frac{4\pi\sigma}{c}\frac{\partial A}{\partial t} \tag{19.14}$$

while 19.13 is a gauge condition equivalent to the Lorentz gauge in the limit $\sigma \to 0$. Equation (19.14) can be solved by standard finite difference techniques, using a separate incident and scattered formalism, i.e., $A = A^{inc} + A^{scat}$.

This vector potential formulation of the lossy dielectric problem allows the vector potential $\vec{A}$ to be time stepped in place of the six field components E_x, E_y, E_z, H_x, H_y, H_z. In the formulation, ϕ is found from $\vec{A}$, and E and H found from $\vec{A}$ and ϕ in separate postprocessing calculations. Because each component of $\vec{A}$ can be solved separately, only one component at a time is treated for a sixfold reduction in the number of components that must be treated. Once A is known, ϕ can be found from the gauge condition, and from A and ϕ the field quantities can be calculated, using (19.3) and (19.4).

The sixfold reduction in the storage requirements is partially offset by the need for two prior time values of A in storage at each cell location, so that $\partial^2 A/\partial t^2$ can be calculated, using the standard difference formula

$$\frac{\partial^2 A}{\partial t^2} \doteq \left[\left(\frac{A^{n+1} - A^n}{\Delta t}\right) - \left(\frac{A^n - A^{n-1}}{\Delta t}\right)\right] / \Delta t$$
$$= \left(A^{n+1} - 2A^n + A^{n-1}\right)/(\Delta t)^2$$

In contrast, the field component formulation requires only a first derivative in time. Therefore, only a single value of the field component a time step earlier must be stored to evaluate

$$\frac{\partial F}{\partial t} \doteq \frac{F^{n+1} - F^n}{\Delta t}$$

where $F = E_i$ or H_i ($i = x,y,z$). The net result is a threefold decrease in storage requirements.

While apparently feasible and offering the advantage of reduced storage requirements this somewhat different approach has never been employed. A similar alternative would be to employ the inhomogeneous wave equations for E or H.

19.4 IMPLICIT SCHEMES

When the term $\partial D/\partial t$ in the Maxwell equations is small and can be ignored, the resulting equations with $\partial D/\partial t = 0$ are parabolic and may be solved in a number of new ways. Brief mention was made of the Frankel-Dufort method in Chapter 16, a method that can be implemented in either explicit or implicit form. In either instance larger time steps than that allowed by the Courant stability condition are possible, albeit with some potential problems with oscillations in the predicted response about the true response. While we have not encountered an implicit scheme that has proven useful for the complete Maxwell equations ($\partial D/\partial t \neq 0$) in hyperbolic form, it is appropriate to discuss other implicit schemes that may be applicable to the Maxwell equations in parabolic form ($\partial D/\partial t = 0$).

Chief among these is the Crank-Nicholson scheme widely used in heat flow analysis, among other applications. We briefly summarize this technique following Potter's[3] treatment. The Crank-Nicholson implicit scheme for parabolic equations is

$$u_j^{n+1} = u_j^n + \frac{\kappa \Delta t}{2\Delta^2}\left(u_{j+1}^{n+1} - 2u_j^{n+1} + u_{j-1}^{n+1}\right)$$

$$+ \frac{\kappa \Delta t}{2\Delta^2}\left(u_{j+1}^{n} - 2u_j^{n} + u_{j-1}^{n}\right)$$

for $\quad \dfrac{\partial u}{\partial t} + \dfrac{\kappa \partial^2 u}{\partial x^2} = 0 \quad$ which is of the form

$$\frac{\partial H}{\partial t} - (\sigma\mu)\frac{\partial^2 H}{\partial x^2} = 0 \quad \text{for the H field when } \partial D/\partial t = 0$$

This equation is always stable.

19.5 HIGH FREQUENCY APPROXIMATION

The Crank-Nicholson and the Frankel-Dufort schemes, as applied to electromagnetic problems where $\partial D/\partial t = 0$, are in effect low frequency approximations, which suggests that there may also be high frequency schemes that are effective for extending FDTD modeling capabilities to higher frequencies. We will examine here a newly developed high frequency approach as an example.

19.5.1 TIME DOMAIN SMYTHE-KIRCHHOFF APERTURE APPROXIMATION AND METHOD OF IMAGES FOR HIGH FREQUENCY FDTD MODELING

The Smythe-Kirchhoff approximation for aperture coupling[4] can be used above aperture cut off to obtain a time domain representation of the field traversing the aperture with little effort, assuming a Gaussian time domain behavior for the incident field. The $J_1(x)/x$ part of the Smythe-Kirchhoff expression

$$E(k) = -jE_0 \frac{e^{jkr}}{r}(\cos\alpha)(kx\hat{u})\frac{J_1(x)}{x}$$

$$\begin{aligned} x &= ka\xi \\ a &= \text{aperture radius} \\ \xi &= \left(\sin^2\theta + \sin^2\alpha - 2\sin\theta\sin\alpha\cos\phi\right)^{1/2} \end{aligned} \tag{19.15}$$

can be expanded in $ka\xi$. Recognizing that $\omega = kc$ and that ω represents a derivative in the time domain, the transmitted time domain field can be written out by inspection as

$$E(t) \cong \frac{a}{r}(\cos\alpha)\left(\frac{a}{c}\right)\left(\bar{k}x\hat{u}\right)$$
$$\frac{1}{2}\left[\frac{d}{dt} + \frac{\sigma^2}{8}\frac{d^3}{dt^3} + \frac{\sigma^3}{192}\frac{d^5}{dt^5} + \frac{\sigma^6}{9216}\frac{d^7}{dt^7}\right]Ae^{-(\sigma t)^2} \tag{19.16}$$

and the derivatives are easily evaluated to yield the time domain field.

This field appears to emanate from a point centered in the aperture and can be imaged many times. Because the imaging is periodic in a rectangular cavity it has proved feasible to easily generate upwards of 1 million images with a few lines of code that also account for polarity effects of the multiple reflections producing the components of each image. A field within the cavity can be produced in under 3 min on a 6 MFLOP machine that accounts for 500,000 images when it is noted that many images arrive nearly coincidentally. In this case 8192 time bins were used so that on an average ~60 images coalesce into a single bin. Producing on the order of 100 field points to illuminate an object within the cavity is a matter of 5 h computer time.

The resulting field can model a field on the order of ten times higher in its frequency limit as compared to what can be achieved with an FDTD model of the entire cavity. It is therefore possible for the aperture rectangular cavity geometry to model response ten times higher in frequency than a brute force FDTD model. Recalling that a tenfold increase in frequency would require ten times finer cells and ten times smaller time steps, or 10,000 times the resources, this is a significant savings and clearly demonstrates the utility of high frequency approximations.

19.6 ACOUSTIC ANALOG/SCALAR EQUIVALENT

This treatment follows Reference 5. It assumes $P = P_0 + p$, where P_0 is the undisturbed pressure and p is the pressure change for the sound wave, and $V = V_0 + \tau$, where V_0 is the undisturbed volume and τ is the volume change. The treatment uses three parts in its development: (1) Newton's second law of motion (f = ma), (2) gas law, and (3) conservation of mass. This treatment leads to a wave equation, but more importantly, two linear coupled first order differential equations

$$\frac{\partial \bar{q}}{\partial t} = \frac{-1}{\rho_o} \nabla p \qquad \rho_o \text{ average density} \tag{19.17}$$

$$\frac{\partial p}{\partial t} = \gamma P_0 \vec{\nabla} \cdot \bar{q} \tag{19.18}$$

These equations are used instead of the Maxwell curl equations for the acoustic analog.

The derivation is approximate and for standard temperature and pressure (STP) air is good to about 110 dB 0.0002 μbar. It assumes Lagrangian cells but simplifies to Eulerian cells by requiring the vector particle velocity $\bar{q}$ to be small enough that the rate of change of momentum of the particles in a cell (moving) can be approximated by the rate of change of momentum at a fixed point $D\bar{q}/Dt \doteq \partial\bar{q}/\partial t$. It also assumes $\rho = \rho_0 + \Delta\rho \approx \rho_0$.

19.6.1 EQUATION OF MOTION

Define the sound pressure change in space by

$$\nabla p = \hat{i}\frac{\partial p}{\partial x} + \hat{j}\frac{\partial p}{\partial y} + \hat{k}\frac{\partial p}{\partial z} \tag{19.19}$$

Forces on a box of particles (accelerating the box in the positive direction) are just

$$F = -\left[\hat{i}\left(\frac{\partial p}{\partial x}\Delta x\right)\Delta y \Delta z + \hat{j}\left(\frac{\partial p}{\partial y}\Delta y\right)\Delta x \Delta z + \hat{k}\left(\frac{\partial p}{\partial z}\Delta z\right)\Delta x \Delta y\right] \tag{19.20}$$

so that

$$\frac{F}{V} = \frac{F}{\Delta x \Delta y \Delta z} = -\nabla p \tag{19.21}$$

Now note that

$$\frac{F}{V} = \frac{1}{V}\left(M\frac{D\vec{q}}{Dt}\right)$$

= time rate of change of the momentum of the cell (acceleration)

Thus,

$$\frac{F}{V} = -\nabla p = \frac{M}{V}\frac{D\vec{q}}{Dt} = \rho'\frac{Dq}{Dt} \tag{19.22}$$

or approximately

$$-\nabla p = \rho_o \frac{\partial \vec{q}}{\partial t} \tag{19.23}$$

This is the first coupled partial differential equation.

19.6.2 GAS LAW

For acoustic frequencies (≤20 KHz) the gas is adiabatic (thermal velocity at 10 KHz is ~1.5 m/s) and PV^γ = constant, where $\gamma = C_p/C_v$, where C is specific heat, C_p is constant pressure, and C_v is constant volume. In differential form

$$\begin{aligned} &\frac{dP}{P} = -\gamma\frac{dV}{V} \text{ and with } P = P_o + p, \\ &V = V_o + \tau,\ P_o >> p,\ V_o >> \tau \end{aligned} \tag{19.24}$$

we obtain

$$\frac{P}{P_o} = \frac{-\gamma\tau}{V_o}$$

The time derivative of this equation gives what will become the second coupled partial differential equation

$$\frac{1}{P_o}\frac{\partial p}{\partial t} = \frac{-\gamma}{V_o}\frac{\partial \tau}{\partial t} \tag{19.25}$$

19.6.3 CONTINUITY EQUATION

To transform 19.25 into the desired form requires the continuity equation

$$\frac{\partial \tau}{\partial t} = V_o \vec{\nabla} \cdot \vec{q}$$

for a Lagrangian cell.

Setting (19.25) to equal (19.26) gives

$$\frac{\partial p}{\partial t} = \gamma P_o \vec{\nabla} \cdot \vec{q} \tag{19.26}$$

the second desired coupled differential equation.

Letting $\vec{E}$ be replaced by $\vec{v}$, i.e., v_x in place of E_x, etc., and letting p, a scalar, occupy the vertex of an "acoustic" Yee cell completes the development of an acoustic FDTD equivalent.

REFERENCES

1. **Taflove, A. and Umashankar, K. R. U.,** Finite-difference time-domain (FD-TD) modeling of electromagnetic wave scattering and interaction problems, *IEEE Ant. Prop. Newsl.,* April 1988, 5.
2. **Holland, R. and Williams, J. W.,** Total-field vs scattered-field finite-difference codes: a comparative assessment, *IEEE Trans. Nucl. Sci.,* 6, 4583, 1983.
3. **Potter, D.,** *Computational Physics,* John Wiley & Sons, New York, 1973.
4. **Jackson, J. D.,** *Classical Electrodynamics,* 2nd ed., John Wiley & Sons, New York, 1975.
5. **Beranek, L.,** *Acoustics,* McGraw-Hill, New York, 1954.

Appendix A

OTHER COORDINATE SYSTEMS AND REDUCED DIMENSIONS

INTRODUCTION

While a fully 3-D Cartesian coordinate FDTD computer code is generally the most useful, there are situations in which a 2-D or even 1-D FDTD code may also be useful. In 2-D problems there is no variation in one of the coordinate directions, usually taken as the z direction in Cartesian coordinates or the ϕ direction in cylindrical or spherical coordinates. If there is no z variation, the scattering structure is infinitely long in the z direction with an arbitrary cross section in the orthogonal directions. This geometry may be modeled in 2-D in either Cartesian or cylindrical coordinates. The Cartesian system is preferred because the spatial differences are uniform over the FDTD space, while in the cylindrical system one side of the cell has $\ell_\phi = r\Delta\phi$ which varies with r. While no actual object is infinitely long, these 2-D results may be quite useful in approximating scattering by a finite length object. For example, scattering by a rectangular plate illuminated by a plane wave normal to an edge can be estimated by considering scattering by an infinitely long strip of the same width. For a discussion of this, with formulas for converting 2-D scattering results to 3-D objects, see Reference 1.

The other quite useful 2-D geometry is one that has no variation in the ϕ variable in cylindrical coordinates. This 2-D geometry can be used to model a wide range of actual geometries with rotational symmetry about the z axis. It is most useful in calculating radiation patterns, impedance, and other parameters for antenna radiation problems with rotational symmetry.[2] Such antennas include coaxially fed monopoles, conical and biconical antennas, lens antennas, and even reflector antennas. These can be directly analyzed with a 2-D cylindrical coordinate FDTD code since not only the geometry but also the radiated fields will have no ϕ dependence.

While body of revolution (BOR) scattering problems have ϕ-independent scattering geometries, the scattered fields are not ϕ-independent, and in general, 3-D cylindrical or spherical coordinate FDTD codes are necessary. However, in special cases coordinate transformations may be used to reduce the dimensionality to 2-D in BOR scattering problems. For example, for a BOR scatterer in cylindrical coordinates, with plane wave incidence along the z axis, by symmetry the fields must vary as $\cos(\phi)$ or $\sin(\phi)$. With this known ϕ variation the Maxwell curl equations can be reduced to 2-D and from this an FDTD code can be developed.[3] However, this approach is not easily generalized to other incidence angles as the ϕ variation of the fields will be more complicated.

An important point to remember is that Cartesian 2-D geometries and ϕ-independent cylindrical 2-D geometries are fundamentally different in an important way. In a 2-D Cartesian geometry the scattered fields are cylindrical waves and vary as $1/\sqrt{r}$ while in a cylindrical coordinate calculation with ϕ independence the radiated (scattered) fields are spherical waves and vary as 1/r, where r is the radial distance from the scatterer or radiator. Thus, in a 2-D BOR antenna calculation a 3-D object with no ϕ variation is located in a 3-D space, while in a 2-D Cartesian calculation a 2-D object is located in a 2-D space. This distinction is important. For example, consider transforming the near zone FDTD fields to the far zone. The far zone fields for a 2-D cylindrical coordinate system BOR radiator can be determined by applying a 3-D transformation with no ϕ variation (Section 7.2). On the other hand, the far zone fields for a 2-D Cartesian calculation require a different far zone transformation derived specifically for 2-D with $1/\sqrt{r}$ field variation (Section 7.3). (A cylindrical coordinate calculation with no z variation is 2-D in the same sense as a Cartesian coordinate calculation with no z variation.)

While the above considerations should be kept in mind, we realize that the FDTD technique can be applied to any orthogonal coordinate system. It is a straightforward exercise to generate linearized FDTD equations for cylindrical and spherical geometries as well as for other orthogonal coordinate systems, ellipsoidal, for example. Our discussion is limited to cylindrical and spherical geometries. One can also consider nonorthogonal coordinate systems with corresponding increases in complexity.

In this appendix we are primarily interested in 2-D FDTD calculations in Cartesian and cylindrical coordinates, because these are the most useful in practical problems. We briefly consider the simplest case, which is 1-D and is exemplified by the telegraphy equations. Even 1-D calculations can be useful in testing new FDTD methods, for example, extensions of FDTD capabilities to special materials.

Even though the applications of 3-D FDTD in spherical and cylindrical coordinates are somewhat limited, we will first derive the FDTD equations for these cases for generality, then specialize them to reduced dimensions. We also will present reduced dimension FDTD equations for Cartesian coordinates in this appendix.

3-D CYLINDRICAL AND SPHERICAL

We are in general concerned with three orthogonal coordinate systems: rectangular, cylindrical, and spherical, and treat them first in 3-D. 3-D FDTD in rectangular coordinates was presented in Chapter 2. The finite difference form of the Maxwell equations for the other two coordinate systems is presented here.

We start in either case with the two curl equations for the scattered fields

$$\nabla \times H^{scat} = \varepsilon \frac{\partial E^{scat}}{\partial t} + \sigma E^{scat} + \left(\varepsilon - \varepsilon_0\right)\frac{\partial E^{inc}}{\partial t} + \sigma E^{inc} \qquad \text{(A.1)}$$

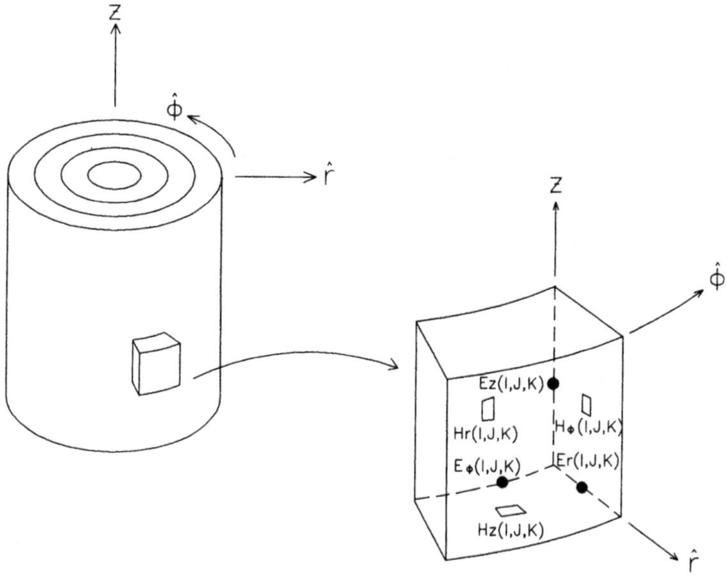

Figure A-1. Cylindrical coordinates.

$$\nabla \times E^{scat} = -\mu \frac{\partial H^{scat}}{\partial t} - \left(\mu - \mu_0\right) \frac{\partial H^{inc}}{\partial t} \tag{A.2}$$

The key is to express the ∇x operation for the particular coordinate system. Thus, we have

$$\nabla \times A = \hat{r}\left(\frac{1}{r}\frac{\partial A_z}{\partial \phi} - \frac{\partial A_\phi}{\partial z}\right) + \hat{\phi}\left(\frac{\partial A_r}{\partial z} - \frac{\partial A_z}{\partial r}\right) + \hat{z}\left(\frac{1}{r}\frac{\partial}{\partial r}\left(rA_\phi\right) - \frac{1}{r}\frac{\partial A_r}{\partial \phi}\right) \tag{A.3}$$

for cylindrical coordinates (Figure A-1), where $\hat{r}, \hat{\phi}$ and $\hat{z}$ are the unit vectors in the r, ϕ, z directions and

$$\begin{aligned} \nabla \times A = \hat{r}\frac{1}{r\sin\theta}&\left[\frac{\partial}{\partial \theta}\left[\sin\theta A_\phi\right] - \frac{\partial A_\phi}{\partial \phi}\right] \\ &+ \hat{\theta}\frac{1}{r}\left[\frac{1}{\sin\theta}\frac{\partial A_r}{\partial \phi} - \frac{\partial}{\partial r}\left[rA_\phi\right]\right] \\ &+ \hat{\phi}\frac{1}{r}\left[\frac{\partial}{\partial r}\left[rA_\theta\right]\frac{\partial A_r}{\partial \theta}\right] \end{aligned} \tag{A.4}$$

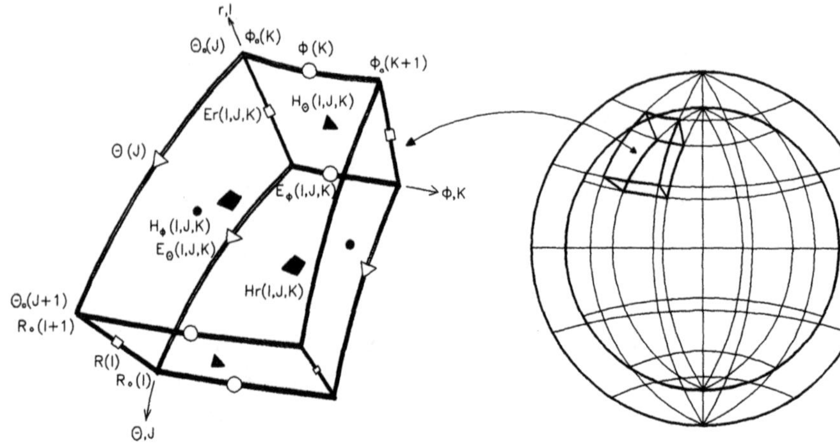

Figure A-2. Spherical coordinates.

for spherical coordinates (Figure A-2), where $\hat{r}, \hat{\phi}$ and $\hat{\theta}$ are the unit vectors in the r, ϕ, and θ directions.

For cylindrical coordinates the Maxwell curl equations in component form are

$$\frac{1}{r}\frac{\partial H_z^{scat}}{\partial \phi} - \frac{\partial H_\phi^{scat}}{\partial z} = \varepsilon\frac{\partial E_r^{scat}}{\partial t} + \sigma E_r^{scat} + \left(\varepsilon - \varepsilon_0\right)\frac{\partial E_r^{inc}}{\partial t} + \sigma E_r^{scat} \quad \text{(A.5a)}$$

$$\frac{\partial H_r^{scat}}{\partial z} - \frac{\partial H_z^{scat}}{\partial r} = \varepsilon\frac{\partial E_\phi^{scat}}{\partial t} + \sigma E_\phi^{scat} + \left(\varepsilon - \varepsilon_0\right)\frac{\partial E^{inc}}{\partial t}\phi + \sigma E_\phi^{scat}$$

$$\frac{1}{r}\frac{\partial}{\partial r}\left(rH_\phi^{scat}\right) - \frac{1}{r}\frac{\partial H_r^{scat}}{\partial \phi} = \varepsilon\frac{\partial H_z^{scat}}{\partial t} \quad \text{(A.5b)}$$

$$+ \sigma E_z^{scat} + \left(\varepsilon - \varepsilon_0\right)\frac{\partial E_z^{inc}}{\partial t} + \sigma E_z^{inc} \quad \text{(A.5c)}$$

$$\frac{1}{r}\frac{\partial E_z^{scat}}{\partial \phi} - \frac{\partial E_\phi^{scat}}{\partial z} = -\mu\frac{\partial H_r^{scat}}{\partial z} - \left(\mu - \mu_0\right)\frac{\partial H_r^{inc}}{\partial t} \quad \text{(A.5d)}$$

$$\frac{\partial E_r^{scat}}{\partial z} - \frac{\partial E_z^{scat}}{\partial r} = -\mu\frac{\partial H_\phi^{scat}}{\partial t} - \left(\mu - \mu_0\right)\frac{\partial H_\phi^{inc}}{\partial t} \quad \text{(A.5e)}$$

$$\frac{1}{r}\frac{\partial}{\partial r}\left(rE_{\phi}^{scat}\right)-\frac{1}{r}\frac{\partial E_{r}^{scat}}{\partial \phi}=-\mu\frac{\partial H_{z}^{scat}}{\partial t}-\left(\mu-\mu_{0}\right)\frac{\partial H_{z}^{inc}}{\partial t} \tag{A.5f}$$

In linearized form, where the derivatives are approximated as differences denoted by Δ, these equations can be recast as

$$\varepsilon\frac{\Delta E_{r}^{scat}}{\Delta t}+\sigma E_{r}^{scat}=-\sigma E_{r}^{inc}-\left(\varepsilon-\varepsilon_{0}\right)\frac{\Delta E_{r}^{inc}}{\Delta t}+\frac{1}{r}\frac{\Delta H_{z}^{scat}}{\Delta \phi}-\frac{\Delta H_{\phi}^{scat}}{\Delta z} \tag{A.6a}$$

$$\varepsilon\frac{\Delta E_{\phi}^{scat}}{\Delta t}+\sigma E_{\phi}^{scat}=-\sigma E_{\phi}^{inc}-\left(\varepsilon-\varepsilon_{0}\right)\frac{\Delta E_{\phi}^{inc}}{\Delta t}+\frac{\Delta H_{r}^{scat}}{\Delta z}-\frac{\Delta H_{z}^{scat}}{\Delta r} \tag{A.6b}$$

$$\varepsilon\frac{\Delta E_{z}^{scat}}{\Delta t}+\sigma E_{z}^{scat}=-\sigma E_{z}^{inc}-\left(\varepsilon-\varepsilon_{0}\right)\frac{\Delta E_{z}^{inc}}{\Delta t}+\frac{1}{r}\frac{\Delta\left(rH_{\phi}^{scat}\right)}{\Delta r}-\frac{\Delta H_{r}^{scat}}{\Delta \phi} \tag{A.6c}$$

$$\mu\frac{\Delta H_{r}^{scat}}{\Delta t}=-\left(\mu-\mu_{0}\right)\frac{\Delta H_{r}^{inc}}{\Delta t}-\frac{1}{r}\frac{\Delta E_{z}^{scat}}{\Delta \phi}+\frac{\Delta E_{\phi}^{scat}}{\Delta z} \tag{A.6d}$$

$$\mu\Delta\frac{H_{\phi}^{scat}}{\Delta t}=-\left(\mu-\mu_{0}\right)\frac{\Delta H_{\phi}^{inc}}{\Delta t}-\frac{\Delta E_{r}^{scat}}{\Delta z}+\frac{\Delta E_{z}^{scat}}{\Delta r} \tag{A.6e}$$

$$\mu\frac{\Delta H_{z}^{scat}}{\Delta t}=-\left(\mu-\mu_{0}\right)\frac{\Delta H_{z}^{inc}}{\Delta t}-\frac{1}{r}\frac{\Delta\left(rE_{\phi}^{scat}\right)}{\Delta r}+\frac{1}{r}\frac{\Delta E_{r}^{scat}}{\Delta \phi} \tag{A.6f}$$

In explicitly finite differenced form they become

$$\begin{aligned}
&\varepsilon\left[E_{r}^{s,\,n+1/2}(I,J,K)-E_{r}^{s,\,n-1/2}(I,J,K)\right]+\sigma\Delta t E_{r}^{s,\,n+1/2}(I,J,K)\\
&=\sigma\Delta t E_{r}^{i,\,n+1/2}(I,J,K)-\left(\varepsilon-\varepsilon_{0}\right)\Delta t\dot{E}_{r}^{i,\,n+1/2}(I,J,K)\\
&-\Delta t\left[\frac{H_{r}^{s,\,n}(I,J,K)-H_{\phi}^{s,\,n}(I,J,K)}{\Delta z}\right]
\end{aligned} \tag{A.7a}$$

$$\begin{aligned}
&\varepsilon\left[E_\phi^{s,\,n+1/2}(I,J,K)-E_\phi^{s,\,n-1/2}(I,J,K)\right]\\
&=-\sigma\Delta t E_\phi^{i,\,n+1/2}(I,J,K)\;-(\varepsilon-\varepsilon_0)\Delta t\dot{E}_\phi^{i,\,n+1/2}(I,J,K)\\
&\quad+\Delta t\left[\left[\frac{H_r^{s,\,n}(I,J,K)-H_r^{s,\,n}(I,J,K)}{\Delta z}\right]\right.\\
&\quad\left.-\left[\frac{H_z^{s,\,n}(I,J,K)-H_z^{s,\,n}(I-1,J,K)}{\Delta r}\right]\right]
\end{aligned} \tag{A.7b}$$

$$\begin{aligned}
&\varepsilon\left[E_z^{s,\,n+1/2}(I,J,K)-E_z^{s,\,n-1/2}(I,J,K)\right]\\
&+\sigma\Delta t E_z^{s,\,n+1/2}(I,J,K)=-\sigma\Delta t E_z^{i,\,n+1/2}(I,J,K)-(\varepsilon-\varepsilon_0)\Delta t\dot{E}_z^{i,\,n+1/2}(I,J,K)\\
&\quad+\frac{\Delta t}{(I\Delta r)}\left[\left[\frac{(I+1/2)\Delta r H_\phi^{s,\,n}(I,J,K)-(I-1/2)\Delta r H_\phi^{s,\,n}(I-1,J,K)}{\Delta r}\right.\right.\\
&\quad\left.\left.-\left[H^{s,\,n}\,\frac{H_r^{s,\,n}(I,J,K)-H_r^{s,\,n}(I,J-1,K)}{\Delta\phi}\right]\right]\right]
\end{aligned} \tag{A.7c}$$

$$\begin{aligned}
&\mu\left[H_r^{s,\,n+1/2}(I,J,K)-H_r^{s,\,n}(I,J,K)\right]=-(\mu-\mu_0)\Delta t\dot{H}_r^{i,\,n}(I,J,K)\\
&\quad-\frac{\Delta t}{I\Delta r}\left[\frac{E_z^{s,\,n+1/2}(I,J+1,K)-E_z^{s,\,n+1/2}(I,J,K)}{\Delta\phi}\right]\\
&\quad+\Delta t\left[\frac{E_\phi^{s,\,n+1/2}(I,J,K+1)-E_\phi^{s,\,n+1/2}(I,J,K)}{\Delta z}\right]
\end{aligned} \tag{A.7d}$$

$$\mu\left[H_\phi^{s,\,n+1}(I,J,K)-H_\phi^{s,\,n}(I,J,K)\right]=-(\mu-\mu_0)\Delta t H_\phi^{i,\,n}(I,J,K)$$
$$-\Delta t\left[\left[\frac{E_r^{s,\,n+1/2}(I,J,K+1)-E_r^{s,\,n+1/2}(I,J,K)}{\Delta z}\right]-\left[\frac{E_z^{s,\,n+1/2}(I+1,J,K)-E_z^{s,\,n+1/2}(I,J,K)}{\Delta r}\right]\right] \tag{A.7e}$$

$$\mu\left[H_z^{s,\,n+1/2}(I,J,K)-H_z^{s,\,n}(I,J,K)\right]=-(\mu-\mu_0)\Delta t H_z^{i,\,n}(I,J,K)$$
$$-\frac{\Delta t}{((I+1/2)\Delta r)}\left[\frac{(I+1)\Delta r E_\phi^{s,\,n+1/2}(I+1,J,K)-I\Delta r E_\phi^{s,\,n+1/2}(I,J,K)}{\Delta r}-\left[\frac{E_r^{s,\,n+1/2}(I,J+1,K)-E_r^{s,\,n+1/2}(I,J,K)}{\Delta\phi}\right]\right] \tag{A.7f}$$

where the I,J,K notation has been explained previously and we assume that the equivalent to the Cartesian Yee cell is used in this coordinate system.

These equations can in turn be solved for the latest value in time of the field components:

$$E_r^{s,\,n+1/2}(I,J,K)=\left(\frac{\varepsilon}{\varepsilon+\sigma\Delta t}\right)E_r^{s,\,n-1/2}(I,J,K)$$
$$-\left(\frac{\sigma\Delta t}{\varepsilon+\sigma\Delta t}\right)E_r^{i,\,n+1/2}(I,J,K)-\left(\frac{(\varepsilon-\varepsilon_0)\Delta t}{\varepsilon+\sigma\Delta t}\right)\dot{E}_r^{i,\,n+1/2}(I,J,K)$$
$$+\left(\frac{\Delta t}{\varepsilon+\sigma\Delta t}\right)\left[\frac{1}{(I+1/2)\Delta r}\left[\frac{H_z^{s,\,n}(I,J,K)-H_z^{s,\,n}(I,J,K-1)}{\Delta\phi}\right]-\left[\frac{H_\phi^{s,\,n}(I,J,K)-H_z^{s,\,n}(I,J,K-1)}{\Delta z}\right]\right] \tag{A.8a}$$

$$E_\phi^{s,\,n+½}(I,J,K)=\left(\frac{\varepsilon}{\varepsilon+\sigma\Delta t}\right)E_\phi^{s,\,n-½}(I,J,K)$$
$$-\left(\frac{\sigma\Delta t}{\varepsilon+\sigma\Delta t}\right)E_\phi^{i,\,n+½}(I,J,K)-\left(\frac{(\varepsilon-\varepsilon_0)\Delta t}{\varepsilon+\sigma\Delta t}\right)\dot{E}_\phi^{i,\,n+½}(I,J,K)$$
$$+\left(\frac{\Delta t}{\varepsilon+\sigma\Delta t}\right)\left[\left[\frac{H_r^{s,\,n}(I,J,K)-H_r^{s,\,n}(I,J,K-1)}{\Delta z}\right]\right. \tag{A.8b}$$
$$\left.-\left[\frac{H_z^{s,\,n}(I,J,K)-H_z^{s,\,n}(I-1,J,K)}{\Delta r}\right]\right]$$

$$E_z^{s,\,n+½}(I,J,K)=\left(\frac{\varepsilon}{\varepsilon+\sigma\Delta t}\right)E_z^{s,\,n-½}(I,J,K)$$
$$-\left(\frac{\sigma\Delta t}{\varepsilon+\sigma\Delta t}\right)E_z^{i,\,n+½}(I,J,K)$$
$$-\left(\frac{(\varepsilon-\varepsilon_0)\Delta t}{\varepsilon+\sigma\Delta t}\right)\dot{E}_z^{i,\,n+½}(I,J,K)+\left(\frac{\Delta t}{\varepsilon+\sigma\Delta t}\right)\left(\frac{1}{I\Delta r}\right)$$
$$\left[\left[\frac{(I+1/2)\Delta r H_\phi^{s,\,n}(I,J,K)-(I+1/2)\Delta r H_\phi^{s,\,n}(I-1,J,K)}{\Delta r}\right]\right. \tag{A.8c}$$
$$\left.-\left[\frac{H_r^{s,\,n}(I,J,K)-H_r^{s,\,n}(I,J-1,K)}{\Delta\phi}\right]\right]$$

$$H_r^{s,\,n+1}(I,J,K)=H_r^{s,\,n}(I,J,K)=\frac{-(\mu-\mu_0)}{\mu}\Delta t\dot{H}_r^{i,\,n}(I,J,K)$$
$$-\frac{\Delta t}{\mu(I\Delta r)}\left[\frac{E_z^{s,\,n+½}(I,J+1,K)-E_z^{s,\,n+½}(I,J,K)}{\Delta\phi}\right] \tag{A.8d}$$
$$+\frac{\Delta t}{\mu}\left[\frac{E_\phi^{s,\,n+½}-E_\phi^{s,\,n+½}(I,J,K)}{\Delta z}\right]$$

$$H_\phi^{s,\,n+1}(I,J,K) = H_\phi^{s,\,n}(I,J,K) - \frac{(\mu-\mu_0)}{\mu}\Delta t\dot{H}_\phi^{i,\,n}(I,J,K)$$
$$-\frac{\Delta t}{\mu}\left[\left[\frac{E_r^{s,\,n+½}(I,J,K+1) - E_r^{s,\,n+½}(I,J,K)}{\Delta z}\right] - \left[\frac{E_z^{s,\,n+½}(I+1,J,K) - E_z^{s,\,n+½}(I,J,K)}{\Delta r}\right]\right] \quad \text{(A.8e)}$$

$$H_z^{s,\,n+1}(I,J,K) = H_z^{s,\,n}(I,J,K) - \frac{(\mu-\mu_0)}{\mu}\Delta t\dot{H}_z^{i,\,n}(I,J,K)$$
$$-\frac{\Delta t}{\mu(I+1/2)\Delta r}\left[\left[\frac{(I+1)\Delta rE_\phi(I+1,J,K) - I\Delta rE_\phi(I,J,K)}{\Delta r}\right] - \left[\frac{E_r^{s,\,n+½}(I,J+1,K) - E_r^{s,\,n+½}(I,J,K)}{\Delta\phi}\right]\right] \quad \text{(A.8f)}$$

A loop can be formed around each component with four neighboring (H,E) components except for the E_z components at r = 0 or $E_z(0,0,K)$. For this component at I = 0, (A.8c) for E_z is singular. Instead of using (A.8c) for these E field components, a loop of H with all the H_ϕ field components immediately about E_z must be evaluated (Figure A-3). In (A.8d) I = 0 is not allowed as there will be no radial component of H_r at r = 0.

A complication that arises in applying cylindrical coordinate FDTD in 3-D is that the cell size in the ϕ dimension decreases with decreasing r. This means that very small time steps may be necessary in order to satisfy the Courant stability criterion, unless the region in the vicinity of r = 0 is filled with perfect conductor or otherwise excluded. This difficulty is removed for 2-D ϕ-independent calculations considered later in this appendix.

For spherical coordinates the same exercise starting with the curl equations[1] leads to

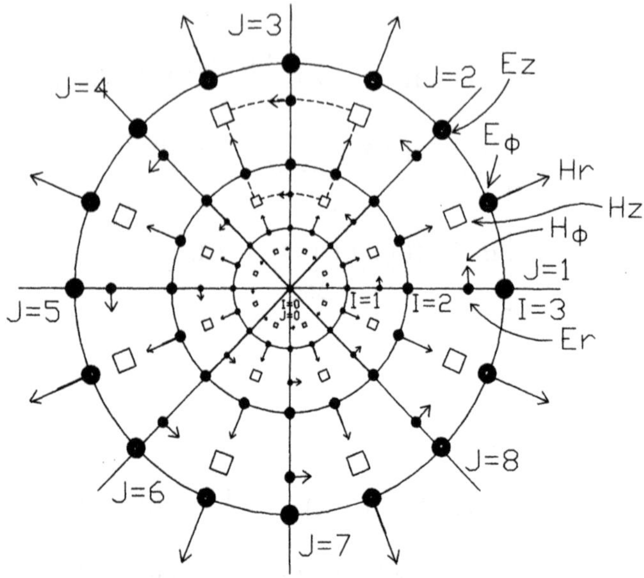

Figure A-3. Field components for evaluating innermost E field component at r = 0.

$$
\begin{aligned}
E_r^{s,\,n+1/2}(I,J,K) &= \left(\frac{\varepsilon}{\varepsilon+\sigma\Delta t}\right)E_r^{s,\,n-1/2}(I,J,K) \\
&-\left(\frac{\sigma\Delta t}{\varepsilon+\sigma\Delta t}\right)E_r^{i,\,n+1/2}(I,J,K) \\
&-\left(\frac{(\varepsilon-\varepsilon_0)\Delta t}{\varepsilon+\sigma\Delta t}\right)\dot{E}_r^{i,\,n+1/2}(I,J,K)+\left(\frac{\Delta t}{\varepsilon+\sigma\Delta t}\right)\left(\frac{1}{R(I)\sin(J\Delta\theta)}\right) \\
&\left[\left[\frac{\sin\theta(J)H_\phi^{s,\,n}(I,J,K)-\sin\theta(J-1)H_\phi^{s,\,n}(I,J-1,K)}{\Delta\theta(J-1)}\right]\right. \\
&\left.-\left[\frac{H_\theta^{s,\,n}(I,J,K)-H_\theta^{s,\,n}(I,J,K-1)}{\Delta\phi(K-1)}\right]\right]
\end{aligned}
\tag{A.9a}
$$

$$
\begin{aligned}
E_\theta^{s,\,n+1/2}(I,J,K) &= \left(\frac{\varepsilon}{\varepsilon+\sigma\Delta t}\right)E_\theta^{s,\,n-1/2}(I,J,K)-\left(\frac{\sigma\Delta t}{\varepsilon+\sigma\Delta t}\right)E_\theta^{i,\,n+1/2}(I,J,K) \\
&-\left(\frac{(\varepsilon-\varepsilon_0)\Delta t}{\varepsilon+\sigma\Delta t}\right)\dot{E}_\theta^{i,\,n+1/2}(I,J,K) \\
&+\left(\frac{\Delta t}{\varepsilon+\sigma\Delta t}\right)\left(\frac{1}{R_0(I)}\right)\left[\frac{H_r^{s,\,n}(I,J,K)-H_r^{s,\,n}(I,J,K-1)}{\sin\theta(J)(\Delta\theta(K-1))}\right. \\
&\left.-\frac{R(I)H_\phi^{s,\,n}(I,J,K)-R(I-1)H_\phi^{s,\,n}(I-1,J,K)}{\Delta R(I-1)}\right]
\end{aligned}
\tag{A.9b}
$$

$$
\begin{aligned}
E_\phi^{s,\,n+1/2}(I,J,K) &= \left(\frac{\varepsilon}{\varepsilon+\sigma\Delta t}\right)E_\phi^{s,\,n-1/2}(I,J,K) \\
&\quad - \left(\frac{\sigma\Delta t}{\varepsilon+\sigma\Delta t}\right)E_\phi^{i,\,n+1/2}(I,J,K) \\
&\quad - \left(\frac{(\varepsilon-\varepsilon_0)\Delta t}{\varepsilon+\sigma\Delta t}\right)\dot{E}_\phi^{i,\,n+1/2}(I,J,K) \\
&\quad + \left(\frac{\Delta t}{\varepsilon+\sigma\Delta t}\right)\left(\frac{1}{R_0(I)}\right)\left[\frac{R(I)H_\theta^{s,\,n}(I,J,K)-R(I-1)H_\theta^{s,\,n}(I-1,J,K)}{\Delta R(I-1)}\right. \\
&\quad \left. - \frac{H_r^{s,\,n}(I,J,K)-H_r^{s,\,n}(I,J-1,K)}{\Delta\theta(J-1)}\right]
\end{aligned}
\tag{A.9c}
$$

$$
\begin{aligned}
H_r^{s,\,n+1}(I,J,K) &= H_r^{s,\,n}(I,J,K) - \frac{(\mu-\mu_0)}{\mu}\Delta t\dot{H}_r^{i,\,n}(I,J,K) \\
&\quad - \frac{\Delta t}{\mu}\frac{1}{R_0(I)\sin\theta(J)} \\
&\quad \times\left[\frac{\sin\theta_0(J+1)E_\phi^{s,\,n+1/2}(I,J+1,K)-\sin\theta_0(J)E_\phi^{s,\,n+1/2}(I,J,K)}{\Delta\theta_0(J)}\right. \\
&\quad \left. - \frac{E_\theta^{s,\,n+1/2}(I,J,K+1)-E_\theta^{s,\,n+1/2}(I,J,K)}{\Delta\varnothing_0(K)}\right]
\end{aligned}
\tag{A.9d}
$$

$$
\begin{aligned}
H_\theta^{s,\,n+1}(I,J,K) &= H_\theta^{s,\,n}(I,J,K) \\
&\quad - \frac{(\mu-\mu_0)}{\mu}\Delta t\dot{H}_\theta^{i,\,n}(I,J,K) - \frac{\Delta t}{\mu}\frac{1}{R(I)} \\
&\quad \times\left[\frac{E_r^{s,\,n+1/2}(I,J,K+1)-E_r^{s,\,n+1/2}(I,J,K)}{\sin\theta_0(J)(\Delta\varnothing_0(K))}\right. \\
&\quad \left. - \frac{R_0(I+1)E_\phi^{s,\,n+1/2}(I+1,J,K)-R_0(I)E_\phi^{s,\,n+1/2}(I,J,K)}{\Delta R_0(I)}\right]
\end{aligned}
\tag{A.9e}
$$

$$
\begin{aligned}
H_\phi^{s,\,n+1}(I,J,K) &= H_\phi^{s,\,n}(I,J,K) \\
&-\frac{(\mu-\mu_0)}{\mu}\Delta t\dot{H}_\phi^{i,\,n}(I,J,K)-\frac{\Delta t}{\mu}\frac{1}{R(I)} \\
&\times\left[\frac{R_0(I+1)E_\theta^{s,\,n+1/2}(I+1,J,K)-R_0(I)E_\theta^{s,\,n+1/2}(I,J,K)}{\Delta R_0(I)}\right. \\
&\left.-\frac{E_r^{s,\,n+1/2}(I,J+1,K)-E_r^{s,\,n+1/2}(I,J,K)}{\Delta\theta_0(J)}\right]
\end{aligned}
\tag{A.9f}
$$

The field components are on a modified Yee cell (Figure A-2) with unit vectors $\hat{r}, \hat{\theta}$ and $\hat{\phi}$ and the corresponding indices I,J,K. Spatial locations are given[4] in terms of positions

$R_0(I)$	$(R_0(1) = 0$ and $\Delta R_0(I) \equiv R_0(I+1) - R_0(I)$, typically $\Delta R_0(I) = \Delta R$ = constant so that $R_0(I) = (I-1)R)$
$\theta_0(J)$	$(\theta_0(1) = 0$ and $\Delta\theta_0(J) = \theta_0(J+1) - \theta_0(J)$, typically $\Delta\theta_0(J) = \Delta\theta$ = constant = 2π/integer so that $\theta_0(J) = (J-1)\Delta\theta)$
$\phi_0(K)$	$(\phi_0(1) = 0$ and $\Delta\phi_0(K) = \phi_0(K+1) - \phi_0(K)$, typically $\Delta\phi_0(K) = \Delta\phi$ = constant = 2π/integer so that $\phi_0(K) = (K-1)\Delta\phi)$

Points lying midway between these locations are given by R(I), θ(J), ϕ(K).

This process breaks down at the origin (r = 0) and some scheme must be employed to span this volume. Additional problems occur where indices overlap as for

$$
E_r^s(I,J,1) = E_r^s\left(I,J,N_{\text{for }\phi=2\pi}\right)
$$

This problem can be readily overcome,[4] but not so the problem with the singularity at r = 0 except by some expedient such as assuming a finite radius perfect conductor at the center of the problem space. While effective, this does limit the utility of the spherical coordinate treatment. The alternative we prefer is to define cells that are not quasi-rectangular at the center, rather ones that are pyramidal so that assuming an E field component at the center of undefined orientation it makes no contribution to the updated neighboring fields in that it lies on a zero length "leg"[4] (Figure A-4).

As with cylindrical coordinates, another limitation in applying spherical coordinates in 3-D is caused by the changing cell size as r is increased. In the ϕ dimension the arc length of the cell side decreases with decreasing r and as the poles (θ = 0, 180) are approached. Because, as discussed in Chapter 3, the

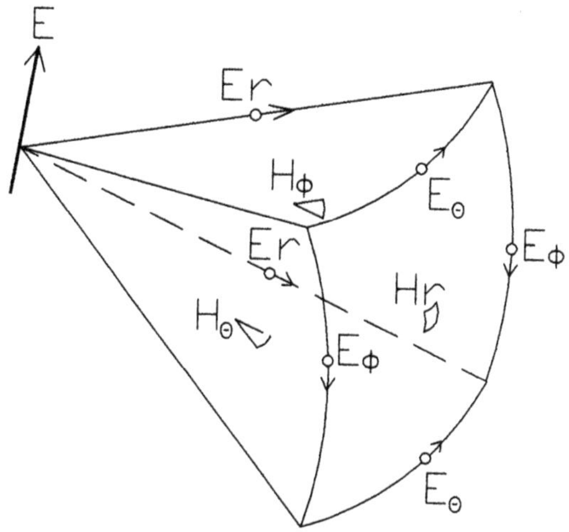

Figure A-4. Vanishing E field contribution at the spherical problem space center.

Courant stability criterion requires that the time step be reduced as the cell size is reduced, the very small cells ($\Delta\phi$ is tending to zero) at the poles of the problem space force the use of very small time steps. These cells cannot be made larger since they are geometrically related to the cells at the largest value of r at $\theta = 90°$ and these largest cells must be much smaller than a wavelength. This difficulty is discussed in Reference 4 and some methods to alleviate it are considered.

We have now formulated FDTD in three different orthogonal 3-D coordinate systems: rectangular (in Chapter 2), cylindrical, and spherical. The FDTD 3-D formulation in the first two coordinate systems can be reduced to usable 2-D formulations for a number of problems. The 2-D rectangular coordinate system formulation applies to any problem with translational symmetry in one direction, i.e., the problem remains unchanged in one direction. The 2-D cylindrical coordinate system formulation is suitable to BOR radiation problems where the problem geometry and radiated fields are invariant with respect to ϕ. By starting with a 3-D coordinate system formulation and orienting the coordinate system so that the problem is invariant in one of the dimensions it is easy to obtain the 2-D formulation that can solve this problem with its invariance or symmetry. This 2-D formulation is far more efficient than a 3-D treatment for this class of problem. In the next sections of this appendix 2-D FDTD equations for Cartesian and cylindrical coordinate systems are given.

2-D CARTESIAN

Consider the problem of scattering by a cylinder which is infinitely long in the z dimension, as discussed in Section 7.3. With no variation in the geometry in the z direction $\partial A/\partial z = 0$ for all A where A is any field component. The Maxwell equations for the six rectangular components (2.33 a-f) become

$$\frac{\partial E_x^{scat}}{\partial t} = \frac{1}{\varepsilon_0}\left(\frac{\partial H_x^{scat}}{\partial y}\right) \qquad \text{(A.10a)}$$

$$\frac{\partial E_y^{scat}}{\partial t} = -\frac{1}{\varepsilon_0}\left(\frac{\partial H_z^{scat}}{\partial x}\right) \qquad \text{(A.10b)}$$

$$\frac{\partial E_z^{scat}}{\partial t} = \frac{1}{\varepsilon_0}\left(\frac{\partial H_y^{scat}}{\partial x} - \frac{\partial H_x^{scat}}{\partial y}\right) \qquad \text{(A.10c)}$$

$$\frac{\partial H_x^{scat}}{\partial t} = -\frac{1}{\mu_0}\left(\frac{\partial E_z^{scat}}{\partial y}\right) \qquad \text{(A.10d)}$$

$$\frac{\partial H_y^{scat}}{\partial t} = \frac{1}{\mu_0}\left(\frac{\partial E_z^{scat}}{\partial x}\right) \qquad \text{(A.10e)}$$

$$\frac{\partial H_z^{scat}}{\partial t} = -\frac{1}{\mu_0}\left(\frac{\partial E_y^{scat}}{\partial x} - \frac{\partial E_x^{scat}}{\partial y}\right) \qquad \text{(A.10f)}$$

Either TE_z or TM_z solutions are possible with $E_z = 0$ or $H_z = 0$, respectively. For TM_z the governing equations are

$$\frac{\partial H_x^{scat}}{\partial t} = -\frac{1}{\mu_0}\left(\frac{\partial E_z^{scat}}{\partial y}\right) \qquad \text{(A.11a)}$$

$$\frac{\partial H_y^{scat}}{\partial t} = \frac{1}{\mu_0}\left(\frac{\partial E_z^{scat}}{\partial x}\right) \qquad \text{(A.11b)}$$

$$\frac{\partial E_z^{scat}}{\partial t} = \frac{1}{\varepsilon_0}\left(\frac{\partial H_y^{scat}}{\partial x} - \frac{\partial H_x^{scat}}{\partial y}\right) \tag{A.11c}$$

and only E_z, H_x, and H_y are required for a solution.

For TE_z fields the governing equations are

$$\frac{\partial E_x^{scat}}{\partial t} = \frac{1}{\varepsilon_0}\left(\frac{\partial H_z^{scat}}{\partial y}\right) \tag{A.12a}$$

$$\frac{\partial E_y^{scat}}{\partial t} = -\frac{1}{\varepsilon_0}\left(\frac{\partial H_z^{scat}}{\partial x}\right) \tag{A.12b}$$

$$\frac{\partial E_z^{scat}}{\partial t} = \frac{1}{\mu_0}\left(\frac{\partial E_x^{scat}}{\partial y} - \frac{\partial E_y^{scat}}{\partial x}\right) \tag{A.12c}$$

and only H_z, E_x, and E_y are required for a solution.

Thus, 2-D Cartesian coordinate problems can be reduced to two separate calculations with only three components each. More importantly the invariance in z reduces either problem to 2-D , x and y, in which either E_z, H_x, and H_y are included for TM_z solutions or H_z, E_x, and E_y for TE_z solutions.

In order to obtain the scattered field formulation in 2-D Cartesian coordinates the 3-D equations given in Chapter 2 are easily modified by eliminating any terms involving differencing in the z dimension, and separating the three components required for either TE_z or TM_z calculations.

The Courant stability criterion for 2-D Cartesian FDTD is given in Section 3.3. The Mur absorbing outer boundary conditions for 2-D Cartesian FDTD can be easily adapted from the 3-D conditions given in Chapter 3 by eliminating terms involving Δz differencing, and are included in the original paper by Mur. A far zone transformation for 2-D Cartesian FDTD is given in Section 7.3 along with example results. Methods for approximating scattering by finite 3-D objects using 2-D scattering results are discussed in Reference 1.

2-D CYLINDRICAL

In a Body of Revolution (BOR) radiation problem there is no variation in ϕ and $\partial A/\partial\phi = 0$ for all field components A. In cylindrical coordinates, then, the FDTD equations (A.5a through f) become

$$\frac{-\partial H_\phi^{scat}}{\partial z} = \frac{\varepsilon \partial E_r^{scat}}{\partial t} + \sigma E_r^{scat} + (\varepsilon - \varepsilon_0)\frac{\partial E_r^{inc}}{\partial t} + \sigma E_r^{inc} \tag{A.13a}$$

$$\frac{\partial H_r^{scat}}{\partial z} - \frac{\partial H_z^{scat}}{\partial r} = \varepsilon \frac{\partial E_\phi^{scat}}{\partial t} + \sigma E_\phi^{scat} + (\varepsilon - \varepsilon_0)\frac{\partial E_\phi^{inc}}{\partial t} + \sigma E_\phi^{inc} \tag{A.13b}$$

$$\frac{1}{r}\frac{\partial}{\partial r}\left(rH_\phi^{scat}\right) = \varepsilon \frac{\partial E_z^{scat}}{\partial t} + \sigma E_z^{scat} + (\varepsilon - \varepsilon_0)\frac{\partial E_z^{inc}}{\partial t} + \sigma E_z^{inc} \tag{A.13c}$$

$$\frac{-\partial E_\phi^{scat}}{\partial z} = -\mu \frac{\partial H_r^{scat}}{\partial t} - (\mu - \mu_0)\frac{\partial H_r^{inc}}{\partial t} \tag{A.13d}$$

$$\frac{\partial E_r^{scat}}{\partial z} - \frac{\partial E_z^{scat}}{\partial r} = \frac{-\mu \partial H_\phi^{scat}}{\partial t} - (\mu - \mu_0)\frac{\partial H_\phi^{inc}}{\partial t} \tag{A.13e}$$

$$\frac{1}{r}\frac{\partial}{\partial r}\left(rE_\phi^{scat}\right) = -\frac{\mu \partial H_z^{scat}}{\partial t} - (\mu - \mu_0)\frac{\partial H_z^{inc}}{\partial t} \tag{A.13f}$$

Here, too, it is possible to obtain two sets of three equations, each of which uniquely describe the behavior of three components each. These sets of equations are

$$\frac{-\partial H_\phi^{scat}}{\partial z} = \frac{\varepsilon \partial E_r^{scat}}{\partial t} + \sigma E_r^{scat} + (\varepsilon - \varepsilon_0)\frac{\partial E_r^{inc}}{\partial t} + \sigma E_r^{inc} \tag{A.14a}$$

$$\frac{1}{r}\frac{\partial}{\partial r}\left(rH_\phi^{scat}\right) = \frac{\varepsilon \partial E_z^{scat}}{\partial t} + \sigma E_z^{scat} + (\varepsilon - \varepsilon_0)\frac{\partial E_z^{inc}}{\partial t} + \sigma E_z^{inc} \tag{A.14b}$$

$$\frac{\partial E_r^{scat}}{\partial z} - \frac{\partial E_z^{scat}}{\partial r} = \frac{-\mu \partial H_\phi}{\partial t} - (\mu - \mu_0)\frac{\partial H_\phi^{inc}}{\partial t} \tag{A.14c}$$

where only H_ϕ and E_r and E_z are required and the remaining three equations where only E_ϕ, H_r, and H_z are required.

The first set of three equations is most useful for modeling antenna-like problems where the BOR is small compared to the shortest wavelength of interest and the incident field can be treated as arriving simultaneously at all parts of the antenna surface (Figure A-5).

As discussed previously one important area of application is antenna radiation and impedance calculations for BOR antennas with cylindrical symmetry.

In Reference 2 results were given for coaxially fed monopole and conical monopole antennas. For these geometries only the H_ϕ, E_r, and E_z field components are required and (A.14) serves as the basis for the FDTD equations. For antenna calculations it is preferable to directly compute total fields (see discussion in Section 3.6), and with our usual notation of t = n Δt, r = I Δr, and z = J Δz, we readily obtain the total field FDTD equations as

$$\begin{aligned} H_\phi^{n+1/2}(I,J) &= H_\phi^{n-1/2}(I,J) \\ &+ \frac{\Delta t}{\mu_0 \Delta r}\left[E_z^n(I+1/2,J) - E_z^n(I+1/2,J)\right] \\ &- \frac{\Delta t}{\mu_0 \Delta z}\left[E_r^n(I,J+1/2) - E_r^n(I,J-1/2)\right] \end{aligned} \qquad \text{(A.15a)}$$

$$\begin{aligned} E_r^{n+1}(I,J-1/2) &= E_r^n(I,J-1/2) \\ &- \frac{\Delta t}{\varepsilon_0 \Delta z}\left[H_\phi^{n+1/2}(I,J) - H_\phi^{n+1/2}(I,J-1)\right] \\ &- \frac{\Delta t}{\mu_0 \Delta z}\left[E_r^n(I,J+1/2) - E_z^n(I,J-1/2)\right] \end{aligned} \qquad \text{(A.15b)}$$

$$\begin{aligned} E_z^{n+1}(I+1/2,J) &= E_z^n(I+1/2,J) \\ &+ \frac{\Delta t}{\varepsilon_0 \Delta r}\frac{1}{(I+1/2)}\left[(I+1)H_\phi^{n+1/2}(I+1,J) - IH_\phi^{n+1/2}(I,J)\right] \\ &- \frac{\Delta t}{\mu_0 \Delta z}\left[E_r^n(I,J+1/2) - E_z^n(I,J-1/2)\right] \end{aligned} \qquad \text{(A.15c)}$$

with the field components located in a Yee grid as shown in Figure A-6. In this set of equations we explicitly locate the field components in the grid with the I,J notation by adding $^1/_2$ as necessary. By locating the field components in this way there is no E_z component at r = 0, so that for all values of I, including I = 0, (A.15c) can be evaluated. In (A.15a) and (A.15b) we constrain I ≥ 1 so that the E_z components necessary to evaluate H_ϕ in (A.15a) and the H_ϕ components needed for E_r are available. This implies that $H_\phi = E_r = 0$ along the z axis at r = 0, a reasonable assumption. We thus exclude from our calculations a cylindrical region along the z axis of radius Δr/2, and locate a z component of electric field on the surface of this cylindrical region.

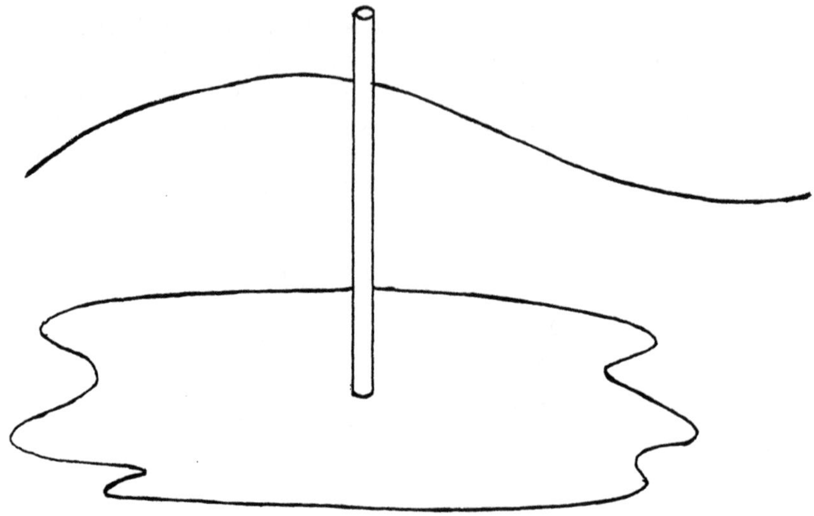

Figure A-5. An antenna-like problem where the BOR is small compared to the shortest wavelength of interest and the incident field can be treated as arriving simultaneously at all parts of the antenna surface.

The Courant stability condition for this calculation is the same as for a 2-D Cartesian grid calculation,

$$c\Delta t \le /\sqrt{\left(\frac{1}{\Delta r}\right)^2 + \left(\frac{1}{\Delta z}\right)^2} \tag{A.16}$$

assuming that the maximum velocity in the FDTD space is the speed of light c, and with Δr and Δz constrained as usual to be much smaller than a wavelength. The 2-D Mur absorbing boundary can be applied directly to this FDTD space, with E_r and E_z components located tangentially on the appropriate free space surfaces.

If we wish to transform our results to the far zone to compute antenna patterns, for example, we must remember that while this is a 2-D calculation the actual geometry and fields are in a 3-D space with no ϕ variation. Therefore, we must apply a 3-D far zone transformation. The required transformation can be adapted from the Cartesian coordinate 3-D transformation given in Section 7.2. This will involve converting this transformation to cylindrical coordinates and integrating the ϕ dependence from the equations.

1-D CARTESIAN AND CYLINDRICAL

Just as 2-D FDTD formulations arise from eliminating variations in 1-D , 1-D formulations arise from eliminating variations in 2-D. We can obtain 1-D

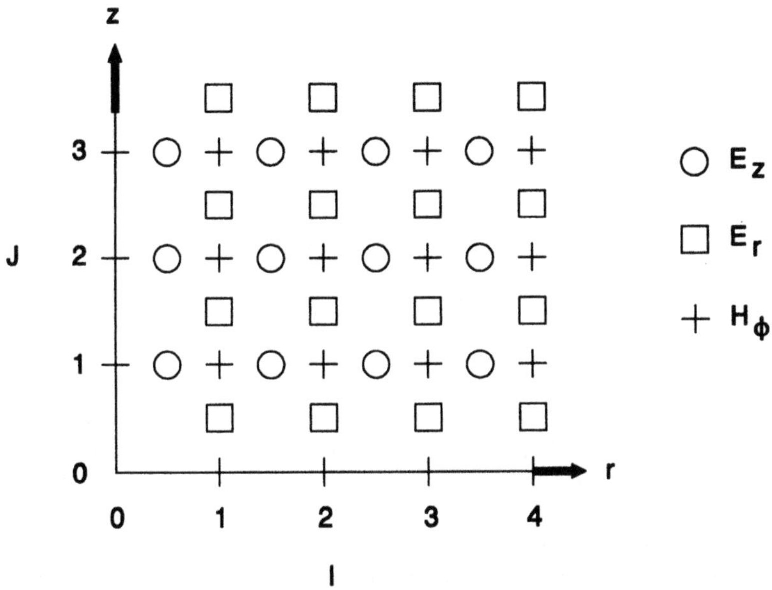

Figure A-6. Total field FDTD with the field components located in a Yee grid.

coordinate system FDTD formulations from the 2-D formulations quite simply by eliminating variations in one additional coordinate. The resulting 1-D formulations will, of course, depend on only a single spatial coordinate, but that coordinate will differ between formulations derived from a rectangular coordinate system vs. a cylindrical coordinate system. The field components will also differ.

In Cartesian coordinates with no variations in two coordinate directions, the problem that can be described is transverse electromagnetic (TEM) plane wave propagation through a planar stratified medium. While this geometry can be handled quite easily by other methods, it still can be useful for testing new FDTD formulations. For example, in this book 1-D tests of FDTD formulations for frequency-dependent materials are presented in Chapter 8. Also, 1-D demonstrations of FDTD for educational purposes have been used with some success.[5]

The TEM solutions can be separated into two parts, depending on the polarization of the fields. If we assume propagation (variation) in the z dimension, then the two sets of TEM solutions in free space are

$$\frac{\partial H_x^{scat}}{\partial t} = \frac{1}{\mu_0}\left(\frac{\partial E_y^{scat}}{\partial z}\right) \tag{A.17a}$$

$$\frac{\partial E_y^{scat}}{\partial t} = \frac{1}{\varepsilon_0}\left(\frac{\partial H_x^{scat}}{\partial z}\right) \tag{A.17b}$$

for E_y polarization and

$$\frac{\partial E_x^{scat}}{\partial t} = \frac{-1}{\varepsilon_0}\left(\frac{\partial H_y^{scat}}{\partial z}\right) \tag{A.18a}$$

$$\frac{\partial H_y^{scat}}{\partial t} = \frac{-1}{\mu_0}\left(\frac{\partial E_x^{scat}}{\partial z}\right) \tag{A.18b}$$

for E_x polarization. Each pair of equations tracks two field components in the x and y direction that vary only in the z direction. Either formulation can be employed for a linearly polarized 1-D problem, both will be needed for other polarizations, circular or elliptical, for example. These equations are the same as the telegraphy equations where V replaces E and I replaces H and L and C appear instead of ε_0 and μ_0.

Now let us consider a 1-D problem in cylindrical coordinates. The 2-D cylindrical representation was obtained by assuming an azimuthal symmetry or invariance in the ϕ direction. The additional symmetry leading to a 1-D formulation chosen here is a translational invariance in the z direction, i.e., $\partial A/\partial z = 0$ for all A where A may be any field component. This will allow modeling the propagation of cylindrical waves through a cylindrically stratified medium.

The 1-D cylindrical FDTD equations arising from the three 2-D equations in H_ϕ, E_r, and E_z are

$$\frac{1}{r}\frac{\partial}{\partial r}\left(rH_\phi^{scat}\right) = \frac{\varepsilon \partial E_z^{scat}}{\partial t} + \sigma E_z^{scat} + \left(\varepsilon - \varepsilon_0\right)\frac{\partial E_z^{inc}}{\partial t} + \sigma E_z^{inc} \tag{A.19a}$$

$$-\frac{\partial E_z^{scat}}{\partial r} = \frac{-\mu \partial H_\phi}{\partial t} - \left(\mu - \mu_0\right)\frac{\partial H_\phi^{inc}}{\partial t} \tag{A.19b}$$

Note that there are only two components, H_ϕ and E_z, and that they vary only in r. Because of the $\frac{1}{r}\frac{\partial}{\partial r}(rH\phi^{scat})$ term the fields have a $1/\sqrt{r}$ variation. This geometry is actually closely related to the 2-D rectangular geometry, because both have no z variation and the fields in both have the same $1/\sqrt{r}$ variation. The field components are also related to the TM_z fields in the Cartesian 2-D geometry, with the E_z field components identical and the cylindrical H_ϕ field component equivalent to the combination of the H_x and H_x components in the Cartesian geometry. In this geometry a cylindrical coordinate absorbing boundary condition would be required, taking into account the $1/\sqrt{r}$ variation in the fields. While it is unlikely that a far zone transformation would be desired for this geometry, one could be adapted from the 2-D Cartesian coordinate far zone transformation of Section 7.3.

REFERENCES

1. **Balanis, C.,** *Advanced Engineering Electromagnetics,* John Wiley & Sons, New York, 1989, 577.
2. **Maloney, J. et al.,** Accurate computation of the radiation from simple antennas using the finite-difference time-domain method, *IEEE Trans. Ant. Prop.,* 38(7), 1181, 1990.
3. **Britt, C.,** Solution of electromagnetic scattering problems using time domain techniques, *IEEE Trans. Ant. Prop.,* 37(9), 1181, 1989.
4. **Holland, R.,** THREDS: a finite-difference time-domain EMP code in 3D spherical coordinates, *IEEE Trans. Nucl. Sci.,* 30(6), 1983.
5. **Luebbers, R., Kunz, K. S., and Chamberlin, K. A.,** An interactive demonstration of electromagnetic wave propagation using time-domain finite differences, *IEEE Trans. Educ.,* 1990.

Appendix B

FORTRAN LISTINGS

DESCRIPTION

This Appendix contains FORTRAN listings for an FDTD computer code. The code is 3-D. It is capable of including perfect conductors and lossy dielectric materials. The IDXXX arrays described in Chapters 2 and 3 are used to describe the geometry.

The example geometry is a sphere composed of lossy dielectric with relative permittivity 4 and conductivity 0.005 S/m. A Gaussian pulse plane wave is incident on the sphere. The β value for the Gaussian pulse (see Chapter 3) is set at 64 to compensate for the shorter wavelength inside the dielectric sphere. The sphere has a radius of eight cells and is centered in a $34 \times 34 \times 34$ cell problem space.

There are four separate listings. The first is the main program, FDTDA.FOR, with contains the actual FDTD calculation code. This program requires the include file COMMONA.FOR. This file contains the arrays and COMMON files that are included in the MAIN section and the subroutines in FDTDA.FOR. The file DIAGS3D.DAT is a diagnostics file that contains the pertinent calculation parameters. Finally, the file NZOUT3D.DAT contains the time domain scattered electric and magnetic fields sampled at two locations (each) in the FDTD space for 256 time steps.

The listings are liberally commented. The basic organization of the program is modular, with the MAIN section of FDTDA.FOR calling subroutines to perform all the actual calculations.

FDTDA.FOR

```
C
      PROGRAM FDTDA
C
C     PENN STATE UNIVERSITY FINITE DIFFERENCE TIME DO-
      MAIN ELECTROMAGNETIC ANALYSIS COMPUTER CODE
C     —VERSION A
C
C     THIS CODE IS A SCATTERED FIELD FORMULATION
C
C     VERSION A:  August 25, 1992
C
C     TO REPORT ANY CODE ERRORS OR FOR ADDITIONAL
C     INFORMATION CONTACT:
C
```

```
C    DR. RAYMOND J. LUEBBERS
C    ELECTRICAL ENGINEERING DEPARTMENT
C    THE PENNSYLVANIA STATE UNIVERSITY
C    UNIVERSITY PARK, PA 16802
C
C    INTERNET:    LU4@PSUVM.PSU.EDU;    BITNET:
C    LU4@PSUVM.BITNET
C
C    THIS VERSION INCLUDES:
C    1.) 1ST ORDER E FIELD OUTER RADIATION BOUNDARY
C        CONDITION    (ORBC)
C    2.) 2ND ORDER E FIELD OUTER RADIATION BOUNDARY
C        CONDITION C     (ORBC)
C    3.) CAPABILITY TO SPECIFY DIRECTION AND POLARIZA-
C        TION OF INCIDENT PLANE WAVE
C    4.) ERROR CHECKING FOR NTYPE AND IOBS, JOBS AND
C        KOBS FOR NEAR ZONE FIELD SAMPLING IN SUBROUTINE
C        DATSAV
C    5.) ERROR CHECKING OF IDONE, IDTWO AND IDTHRE
C        VALUES
C
C
C    COMMONA.FOR:
C    NX, NY, NZ DETERMINE NUMBER OF CELLS IN PROBLEM
C    SPACE NTEST IS NUMBER OF QUANTITIES SAMPLED AND
C    WRITTEN VS TIME TIME VARIABLE IS NSTOP.
C
C    THE QUANTITIES TO BE SAMPLED (NEAR ZONE COMPUTA-
C    TIONS) ARE TO BE SET IN SUBROUTINE DATSAV
C
C    DEFINE OUTPUT FILES
C
C    DIAGS3D.DAT   =>   DIAGNOSTICS OF SOME SETUP
C    PARAMETERS
C    NZOUT3D.DAT  =>  NEAR-ZONE FIELDS OR CURRENTS AS
C    DEFINED IN DATSAV
C
C    WARNING:  PLEASE READ THE FOLLOWING COMMENTS
C                   REGARDING THE INCLUDE STATEMENT!
C
C    INCLUDE COMMON FILE (STATEMENT APPEARS IN EVERY
C    FUNCTION SUBPROGRAM AND SUBROUTINE). THIS IN
C    CLUDE STATEMENT IS APPROPRIATE FOR MOST MACHINES
C    (I.E. VAX,  SILICON GRAPHICS, 386/486 PC'S WITH LAHEY
C    COMPILER).  THE WATFOR (PC WATFOR) VERSION OF THE
```

```
C    INCLUDE STATEMENT  IS DIFFERENT AND IS INCLUDED
C    IMMEDIATELY AFTER THE NORMAL INCLUDE STATEMENT.
C    IF USING THE WATFOR COMPILER, DELETE THE NORMAL
C    INCLUDE STATEMENTS OR COMMENT THEM OUT. THE
C    INCLUDE STATEMENT FOR IBM VS FORTRAN IS DIFFERENT
C    ALSO. THE INCLUDE STATEMENT FOR AN IBM VS FORTRAN
C    COMPILER MUST BE CHANGED TO:  INCLUDE 'COMMONA
C    FORTRAN A1' THE INCLUDE STATEMENT MUST BE
C    CHANGED IN EVERY SUBROUTINE AND FUNCTION SUBPRO
C    GRAM AND IS MOST EFFICIENTLY DONE USING A GLOBAL
C    SEARCH AND REPLACE WITH AN EDITOR.  IF YOU ARE
C    USING THE WATFOR COMPILER, MAKE SURE THE COMMON
C    FILE HAS EXTENSION '.FOR'.
C
     INCLUDE 'COMMONA.FOR'
C$INCLUDE COMMONA
C
C    OPEN DATA FILES.  THE OPEN STATEMENTS SHOWN BE
C    LOW ARE APPROPRIATE FOR MOST MACHINES (I.E. VAX,
C    SILICON GRAPHICS, 386/486 PC'S). HOWEVER, THE IBM
C    VERSION OF THE OPEN STATEMENTS ARE DENOTED BY
C    LINES WITH C###. UNCOMMENTAPPROPRIATE LINES FOR
C    IBM VERSION AND COMMENT OUT OTHER LINES.
C
    OPEN(UNIT=17,FILE='DIAGS3D.DAT',STATUS='UNKNOWN')
C###      OPEN (UNIT=17,FILE='/DIAGS3D DATA E1',
                 STATUS='UNKNOWN')
     OPEN(UNIT=10,FILE='NZOUT3D.DAT',STATUS='UNKNOWN')
C###       OPEN (UNIT=10,FILE='/NZOUT3D DATA E1',
                 STATUS='UNKNOWN')
C
C    ZERO PARAMETERS, GENERATE PROBLEM SPACE, INTER
C    ACTION OBJECT, AND EXCITATION
C
    CALL ZERO
    CALL BUILD
    CALL SETUP
C
C   ***********************************************************
C
C    MAIN LOOP FOR FIELD COMPUTATIONS AND DATA SAVING
C
C   ***********************************************************
    T=0.0
    DO 100 N=1,NSTOP
```

```
C
    WRITE (*,*) N
C
C     ADVANCE SCATTERED ELECTRIC FIELD
C
      CALL EXSFLD
      CALL EYSFLD
      CALL EZSFLD
C
C     APPLY RADIATION BC (SECOND ORDER)
C
      CALL RADEYX
      CALL RADEZX
      CALL RADEZY
      CALL RADEXY
      CALL RADEXZ
      CALL RADEYZ
C
C     ADVANCE TIME BY 1/2 TIME STEP
C
      T=T+DT/2.
C
C     ADVANCE SCATTERED MAGNETIC FIELD
C
      CALL HXSFLD
      CALL HYSFLD
      CALL HZSFLD
C
C     ADVANCE TIME ANOTHER 1/2 STEP
C
      T=T+DT/2.
C
C     SAMPLE FIELDS IN SPACE AND WRITE TO DISK
C
      CALL DATSAV
C
 100  CONTINUE
    T=NSTOP*DT
    WRITE (17,200) T,NSTOP
  200 FORMAT(T2,' EXIT TIME= ',E14.7,' SECONDS,  AT TIME
      STEP', I6,/)
C
C    CLOSE DATA FILES
C
    CLOSE (UNIT=10)
```

```
      CLOSE (UNIT=17)
      STOP
      END
C     ********************************************************
      SUBROUTINE BUILD
C
      INCLUDE 'COMMONA.FOR'
C$INCLUDE COMMONA
C
C     THIS SUBROUTINE IS USED TO DEFINE THE SCATTERING
C     OBJECT WITHIN THE FDTD SOLUTION SPACE. USER MUST
C     SPECIFY IDONE, IDTWO AND IDTHRE AT DIFFERENT CELL
C     LOCATIONS TO DEFINE THE SCATTERING OBJECT. SEE THE
C     YEE PAPER (IEEE TRANS. ON AP, MAY 1966) FOR A DE-
C     SCRIPTION OF THE FDTD ALGORITHM AND THE LOCATION
C     OF FIELD COMPONENTS.
C
C     GEOMETRY DEFINITION
C
C     IDONE, IDTWO, AND IDTHRE ARE USED TO SPECIFY MATE
C     RIAL IN CELL I,J,K. IDONE DETERMINES MATERIAL FOR X
C     COMPONENTS OF E IDTWO FOR Y COMPONENTS, IDTHRE
C     FOR Z COMPONENTS THUS ANISOTROPIC MATERIALS WITH
C     DIAGONAL TENSORS CAN BE MODELED
C
C     SET IDONE,IDTWO, AND/OR IDTHRE FOR EACH I,J,K CELL =
C             0      FOR FREE SPACE
C             1      FOR PEC
C             2-9    FOR LOSSY DIELECTRICS
C
C     SUBROUTINE DCUBE BUILDS A CUBE OF DIELECTRIC
C     MATERIAL BY SETTING IDONE, IDTWO, IDTHRE TO THE
C     SAME MATERIAL TYPE. THE MATERIAL TYPE IS SPECIFIED
C     BY MTYPE. SPECIFY THE STARTING CELL (LOWER LEFT
C     CORNER (I.E. MINIMUM I,J,K VALUES) AND SPECIFY THE
C     CELL WIDTH IN EACH DIRECTION (USE THE NUMBER OF
C     CELLS IN EACH DIRECTION). USE NZWIDE=0 FOR A INFI-
C     NITELY THIN PLATE IN THE XY PLANE. FOR PEC PLATE
C     USE MTYPE=1. ISTART, JSTART, KSTART ARE USED TO
C     DEFINE THE STARTING CELL AND NXWIDE, NYWIDE AND
C     NZWIDE EACH SPECIFY THE OBJECT WIDTH IN CELLS IN
C     THE X, Y AND Z DIRECTIONS. INDIVIDUAL IDONE, TWO OR
C     THRE COMPONENTS CAN  BE SET MANUALLY FOR WIRES,
C     ETC. DCUBE DOES NOT WORK FOR WIRES (I.E. NXWIDE=0
C     AND NYWIDE=0 FOR EXAMPLE)!
```

```
C
C  Build sphere with center at (SC,SC,SC) and radius RA.
      MTYPE=2
      RA=8.2
      SC=17.5
      DO 100 I=1,NX
        DO 200 J=1,NY
          DO 300 K=1,NZ
            R=SQRT((I-SC)**2+(J-SC)**2+(K-SC)**2)
            IF (R.LT.RA) CALL DCUBE (I,J,K,1,1,1,MTYPE)
 300      CONTINUE
 200    CONTINUE
 100  CONTINUE
C
C     THE FOLLOWING SECTION OF CODE IS USED TO CHECK IF
C     THE USER HAS SPECIFIED THE PROPER MATERIAL TYPES
C     TO THE PROPER IDXXX ARRAYS.
C
C     FOR MATERIAL TYPE = ?
C       0 FOR FREE SPACE              USE  IDONE-IDTHRE ARRAYS
C       1 FOR PEC                     USE  IDONE-IDTHRE ARRAYS
C       2-9 FOR LOSSY DIELECTRICS USE  IDONE-IDTHRE ARRAYS
C
      DO 1000 K=1,NZ
        DO 900 J=1,NY
          DO 800 I=1,NX
            IF((IDONE(I,J,K).GE.10).OR.(IDTWO(I,J,K).GE.10).OR.
     $(IDTHRE(I,J,K).GE.10)) THEN
              WRITE (17,*)'ERROR OCCURED. ILLEGAL VALUE FOR'
              WRITE (17,*)'DIELECTRIC TYPE (IDONE-IDTHRE) '
              WRITE (17,*)'AT LOCATION:',I,',',J,',',K
              WRITE (17,*)'EXECUTION HALTED.'
              STOP
            ENDIF
 800      CONTINUE
 900    CONTINUE
 1000 CONTINUE
      RETURN
      END
C     *****************************************************
      SUBROUTINE DCUBE
(ISTART,JSTART,KSTART,NXWIDE,NYWIDE,NZWIDE,MTYPE)
C
      INCLUDE 'COMMONA.FOR'
C$INCLUDE COMMONA
```

```
C
C   THIS SUBROUTINE SETS ALL TWELVE IDXXX COMPONENTS
C   FOR ONE CUBE TO THE SAME MATERIAL TYPE SPECIFIED
C   BY MTYPE. IF NXWIDE, NYWIDE, OR NZWIDE=0, THEN
C   ONLY 4 IDXXX ARRAY COMPONENTS WILL BE SET CORRE-
C   SPONDING TO AN INFINITELY THIN PLATE. THE SUB-
C   ROUTINE IS MOST USEFUL CONSTRUCTING OBJECTS
C   WITH MANY CELLS OF THE SAME MATERIAL (I.E. CUBES,
C   PEC PLATES, ETC.). THIS SUBROUTINE DOES NOT AUTO-
C   MATICALLY DO WIRES!
C
      IMAX=ISTART+NXWIDE-1
      JMAX=JSTART+NYWIDE-1
      KMAX=KSTART+NZWIDE-1
C
      IF (NXWIDE.EQ.0) THEN
        DO 20 K=KSTART,KMAX
          DO 10 J=JSTART,JMAX
            IDTWO(ISTART,J,K)=MTYPE
            IDTWO(ISTART,J,K+1)=MTYPE
            IDTHRE(ISTART,J,K)=MTYPE
            IDTHRE(ISTART,J+1,K)=MTYPE
 10       CONTINUE
 20     CONTINUE
      ELSEIF (NYWIDE.EQ.0) THEN
        DO 40 K=KSTART,KMAX
          DO 30 I=ISTART,IMAX
            IDONE(I,JSTART,K)=MTYPE
            IDONE(I,JSTART,K+1)=MTYPE
            IDTHRE(I,JSTART,K)=MTYPE
            IDTHRE(I+1,JSTART,K)=MTYPE
 30       CONTINUE
 40     CONTINUE
      ELSEIF (NZWIDE.EQ.0) THEN
        DO 60 J=JSTART,JMAX
          DO 50 I=ISTART,IMAX
            IDONE(I,J,KSTART)=MTYPE
            IDONE(I,J+1,KSTART)=MTYPE
            IDTWO(I,J,KSTART)=MTYPE
            IDTWO(I+1,J,KSTART)=MTYPE
 50       CONTINUE
 60     CONTINUE
      ELSE
        DO 90 K=KSTART,KMAX
          DO 80 J=JSTART,JMAX
```

```
          DO 70 I=ISTART,IMAX
            IDONE(I,J,K)=MTYPE
            IDONE(I,J,K+1)=MTYPE
            IDONE(I,J+1,K+1)=MTYPE
            IDONE(I,J+1,K)=MTYPE
            IDTWO(I,J,K)=MTYPE
            IDTWO(I+1,J,K)=MTYPE
            IDTWO(I+1,J,K+1)=MTYPE
            IDTWO(I,J,K+1)=MTYPE
            IDTHRE(I,J,K)=MTYPE
            IDTHRE(I+1,J,K)=MTYPE
            IDTHRE(I+1,J+1,K)=MTYPE
            IDTHRE(I,J+1,K)=MTYPE
 70       CONTINUE
 80     CONTINUE
 90   CONTINUE
      ENDIF
      RETURN
      END
C     ***************************************************
      SUBROUTINE SETUP
C
      INCLUDE 'COMMONA.FOR'
C$INCLUDE COMMONA
C
C     THIS SUBROUTINE INITIALIZES THE COMPUTATIONS
C
C     DEFINE PI AND C
C
      C=1.0/SQRT(XMU0*EPS0)
      PI=4.0*ATAN(1.0)
C
C     CALCULATE DT—THE MAXIMUM TIME STEP ALLOWED BY
THE
C     COURANT STABILITY CONDITION
C
      DTXI=C/DELX
      DTYI=C/DELY
      DTZI=C/DELZ
C
      DT=1./SQRT(DTXI**2+DTYI**2+DTZI**2)
C
C     PARAMETER ALPHA IS THE DECAY RATE DETERMINED BY
C     BETA.
```

```
C
      ALPHA=(1./(BETA*DT/4.0))**2
C
      BETADT =  BETA*DT
      PERIOD = 2.0*BETADT
C
C     SET OFFSET FOR COMPUTING INCIDENT FIELDS
C
      OFF=1.0
C
C     THE FOLLOWING LINES ARE FOR SMOOTH COSINE INCI-
C     DENT FUNCTION
C
      W1 = 2.0*PI/PERIOD
      W2 = 2.0*W1
      W3 = 3.0*W1
C
C     FIND DIRECTION COSINES FOR INCIDENT FIELD
C
      COSTH=COS(PI*THINC/180.)
      SINTH=SIN(PI*THINC/180.)
      COSPH=COS(PI*PHINC/180.)
      SINPH=SIN(PI*PHINC/180.)
C
C     FIND AMPLITUDE OF INCIDENT FIELD COMPONENTS
C
      AMPX=AMP*(ETHINC*COSTH*COSPH-EPHINC*SINPH)
      AMPY=AMP*(ETHINC*COSTH*SINPH+EPHINC*COSPH)
      AMPZ=AMP*(-ETHINC*SINTH)
C
C     FIND RELATIVE SPATIAL DELAY FOR X, Y, Z CELL DIS
C     PLACEMENT
C
      XDISP=-COSPH*SINTH
      YDISP=-SINPH*SINTH
      ZDISP=-COSTH
C
C     DEFINE CONSTITUTIVE PARAMETERS—EPS,SIGMA
C     THIS CODE ASSUMES THAT THE DIELECTRIC MATERIALS
C     ALL HAVE A PERMEABILITY OF MU0.  IF YOU NEED A
C     MATERIAL WITH MAGNETIC PROPERTIES, USE ANOTHER
C     FDTD CODE SUCH AS FDTDC, FDTDD, FDTDG OR FDTDH.
C
C     THESE CORRESPOND TO MATERIAL TYPES IN IDONE,
```

```
C     IDTWO, AND IDTHRE ARRAYS.  SEE COMMENTS IN SUB-
C     ROUTINE BUILD
C
C     VALID CASES:  IDXXX(I,J,K)  = 0 FOR FREE SPACE IN CELL
C                                   I,J,K
C                                 = 1 FOR PEC
C                                 = 2-9 FOR LOSSY DIELECTRICS
C     FOR PARTICULAR COMPONENTS OF FIELDS AS DETER
      MINED BY XXX
C
C     IF IDONE (10,10,10) = 3 THEN THE USER MUST SPECIFY AN
C     EPSILON, MU AND SIGMA FOR MATERIAL TYPE 3.  THAT IS,
C     EPS(3)=EPSILON FOR MATERIAL TYPE 3 AND SIGMA(3)=
C     CONDUCTIVITY FOR MATERIAL TYPE 3. ALL PARAMETERS
C     ARE DEFINED IN MKS UNITS.
C
      DO 10 I=1,9
        EPS(I)=EPS0
        SIGMA(I)=0.0
 10   CONTINUE
C
C     DEFINE EPS AND SIGMA FOR EACH MATERIAL HERE
C
      EPS(2)=4.0*EPS0
      SIGMA(2)=0.005
C
C     GENERATE MULTIPLICATIVE CONSTANTS FOR FIELD
C     UPDATE EQUATIONS
C
C     FREE SPACE
C
      DTEDX=DT/(EPS0*DELX)
      DTEDY=DT/(EPS0*DELY)
      DTEDZ=DT/(EPS0*DELZ)
      DTMDX=DT/(XMU0*DELX)
      DTMDY=DT/(XMU0*DELY)
      DTMDZ=DT/(XMU0*DELZ)
C
C     LOSSY DIELECTRICS
C
      DO 20 I=2,9
        ESCTC(I)=EPS(I)/(EPS(I)+SIGMA(I) *DT)
        EINCC(I)=SIGMA(I) *DT/(EPS(I)+SIGMA(I) *DT)
        EDEVCN(I)=DT*(EPS(I)-EPS0)/(EPS(I)+SIGMA(I) *DT)
        ECRLX(I)=DT/((EPS(I)+SIGMA(I) *DT)*DELX)
```

```
      ECRLY(I)=DT/((EPS(I)+SIGMA(I) *DT)*DELY)
      ECRLZ(I)=DT/((EPS(I)+SIGMA(I) *DT)*DELZ)
 20   CONTINUE
C
C     FIND MAXIMUM SPATIAL DELAY TO MAKE SURE PULSE
C     PROPAGATES INTO SPACE PROPERLY.
C
      DELAY=0.0
      IF (XDISP.LT.0.) DELAY=DELAY-XDISP*NX1*DELX
      IF (YDISP.LT.0.) DELAY=DELAY-YDISP*NY1*DELY
      IF (ZDISP.LT.0.) DELAY=DELAY-ZDISP*NZ1*DELZ
C
C     COMPUTE OUTER RADIATION BOUNDARY CONDITION
C     (ORBC) CONSTANTS
C
      CXD=(C*DT-DELX)/(C*DT+DELX)
      CYD=(C*DT-DELY)/(C*DT+DELY)
      CZD=(C*DT-DELZ)/(C*DT+DELZ)
C
      CXU=CXD
      CYU=CYD
      CZU=CZD
C
C     COMPUTE 2ND ORDER ORBC CONSTANTS
C
      CXX=2.*DELX/(C*DT+DELX)
      CYY=2.*DELY/(C*DT+DELY)
      CZZ=2.*DELZ/(C*DT+DELZ)
C
      CXFYD=DELX*C*DT*C*DT/(2.*DELY*DELY*(C*DT+DELX))
      CXFZD=DELX*C*DT*C*DT/(2.*DELZ*DELZ*(C*DT+DELX))
      CYFZD=DELY*C*DT*C*DT/(2.*DELZ*DELZ*(C*DT+DELY))
      CYFXD=DELY*C*DT*C*DT/(2.*DELX*DELX*(C*DT+DELY))
      CZFXD=DELZ*C*DT*C*DT/(2.*DELX*DELX*(C*DT+DELZ))
      CZFYD=DELZ*C*DT*C*DT/(2.*DELY*DELY*(C*DT+DELZ))
C
C     WRITE SETUP DATA TO FILE DIAGS3D.DAT
C
      WRITE (17,400) NX,NY,NZ
      WRITE (17,500) DELX,DELY,DELZ
      WRITE (17,600) DT,NSTOP
      WRITE (17,700) AMP,ALPHA,BETA
      WRITE (17,800) ETHINC,EPHINC
      WRITE (17,900) THINC,PHINC
      WRITE (17,1000) 'THE INCIDENT EX AMPLITUDE = ',
```

```
      AMPX,'V/M'
      WRITE (17,1000) 'THE INCIDENT EY AMPLITUDE = ',
      AMPY,'V/M'
      WRITE (17,1000) 'THE INCIDENT EZ AMPLITUDE = ',
      AMPZ,'V/M'
      WRITE (17,1100) 'RELATIVE SPATIAL DELAY = ',DELAY
C
 100  FORMAT (T2,A26,/)
 200  FORMAT (T2,A25,/)
 300  FORMAT (T2,A34,/)
 400  FORMAT (T2, 'THE PROBLEM SPACE SIZE IS', I4,' BY',I4,'
        BY',I4,
     $' CELLS IN THE X, Y, Z DIRECTIONS',/)
 500  FORMAT (T2,'CELL SIZE: DELX=',F10.6,' DELY=',F10.6,',
        DELZ=',
     $F10.6, ' METERS',/)
 600  FORMAT (T2,'TIME STEP IS ',E12.6,' SECONDS, MAXIMUM
        OF',I6,
     $' TIME STEPS',/)
 700  FORMAT (T2,'INCIDENT GAUSSIAN PULSE
        AMPLITUDE=',F6.0,' V/M',
     $/,T2,'DECAY FACTOR ALPHA=',E12.3,/,T2,'WIDTH
        BETA=',F6.0,/)
 800  FORMAT (T2,'INCIDENT PLANE WAVE POLARIZATION:',/,T2,
     $ 'RELATIVE ELECTRIC FIELD THETA COMPONENT=',F4.1,/,T2,
     $ 'RELATIVE ELECTRIC FIELD PHI COMPONENT=', F4.1,/)
 900  FORMAT (T2,'PLANE WAVE INCIDENT FROM THETA=',F8.2,'
        PHI=',F8.2,
     $' DEGREES',/)
 1000 FORMAT (T2,A28,F12.5,2X,A3)
 1100 FORMAT (/,T2,A25,F10.7,/)
      RETURN
      END
C     ***********************************************************
      SUBROUTINE EXSFLD
C
      INCLUDE 'COMMONA.FOR'
C$INCLUDE COMMONA
C
C     THIS SUBROUTINE UPDATES THE EX SCATTERED FIELD
C
      DO 30 K=2,NZ1
        KK = K
        DO 20 J=2,NY1
          JJ = J
```

```
        DO 10 I=1,NX1
C
C          DETERMINE MATERIAL TYPE
C
          IF(IDONE(I,J,K).EQ.0) GO TO 100
          IF(IDONE(I,J,K).EQ.1) GO TO 200
          GO TO 300
C
C          FREE SPACE
C
 100       EXS(I,J,K)=EXS(I,J,K)+(HZS(I,J,K)-HZS(I,J-1,K))*DTEDY
    $                    -(HYS(I,J,K)-HYS(I,J,K-1))*DTEDZ
C
          GO TO 10
C
C          PERFECT CONDUCTOR
C
 200       II = I
          EXS(I,J,K)=-EXI(II,JJ,KK)
C
          GO TO 10
C
C          LOSSY DIELECTRIC
C
 300       II = I
          EXS(I,J,K)=EXS(I,J,K)*ESCTC(IDONE(I,J,K))
    $     -EINCC( IDONE(I,J,K))*EXI(II,JJ,KK)
    $     -EDEVCN(IDONE(I,J,K))*DEXI(II,JJ,KK)
    $     +(HZS(I,J,K)-HZS(I,J-1,K))*ECRLY(IDONE(I,J,K))
    $     -(HYS(I,J,K)-HYS(I,J,K-1))*ECRLZ(IDONE(I,J,K))
C
 10       CONTINUE
 20     CONTINUE
 30   CONTINUE
C
      RETURN
      END
C     ******************************************************
      SUBROUTINE EYSFLD
C
      INCLUDE 'COMMONA.FOR'
C$INCLUDE COMMONA
C
C     THIS SUBROUTINE UPDATES THE EY SCATTERED FIELD
C     COMPONENTS
```

```
C
      DO 30 K=2,NZ1
        KK = K
        DO 20 J=1,NY1
          JJ = J
          DO 10 I=2,NX1
C
C           DETERMINE MATERIAL TYPE
C
            IF(IDTWO(I,J,K).EQ.0) GO TO 100
            IF(IDTWO(I,J,K).EQ.1) GO TO 200
            GO TO 300
C
C           FREE SPACE
C
 100        EYS(I,J,K)=EYS(I,J,K)+(HXS(I,J,K)-HXS(I,J,K-1))*DTEDZ
     $                   -(HZS(I,J,K)-HZS(I-1,J,K))*DTEDX
C
            GO TO 10
C
C           PERFECT CONDUCTOR
C
 200        II = I
            EYS(I,J,K)=-EYI(II,JJ,KK)
C
            GO TO 10
C
C           LOSSY DIELECTRIC
C
 300        II = I
            EYS(I,J,K)=EYS(I,J,K)*ESCTC(IDTWO(I,J,K))
     $      -EINCC( IDTWO(I,J,K))*EYI(II,JJ,KK)
     $      -EDEVCN(IDTWO(I,J,K))*DEYI(II,JJ,KK)
     $      +(HXS(I,J,K)-HXS(I,J,K-1))*ECRLZ(IDTWO(I,J,K))
     $      -(HZS(I,J,K)-HZS(I-1,J,K))*ECRLX(IDTWO(I,J,K))
C
  10      CONTINUE
  20    CONTINUE
  30  CONTINUE
C
      RETURN
      END
C     ***********************************************************
      SUBROUTINE EZSFLD
C
```

```
      INCLUDE 'COMMONA.FOR'
C$INCLUDE COMMONA
C
C   THIS SUBROUTINE UPDATES THE EZ SCATTERED FIELD
C   COMPONENTS
C
      DO 30 K=1,NZ1
        KK = K
        DO 20 J=2,NY1
          JJ = J
          DO 10 I=2,NX1
C
C           DETERMINE MATERIAL TYPE
C
            IF(IDTHRE(I,J,K).EQ.0) GO TO 100
            IF(IDTHRE(I,J,K).EQ.1) GO TO 200
            GO TO 300
C
C           FREE SPACE
C
 100        EZS(I,J,K)=EZS(I,J,K)+(HYS(I,J,K)-HYS(I-1,J,K))*DTEDX
     $                       -(HXS(I,J,K)-HXS(I,J-1,K))*DTEDY
C
            GO TO 10
C
C           PERFECT CONDUCTOR
C
 200        II = I
            EZS(I,J,K)=-EZI(II,JJ,KK)
C
            GO TO 10
C
C           LOSSY DIELECTRIC
C
 300        II = I
            EZS(I,J,K)=EZS(I,J,K)*ESCTC(IDTHRE(I,J,K))
     $      -EINCC( IDTHRE(I,J,K))*EZI(II,JJ,KK)
     $      -EDEVCN(IDTHRE(I,J,K))*DEZI(II,JJ,KK)
     $      +(HYS(I,J,K)-HYS(I-1,J,K))*ECRLX(IDTHRE(I,J,K))
     $      -(HXS(I,J,K)-HXS(I,J-1,K))*ECRLY(IDTHRE(I,J,K))
C
 10       CONTINUE
 20     CONTINUE
 30   CONTINUE
C
```

```
      RETURN
      END
C     ****************************************************
      SUBROUTINE RADEZX
C
      INCLUDE 'COMMONA.FOR'
C$INCLUDE COMMONA
C
C     DO EDGES WITH FIRST ORDER ORBC
C
      DO 10 K=1,NZ1
        J=2
        EZS(1,J,K)=EZSX1(2,J,K)+CXD*(EZS(2,J,K)-EZSX1(1,J,K))
        EZS(NX,J,K)=EZSX1(3,J,K)+CXU*(EZS(NX1,J,K)-EZSX1(4,J,K))
        J=NY1
        EZS(1,J,K)=EZSX1(2,J,K)+CXD*(EZS(2,J,K)-EZSX1(1,J,K))
        EZS(NX,J,K)=EZSX1(3,J,K)+CXU*(EZS(NX1,J,K)-EZSX1(4,J,K))
 10   CONTINUE
C
      DO 20 J=3,NY1-1
        K=1
        EZS(1,J,K)=EZSX1(2,J,K)+CXD*(EZS(2,J,K)-EZSX1(1,J,K))
        EZS(NX,J,K)=EZSX1(3,J,K)+CXU*(EZS(NX1,J,K)-EZSX1(4,J,K))
        K=NZ1
        EZS(1,J,K)=EZSX1(2,J,K)+CXD*(EZS(2,J,K)-EZSX1(1,J,K))
        EZS(NX,J,K)=EZSX1(3,J,K)+CXU*(EZS(NX1,J,K)-EZSX1(4,J,K))
 20   CONTINUE
C
C     NOW DO 2ND ORDER ORBC ON REMAINING PORTIONS OF
C     FACES
C
      DO 40 K=2,NZ1-1
        DO 30 J=3,NY1-1
          EZS(1,J,K)=-EZSX2(2,J,K)+CXD*(EZS(2,J,K)+EZSX2(1,J,K))
     $    +CXX*(EZSX1(1,J,K)+EZSX1(2,J,K))
     $    +CXFYD*(EZSX1(1,J+1,K)-2.*EZSX1(1,J,K)+EZSX1(1,J-1,K)
     $    +EZSX1(2,J+1,K)-2.*EZSX1(2,J,K)+EZSX1(2,J-1,K))
     $    +CXFZD*(EZSX1(1,J,K+1)-2.*EZSX1(1,J,K)+EZSX1(1,J,K-1)
     $    +EZSX1(2,J,K+1)-2.*EZSX1(2,J,K)+EZSX1(2,J,K-1))
          EZS(NX,J,K)=-EZSX2(3,J,K)+CXD*(EZS(NX1,J,K)+EZSX2(4,J,K))
     $    +CXX*(EZSX1(4,J,K)+EZSX1(3,J,K))
     $    +CXFYD*(EZSX1(4,J+1,K)-2.*EZSX1(4,J,K)+EZSX1(4,J-1,K)
     $    +EZSX1(3,J+1,K)-2.*EZSX1(3,J,K)+EZSX1(3,J-1,K))
     $    +CXFZD*(EZSX1(4,J,K+1)-2.*EZSX1(4,J,K)+EZSX1(4,J,K-1)
     $    +EZSX1(3,J,K+1)-2.*EZSX1(3,J,K)+EZSX1(3,J,K-1))
```

```
 30     CONTINUE
 40   CONTINUE
C
C     NOW SAVE PAST VALUES
C
      DO 60 K=1,NZ1
        DO 50 J=2,NY1
          EZSX2(1,J,K)=EZSX1(1,J,K)
          EZSX2(2,J,K)=EZSX1(2,J,K)
          EZSX2(3,J,K)=EZSX1(3,J,K)
          EZSX2(4,J,K)=EZSX1(4,J,K)
          EZSX1(1,J,K)=EZS(1,J,K)
          EZSX1(2,J,K)=EZS(2,J,K)
          EZSX1(3,J,K)=EZS(NX1,J,K)
          EZSX1(4,J,K)=EZS(NX,J,K)
 50     CONTINUE
 60   CONTINUE
C
      RETURN
      END
C     ******************************************************
      SUBROUTINE RADEYX
C
      INCLUDE 'COMMONA.FOR'
C$INCLUDE COMMONA
C
C     DO EDGES WITH FIRST ORDER ORBC
C
      DO 10 K=2,NZ1
        J=1
        EYS(1,J,K)=EYSX1(2,J,K)+CXD*(EYS(2,J,K)-EYSX1(1,J,K))
        EYS(NX,J,K)=EYSX1(3,J,K)+CXU*(EYS(NX1,J,K)-EYSX1(4,J,K))
        J=NY1
        EYS(1,J,K)=EYSX1(2,J,K)+CXD*(EYS(2,J,K)-EYSX1(1,J,K))
        EYS(NX,J,K)=EYSX1(3,J,K)+CXU*(EYS(NX1,J,K)-EYSX1(4,J,K))
 10   CONTINUE
C
      DO 20 J=2,NY1-1
        K=2
        EYS(1,J,K)=EYSX1(2,J,K)+CXD*(EYS(2,J,K)-EYSX1(1,J,K))
        EYS(NX,J,K)=EYSX1(3,J,K)+CXU*(EYS(NX1,J,K)-EYSX1(4,J,K))
        K=NZ1
        EYS(1,J,K)=EYSX1(2,J,K)+CXD*(EYS(2,J,K)-EYSX1(1,J,K))
        EYS(NX,J,K)=EYSX1(3,J,K)+CXU*(EYS(NX1,J,K)-EYSX1(4,J,K))
 20   CONTINUE
```

```
C
C    NOW DO 2ND ORDER ORBC ON REMAINING PORTIONS OF
C    FACES
C
      DO 40 K=3,NZ1-1
        DO 30 J=2,NY1-1
          EYS(1,J,K)=-EYSX2(2,J,K)+CXD*(EYS(2,J,K)+EYSX2(1,J,K))
     $    +CXX*(EYSX1(1,J,K)+EYSX1(2,J,K))
     $    +CXFYD*(EYSX1(1,J+1,K)-2.*EYSX1(1,J,K)+EYSX1(1,J-1,K)
     $    +EYSX1(2,J+1,K)-2.*EYSX1(2,J,K)+EYSX1(2,J-1,K))
     $    +CXFZD*(EYSX1(1,J,K+1)-2.*EYSX1(1,J,K)+EYSX1(1,J,K-1)
     $    +EYSX1(2,J,K+1)-2.*EYSX1(2,J,K)+EYSX1(2,J,K-1))
          EYS(NX,J,K)=-EYSX2(3,J,K)+CXD*(EYS (NX1, J,K)
           +EYSX2(4,J,K))
     $    +CXX*(EYSX1(4,J,K)+EYSX1(3,J,K))
     $    +CXFYD*(EYSX1(4,J+1,K)-2.*EYSX1(4,J,K)+EYSX1(4,J-1,K)
     $    +EYSX1(3,J+1,K)-2.*EYSX1(3,J,K)+EYSX1(3,J-1,K))
     $    +CXFZD*(EYSX1(4,J,K+1)-2.*EYSX1(4,J,K)+EYSX1(4,J,K-1)
     $    +EYSX1(3,J,K+1)-2.*EYSX1(3,J,K)+EYSX1(3,J,K-1))
 30     CONTINUE
 40   CONTINUE
C
C    NOW SAVE PAST VALUES
C
      DO 60 K=2,NZ1
        DO 50 J=1,NY1
          EYSX2(1,J,K)=EYSX1(1,J,K)
          EYSX2(2,J,K)=EYSX1(2,J,K)
          EYSX2(3,J,K)=EYSX1(3,J,K)
          EYSX2(4,J,K)=EYSX1(4,J,K)
          EYSX1(1,J,K)=EYS(1,J,K)
          EYSX1(2,J,K)=EYS(2,J,K)
          EYSX1(3,J,K)=EYS(NX1,J,K)
          EYSX1(4,J,K)=EYS(NX,J,K)
 50     CONTINUE
 60   CONTINUE
C
      RETURN
      END
C    ********************************************************
      SUBROUTINE RADEZY
C
      INCLUDE 'COMMONA.FOR'
C$INCLUDE COMMONA
C
```

```
C     DO EDGES WITH FIRST ORDER ORBC
C
      DO 10 K=1,NZ1
        I=2
        EZS(I,1,K)=EZSY1(I,2,K)+CYD*(EZS(I,2,K)-EZSY1(I,1,K))
        EZS(I,NY,K)=EZSY1(I,3,K)+CYD*(EZS(I,NY1,K)-EZSY1(I,4,K))
        I=NX1
        EZS(I,1,K)=EZSY1(I,2,K)+CYD*(EZS(I,2,K)-EZSY1(I,1,K))
        EZS(I,NY,K)=EZSY1(I,3,K)+CYD*(EZS(I,NY1,K)-EZSY1(I,4,K))
 10   CONTINUE
C
      DO 20 I=3,NX1-1
        K=1
        EZS(I,1,K)=EZSY1(I,2,K)+CYD*(EZS(I,2,K)-EZSY1(I,1,K))
        EZS(I,NY,K)=EZSY1(I,3,K)+CYD*(EZS(I,NY1,K)-EZSY1(I,4,K))
        K=NZ1
        EZS(I,1,K)=EZSY1(I,2,K)+CYD*(EZS(I,2,K)-EZSY1(I,1,K))
        EZS(I,NY,K)=EZSY1(I,3,K)+CYD*(EZS(I,NY1,K)-EZSY1(I,4,K))
 20   CONTINUE
C
C     NOW DO 2ND ORDER ORBC ON REMAINING PORTIONS OF
C     FACES
C
      DO 40 K=2,NZ1-1
        DO 30 I=3,NX1-1
          EZS(I,1,K)=-EZSY2(I,2,K)+CYD*(EZS(I,2,K)+EZSY2(I,1,K))
     $    +CYY*(EZSY1(I,1,K)+EZSY1(I,2,K))
     $    +CYFXD*(EZSY1(I+1,1,K)-2.*EZSY1(I,1,K)+EZSY1(I-1,1,K)
     $    +EZSY1(I+1,2,K)-2.*EZSY1(I,2,K)+EZSY1(I-1,2,K))
     $    +CYFZD*(EZSY1(I,1,K+1)-2.*EZSY1(I,1,K)+EZSY1(I,1,K-1)
     $    +EZSY1(I,2,K+1)-2.*EZSY1(I,2,K)+EZSY1(I,2,K-1))
          EZS(I,NY,K)=-EZSY2(I,3,K)+CYD*(EZS(I,NY1,K)+EZSY2(I,4,K))
     $    +CYY*(EZSY1(I,4,K)+EZSY1(I,3,K))
     $    +CYFXD*(EZSY1(I+1,4,K)-2.*EZSY1(I,4,K)+EZSY1(I-1,4,K)
     $    +EZSY1(I+1,3,K)-2.*EZSY1(I,3,K)+EZSY1(I-1,3,K))
     $    +CYFZD*(EZSY1(I,4,K+1)-2.*EZSY1(I,4,K)+EZSY1(I,4,K-1)
     $    +EZSY1(I,3,K+1)-2.*EZSY1(I,3,K)+EZSY1(I,3,K-1))
 30     CONTINUE
 40   CONTINUE
C
C     NOW SAVE PAST VALUES
C
      DO 60 K=1,NZ1
        DO 50 I=2,NX1
          EZSY2(I,1,K)=EZSY1(I,1,K)
```

```
        EZSY2(I,2,K)=EZSY1(I,2,K)
        EZSY2(I,3,K)=EZSY1(I,3,K)
        EZSY2(I,4,K)=EZSY1(I,4,K)
        EZSY1(I,1,K)=EZS(I,1,K)
        EZSY1(I,2,K)=EZS(I,2,K)
        EZSY1(I,3,K)=EZS(I,NY1,K)
        EZSY1(I,4,K)=EZS(I,NY,K)
 50    CONTINUE
 60   CONTINUE
C
      RETURN
      END
C     *****************************************************
      SUBROUTINE RADEXY
C
      INCLUDE 'COMMONA.FOR'
C$INCLUDE COMMONA
C
C     DO EDGES WITH FIRST ORDER ORBC
C
      DO 10 K=2,NZ1
       I=1
       EXS(I,1,K)=EXSY1(I,2,K)+CYD*(EXS(I,2,K)-EXSY1(I,1,K))
       EXS(I,NY,K)=EXSY1(I,3,K)+CYD*(EXS(I,NY1,K)-EXSY1(I,4,K))
       I=NX1
       EXS(I,1,K)=EXSY1(I,2,K)+CYD*(EXS(I,2,K)-EXSY1(I,1,K))
       EXS(I,NY,K)=EXSY1(I,3,K)+CYD*(EXS(I,NY1,K)-EXSY1(I,4,K))
 10   CONTINUE
C
      DO 20 I=2,NX1-1
       K=2
       EXS(I,1,K)=EXSY1(I,2,K)+CYD*(EXS(I,2,K)-EXSY1(I,1,K))
       EXS(I,NY,K)=EXSY1(I,3,K)+CYD*(EXS(I,NY1,K)-EXSY1(I,4,K))
       K=NZ1
       EXS(I,1,K)=EXSY1(I,2,K)+CYD*(EXS(I,2,K)-EXSY1(I,1,K))
       EXS(I,NY,K)=EXSY1(I,3,K)+CYD*(EXS(I,NY1,K)-EXSY1(I,4,K))
 20   CONTINUE
C
C     NOW DO 2ND ORDER ORBC ON REMAINING PORTIONS OF
C     FACES
C
      DO 40 K=3,NZ1-1
       DO 30 I=2,NX1-1
        EXS(I,1,K)=-EXSY2(I,2,K)+CYD*(EXS(I,2,K)+EXSY2(I,1,K))
     $    +CYY*(EXSY1(I,1,K)+EXSY1(I,2,K))
```

```
     $   +CYFXD*(EXSY1(I+1,1,K)-2.*EXSY1(I,1,K)+EXSY1(I-1,1,K)
     $   +EXSY1(I+1,2,K)-2.*EXSY1(I,2,K)+EXSY1(I-1,2,K))
     $   +CYFZD*(EXSY1(I,1,K+1)-2.*EXSY1(I,1,K)+EXSY1(I,1,K-1)
     $   +EXSY1(I,2,K+1)-2.*EXSY1(I,2,K)+EXSY1(I,2,K-1))
         EXS(I,NY,K)=-EXSY2(I,3,K)+CYD*(EXS(I,NY1,K)+EXSY2(I,4,K))
     $   +CYY*(EXSY1(I,4,K)+EXSY1(I,3,K))
     $   +CYFXD*(EXSY1(I+1,4,K)-2.*EXSY1(I,4,K)+EXSY1(I-1,4,K)
     $   +EXSY1(I+1,3,K)-2.*EXSY1(I,3,K)+EXSY1(I-1,3,K))
     $   +CYFZD*(EXSY1(I,4,K+1)-2.*EXSY1(I,4,K)+EXSY1(I,4,K-1)
     $   +EXSY1(I,3,K+1)-2.*EXSY1(I,3,K)+EXSY1(I,3,K-1))
 30    CONTINUE
 40  CONTINUE
C
C    NOW SAVE PAST VALUES
C
      DO 60 K=2,NZ1
       DO 50 I=1,NX1
         EXSY2(I,1,K)=EXSY1(I,1,K)
         EXSY2(I,2,K)=EXSY1(I,2,K)
         EXSY2(I,3,K)=EXSY1(I,3,K)
         EXSY2(I,4,K)=EXSY1(I,4,K)
         EXSY1(I,1,K)=EXS(I,1,K)
         EXSY1(I,2,K)=EXS(I,2,K)
         EXSY1(I,3,K)=EXS(I,NY1,K)
         EXSY1(I,4,K)=EXS(I,NY,K)
 50    CONTINUE
 60  CONTINUE
C
      RETURN
      END
C    *****************************************************
      SUBROUTINE RADEXZ
C
      INCLUDE 'COMMONA.FOR'
C$INCLUDE COMMONA
C
C    DO EDGES WITH FIRST ORDER ORBC
C
      DO 10 J=2,NY1
       I=1
       EXS(I,J,1)=EXSZ1(I,J,2)+CZD*(EXS(I,J,2)-EXSZ1(I,J,1))
       EXS(I,J,NZ)=EXSZ1(I,J,3)+CZD*(EXS(I,J,NZ1)-EXSZ1(I,J,4))
       I=NX1
       EXS(I,J,1)=EXSZ1(I,J,2)+CZD*(EXS(I,J,2)-EXSZ1(I,J,1))
       EXS(I,J,NZ)=EXSZ1(I,J,3)+CZD*(EXS(I,J,NZ1)-EXSZ1(I,J,4))
```

```
 10 CONTINUE
C
      DO 20 I=2,NX1-1
       J=2
       EXS(I,J,1)=EXSZ1(I,J,2)+CZD*(EXS(I,J,2)-EXSZ1(I,J,1))
       EXS(I,J,NZ)=EXSZ1(I,J,3)+CZD*(EXS(I,J,NZ1)-EXSZ1(I,J,4))
       J=NY1
       EXS(I,J,1)=EXSZ1(I,J,2)+CZD*(EXS(I,J,2)-EXSZ1(I,J,1))
       EXS(I,J,NZ)=EXSZ1(I,J,3)+CZD*(EXS(I,J,NZ1)-EXSZ1(I,J,4))
 20 CONTINUE
C
C   NOW DO 2ND ORDER ORBC ON REMAINING PORTIONS OF
C   FACES
C
      DO 40 J=3,NY1-1
       DO 30 I=2,NX1-1
        EXS(I,J,1)=-EXSZ2(I,J,2)+CZD*(EXS(I,J,2)+EXSZ2(I,J,1))
     $  +CZZ*(EXSZ1(I,J,1)+EXSZ1(I,J,2))
     $  +CZFXD*(EXSZ1(I+1,J,1)-2.*EXSZ1(I,J,1)+EXSZ1(I-1,J,1)
     $  +EXSZ1(I+1,J,2)-2.*EXSZ1(I,J,2)+EXSZ1(I-1,J,2))
     $  +CZFYD*(EXSZ1(I,J+1,1)-2.*EXSZ1(I,J,1)+EXSZ1(I,J-1,1)
     $  +EXSZ1(I,J+1,2)-2.*EXSZ1(I,J,2)+EXSZ1(I,J-1,2))
        EXS(I,J,NZ)=-EXSZ2(I,J,3)+CZD*(EXS(I,J,NZ1)+EXSZ2(I,J,4))
     $  +CZZ*(EXSZ1(I,J,4)+EXSZ1(I,J,3))
     $  +CZFXD*(EXSZ1(I+1,J,4)-2.*EXSZ1(I,J,4)+EXSZ1(I-1,J,4)
     $  +EXSZ1(I+1,J,3)-2.*EXSZ1(I,J,3)+EXSZ1(I-1,J,3))
     $  +CZFYD*(EXSZ1(I,J+1,4)-2.*EXSZ1(I,J,4)+EXSZ1(I,J-1,4)
     $  +EXSZ1(I,J+1,3)-2.*EXSZ1(I,J,3)+EXSZ1(I,J-1,3))
 30    CONTINUE
 40 CONTINUE
C
C   NOW SAVE PAST VALUES
C
      DO 60 J=2,NY1
       DO 50 I=1,NX1
        EXSZ2(I,J,1)=EXSZ1(I,J,1)
        EXSZ2(I,J,2)=EXSZ1(I,J,2)
        EXSZ2(I,J,3)=EXSZ1(I,J,3)
        EXSZ2(I,J,4)=EXSZ1(I,J,4)
        EXSZ1(I,J,1)=EXS(I,J,1)
        EXSZ1(I,J,2)=EXS(I,J,2)
        EXSZ1(I,J,3)=EXS(I,J,NZ1)
        EXSZ1(I,J,4)=EXS(I,J,NZ)
 50    CONTINUE
 60 CONTINUE
```

```
C
      RETURN
      END
C     ****************************************************
      SUBROUTINE RADEYZ
C
      INCLUDE 'COMMONA.FOR'
C$INCLUDE COMMONA
C
C     DO EDGES WITH FIRST ORDER ORBC
C
      DO 10 J=1,NY1
        I=2
        EYS(I,J,1)=EYSZ1(I,J,2)+CZD*(EYS(I,J,2)-EYSZ1(I,J,1))
        EYS(I,J,NZ)=EYSZ1(I,J,3)+CZD*(EYS(I,J,NZ1)-EYSZ1(I,J,4))
        I=NX1
        EYS(I,J,1)=EYSZ1(I,J,2)+CZD*(EYS(I,J,2)-EYSZ1(I,J,1))
        EYS(I,J,NZ)=EYSZ1(I,J,3)+CZD*(EYS(I,J,NZ1)-EYSZ1(I,J,4))
 10   CONTINUE
C
      DO 20 I=3,NX1-1
        J=1
        EYS(I,J,1)=EYSZ1(I,J,2)+CZD*(EYS(I,J,2)-EYSZ1(I,J,1))
        EYS(I,J,NZ)=EYSZ1(I,J,3)+CZD*(EYS(I,J,NZ1)-EYSZ1(I,J,4))
        J=NY1
        EYS(I,J,1)=EYSZ1(I,J,2)+CZD*(EYS(I,J,2)-EYSZ1(I,J,1))
        EYS(I,J,NZ)=EYSZ1(I,J,3)+CZD*(EYS(I,J,NZ1)-EYSZ1(I,J,4))
 20   CONTINUE
C
C     NOW DO 2ND ORDER ORBC ON REMAINING PORTIONS OF
C     FACES
C
      DO 40 J=2,NY1-1
        DO 30 I=3,NX1-1
          EYS(I,J,1)=-EYSZ2(I,J,2)+CZD*(EYS(I,J,2)+EYSZ2(I,J,1))
     $    +CZZ*(EYSZ1(I,J,1)+EYSZ1(I,J,2))
     $    +CZFXD*(EYSZ1(I+1,J,1)-2.*EYSZ1(I,J,1)+EYSZ1(I-1,J,1)
     $    +EYSZ1(I+1,J,2)-2.*EYSZ1(I,J,2)+EYSZ1(I-1,J,2))
     $    +CZFYD*(EYSZ1(I,J+1,1)-2.*EYSZ1(I,J,1)+EYSZ1(I,J-1,1)
     $    +EYSZ1(I,J+1,2)-2.*EYSZ1(I,J,2)+EYSZ1(I,J-1,2))
          EYS(I,J,NZ)=-EYSZ2(I,J,3)+CZD*(EYS(I,J,NZ1)+EYSZ2(I,J,4))
     $    +CZZ*(EYSZ1(I,J,4)+EYSZ1(I,J,3))
     $    +CZFXD*(EYSZ1(I+1,J,4)-2.*EYSZ1(I,J,4)+EYSZ1(I-1,J,4)
     $    +EYSZ1(I+1,J,3)-2.*EYSZ1(I,J,3)+EYSZ1(I-1,J,3))
     $    +CZFYD*(EYSZ1(I,J+1,4)-2.*EYSZ1(I,J,4)+EYSZ1(I,J-1,4)
```

```
   $   +EYSZ1(I,J+1,3)-2.*EYSZ1(I,J,3)+EYSZ1(I,J-1,3))
 30     CONTINUE
 40   CONTINUE
C
C     NOW SAVE PAST VALUES
C
      DO 60 J=1,NY1
        DO 50 I=2,NX1
          EYSZ2(I,J,1)=EYSZ1(I,J,1)
          EYSZ2(I,J,2)=EYSZ1(I,J,2)
          EYSZ2(I,J,3)=EYSZ1(I,J,3)
          EYSZ2(I,J,4)=EYSZ1(I,J,4)
          EYSZ1(I,J,1)=EYS(I,J,1)
          EYSZ1(I,J,2)=EYS(I,J,2)
          EYSZ1(I,J,3)=EYS(I,J,NZ1)
          EYSZ1(I,J,4)=EYS(I,J,NZ)
 50     CONTINUE
 60   CONTINUE
C
      RETURN
      END
C
***********************************************************************
      SUBROUTINE HXSFLD
C
      INCLUDE 'COMMONA.FOR'
C$INCLUDE COMMONA
C
C     THIS SUBROUTINE UPDATES THE HX SCATTERED FIELD
C     COMPONENTS
C
      DO 30 K=1,NZ1
        DO 20 J=1,NY1
          DO 10 I=2,NX1
            HXS(I,J,K)=HXS(I,J,K)-(EZS(I,J+1,K)-EZS(I,J,K))*DTMDY
     $                      +(EYS(I,J,K+1)-EYS(I,J,K))*DTMDZ
 10       CONTINUE
 20     CONTINUE
 30   CONTINUE
C
      RETURN
      END
C     ****************************************************************
      SUBROUTINE HYSFLD
C
```

```
      INCLUDE 'COMMONA.FOR'
C$INCLUDE COMMONA
C
C    THIS SUBROUTINE UPDATES THE HY SCATTERED FIELD
C    COMPONENTS
C
      DO 30 K=1,NZ1
        DO 20 J=2,NY1
          DO 10 I=1,NX1
            HYS(I,J,K)=HYS(I,J,K)-(EXS(I,J,K+1)-EXS(I,J,K))*DTMDZ
     $                    +(EZS(I+1,J,K)-EZS(I,J,K))*DTMDX
 10       CONTINUE
 20     CONTINUE
 30   CONTINUE
C
      RETURN
      END
C    ******************************************************
      SUBROUTINE HZSFLD
C
      INCLUDE 'COMMONA.FOR'
C$INCLUDE COMMONA
C
C    THIS SUBROUTINE UPDATES THE HZ SCATTERED FIELD
C    COMPONENTS
C
      DO 30 K=2,NZ1
        DO 20 J=1,NY1
          DO 10 I=1,NX1
            HZS(I,J,K)=HZS(I,J,K)-(EYS(I+1,J,K)-EYS(I,J,K))*DTMDX
     $                    +(EXS(I,J+1,K)-EXS(I,J,K))*DTMDY
 10       CONTINUE
 20     CONTINUE
 30   CONTINUE
C
      RETURN
      END
C
***********************************************************
      SUBROUTINE DATSAV
C
      INCLUDE 'COMMONA.FOR'
C$INCLUDE COMMONA
C
C -------------------------------------------------------
```

```
C     To completely specifiy a sampling point, the user must specify
C       4 parameters: NTYPE, IOBS, JOBS, and KOBS.
C       NTYPE for point L is specified by NTYPE(L).
C         NTYPE controls what quantities are sampled as follows:
C           1 = EXS (scattered x-component of electric field)
C           2 = EYS (scattered y-component of electric field)
C           3 = EZS (scattered z-component of electric field)
C           4 = HXS (scattered x-component of magnetic field)
C           5 = HYS (scattered y-component of magnetic field)
C           6 = HZS (scattered z-component of magnetic field)
C           7 = IX (x-component of current through rectangular loop of H)
C           8 = IY (y-component of current through rectangular loop of H)
C           9 = IZ (z-component of current through rectangular loop of H)
C       IOBS(L), JOBS(L) AND KOBS(L) specify the I, J and K coordi-
C         nates of the point L. Thus for point 1, the user would specify
C         NTYPE(1), IOBS(1), JOBS(1) and KOBS(1). For point 2, the
C         user would specify NTYPE(2), IOBS(2), JOBS(2), KOBS(2).
C         The exact locations of a component within a cell follows that
C         of the Yee cell.
C       As an example, the following 4 lines save EYS at the point
C         (33,26,22):
C           NTYPE(1)=2
C           IOBS(1)=33
C           JOBS(1)=26
C           KOBS(1)=22
C
C     To sample total field, issue a statement to add in the appropriate
C       incident field. These total field save statements should be added
C       directly after the "3000 CONTINUE" statement near the end of this
C       SUBPROGRAM to override any previous assignment to
C       STORE(NPT). Make certain to back up the incident field by 1 time
C     step (DT) for sampling total E fields and by 1/2 time step (DT/2.0)
C     for sampling total H fields. For example, to store total EX field at
C     sample point 3, issue the following 3 statements:
C         T=T-DT
C         STORE(3)=EXS(IOBS(3),JOBS(3),KOBS(3))+
C   2       EXI(IOBS(3),JOBS(3),KOBS(3))
C         T=T+DT
C-----------------------------------------------------------------------
C
C  For the first time through (N=1), do some initializations...
      IF (N.NE.1) GO TO 10
C
C   User-defined test point cell location(s). There should be NTEST
C     occurences of NTYPE and [I,J,K]OBS defined. Unless total fields
```

```
C     are desired, this is probably the only part of this SUBROUTINE
C     the user will need to change.
C
C
      NTYPE(1)=1
      NTYPE(2)=1
      NTYPE(3)=5
      NTYPE(4)=5
C
C     DEFINE NTEST TEST POINT CELL LOCATION(S) HERE
C
      IOBS(1)=17
      JOBS(1)=18
      KOBS(1)=25
      IOBS(2)=17
      JOBS(2)=18
      KOBS(2)=18
      IOBS(3)=17
      JOBS(3)=18
      KOBS(3)=24
      IOBS(4)=17
      JOBS(4)=18
      KOBS(4)=17
C
C   Initialize some variables.
      DO 5 II=1,NTEST
        STORE (II)=0.
 5    CONTINUE
      NPTS=NTEST
C
C   Print header info and check [I,J,K]OBS to see that they are in
 C  range. Write header to data file and to DIAGS3D.DAT.
      WRITE (10,1400) DELX,DELY,DELZ,DT,NSTOP,NPTS
      WRITE (17,1500) NPTS
 1400 FORMAT (T2,E12.5,2X,E12.5,2X,E12.5,3X,E14.7,3X,I6,3X,I3)
 1500 FORMAT (T2,'QUANTITIES SAMPLED AND SAVED AT',I4, '
       LOCATIONS',/)
      DO 1700 NPT = 1,NPTS
        WRITE (17,1800) NPT, NTYPE (NPT), IOBS (NPT), JOBS (NPT),
       KOBS (NPT)
        IF (IOBS(NPT).GT.NX) GO TO 1600
        IF (IOBS(NPT).LT.1) GO TO 1600
        IF (JOBS(NPT).GT.NY) GO TO 1600
        IF (JOBS(NPT).LT.1) GO TO 1600
        IF (KOBS(NPT).GT.NZ) GO TO 1600
```

```
      IF (KOBS(NPT).LT.1) GO TO 1600
      GO TO 1700
 1600 WRITE (17,*) 'ERROR IN IOBS, JOBS OR KOBS FOR SAMP-
      LING'
      WRITE (17,*) 'POINT ',NPT
      WRITE (17,*) 'EXECUTION HALTED.'
      CLOSE (UNIT=10)
      CLOSE (UNIT=17)
      CLOSE (UNIT=30)
      STOP
 1700 CONTINUE
 1800 FORMAT (T2,'SAMPLE:',I4,', NTYPE=',I4,', SAMPLED AT
     CELL I=',
    2 I4,', J=',I4,', K=',I4,/)
C
C  Check for ERROR in NTYPE specification.
     DO 20 NPT=1,NTEST
      IF ((NTYPE(NPT).GE.10).OR.(NTYPE(NPTS).LE.0)) THEN
        WRITE (17,*) 'ERROR IN NYPE FOR SAMPLING POINT ',NPT
        WRITE (17,*) 'EXECUTION HALTED.'
        CLOSE (UNIT=10)
        CLOSE (UNIT=17)
        CLOSE (UNIT=30)
        STOP
      ENDIF
 20  CONTINUE
C  Finished with initialization.
C
C  Cycle through all NTEST sample locations.
 10  DO 3000 NPT=1,NTEST
C
C  Put desired quantity for this sample location into STORE(NPT).
     I=IOBS(NPT)
     J=JOBS(NPT)
     K=KOBS(NPT)
C   Scattered fields
     IF (NTYPE(NPT).EQ.1) STORE(NPT)=EXS(I,J,K)
     IF (NTYPE(NPT).EQ.2) STORE(NPT)=EYS(I,J,K)
     IF (NTYPE(NPT).EQ.3) STORE(NPT)=EZS(I,J,K)
     IF (NTYPE(NPT).EQ.4) STORE(NPT)=HXS(I,J,K)
     IF (NTYPE(NPT).EQ.5) STORE(NPT)=HYS(I,J,K)
     IF (NTYPE(NPT).EQ.6) STORE(NPT)=HZS(I,J,K)
C   Current loops
C    X-directed current
     IF (NTYPE(NPT).EQ.7) THEN
```

```
        STORE(NPT)=0.
        DO 711 KK=K,K+1
          DO 710 JJ=J,J+1
            STORE(NPT)=STORE(NPT)+(-HYS(I,JJ,KK)+HYS(I,JJ,KK-1))
     2      *DELY+ (HZS(I,JJ,KK)-HZS(I,JJ-1,KK))*DELZ
 710      CONTINUE
 711    CONTINUE
      ENDIF
C     Y-directed current
      IF (NTYPE(NPT).EQ.8) THEN
        STORE(NPT)=0.
        DO 811 KK=K,K+1
          DO 810 II=I,I+1
            STORE(NPT)=STORE(NPT)+(-HZS(II,J,KK)+HZS
            (II-1,J,KK))*DELZ+
     2      (HXS(II,J,KK)-HXS(II,J,KK-1))*DELX
 810      CONTINUE
 811    CONTINUE
      ENDIF
C     Z-directed current
      IF (NTYPE(NPT).EQ.9) THEN
        STORE(NPT)=0.
        DO 911 JJ=J,J+1
          DO 910 II=I,I+1
            STORE(NPT)=STORE(NPT)+(-HXS(II,JJ,K)+HXS
      (II,JJ-1,K))*DELX+
     2      (HYS(II,JJ,K)-HYS(II-1,JJ,K))*DELY
 910      CONTINUE
 911    CONTINUE
      ENDIF
C
 3000 CONTINUE
C
C    If total fields are desired, place total field save statements
C     here.  Example for saving total EX field at sample point 3:
C     T=T-DT
C     STORE(3)=EXS(IOBS(3),JOBS(3),KOBS(3))+
C    2  EXI(IOBS(3),JOBS(3),KOBS(3))
C     T=T+DT
C
C    Write sampled values to disk
      WRITE (10,*) (STORE(II),II=1,NPTS)
C
      RETURN
      END
```

```
C
*****************************************************************
      FUNCTION EXI(I,J,K)
C
      INCLUDE 'COMMONA.FOR'
C$INCLUDE COMMONA
C
C     THIS FUNCTION COMPUTES THE X COMPONENT OF THE
C     INCIDENT ELECTRIC FIELD
C
      DIST=((I-1)*DELX+0.5*DELX*OFF)*XDISP+((J-1)
       *DELY)*YDISP+
     $((K-1)*DELZ)*ZDISP + DELAY
      EXI = AMPX*SOURCE(DIST)
      RETURN
      END
C
*****************************************************************
      FUNCTION EYI(I,J,K)
C
      INCLUDE 'COMMONA.FOR'
C$INCLUDE COMMONA
C
C     THIS FUNCTION COMPUTES THE Y COMPONENT OF THE
C     INCIDENT ELECTRIC FIELD
C
      DIST=((I-1)*DELX)*XDISP+((J-1)*DELY+0.5*DELY*OFF)*
     $YDISP+((K-1)*DELZ)*ZDISP + DELAY
      EYI = AMPY*SOURCE(DIST)
      RETURN
      END
C
*************************************************************
      FUNCTION EZI(I,J,K)
C
      INCLUDE 'COMMONA.FOR'
C$INCLUDE COMMONA
C
C     THIS FUNCTION COMPUTES THE Z COMPONENT OF THE
INCIDENT ELECTRIC FIELD
C
      DIST=((I-1)*DELX)*XDISP+((J-1)*DELY)*YDISP+((K-1)*DELZ+
     $0.5*DELZ*OFF)*ZDISP + DELAY
      EZI = AMPZ*SOURCE(DIST)
      RETURN
```

```
      END
C     ****************************************************
      FUNCTION SOURCE(DIST)
C
      INCLUDE 'COMMONA.FOR'
C$INCLUDE COMMONA
C
C     THIS FUNCTION DEFINES THE FUNCTIONAL FORM OF THE
C     INCIDENT FIELD
C
      SOURCE = 0.0
C
C     EXECUTE FOLLOWING FOR INCIDENT E FIELD
C
      TAU=T-DIST/C
      IF(TAU.LT.0.0) GO TO 10
      IF (TAU.GT.PERIOD) GO TO 10
C
C
      SOURCE=EXP(-ALPHA*((TAU-BETADT)**2))
C
C
 10   RETURN
      END
C
*************************************************************
      FUNCTION DEXI(I,J,K)
C
      INCLUDE 'COMMONA.FOR'
C$INCLUDE COMMONA
C
C     TIME DERIVATIVE OF INCIDENT EX FIELD FOR DIELEC
C     TRICS
C
       DIST=((I-1)*DELX+0.5*DELX*OFF)*XDISP+((J-1)*DELY)*YDISP+
      $((K-1)*DELZ)*ZDISP + DELAY
      DEXI=AMPX*DSRCE(DIST)
C
      RETURN
      END
C
*************************************************************
      FUNCTION DEYI(I,J,K)
C
```

```
      INCLUDE 'COMMONA.FOR'
C$INCLUDE COMMONA
C
C     TIME DERIVATIVE OF INCIDENT EY FIELD FOR DIELEC-
C     TRICS
C
      DIST=((I-1)*DELX)*XDISP+((J-1)*DELY+0.5*DELY*OFF)*YDISP+
     $((K-1)*DELZ)*ZDISP + DELAY
      DEYI=AMPY*DSRCE(DIST)
      RETURN
      END
C     ********************************************************
      FUNCTION DEZI(I,J,K)
C
      INCLUDE 'COMMONA.FOR'
C$INCLUDE COMMONA
C
C     TIME DERIVATIVE OF INCIDENT EZ FIELD FOR DIELEC-
C        TRICS
C
      DIST=((I-1)*DELX)*XDISP+((J-1)*DELY)*YDISP+((K-1)*DELZ+
     $0.5*DELZ*OFF)*ZDISP + DELAY
      DEZI=AMPZ*DSRCE(DIST)
      RETURN
      END
C
**************************************************************
      FUNCTION DSRCE(DIST)
C
      INCLUDE 'COMMONA.FOR'
C$INCLUDE COMMONA
C
C     TIME DERIVATIVE OF INCIDENT FIELD FOR DIELECTRICS
C
      DSRCE=0.0
C
      TAU=T-DIST/C
      IF(TAU.LT.0.0) GO TO 10
      IF (TAU.GT.PERIOD) GO TO 10
C
       DSRCE=EXP(-ALPHA*((TAU-BETADT)**2))*(-2.*ALPHA*(TAU-
       BETADT))
C
C
 10   RETURN
```

```
      END
C
**************************************************************
      SUBROUTINE ZERO
C
      INCLUDE 'COMMONA.FOR'
C$INCLUDE COMMONA
C
C     THIS SUBROUTINE INITIALIZES VARIOUS ARRAYS AND
C     CONSTANTS TO ZERO.
C
      T=0.0
      DO 30 K=1,NZ
        DO 20 J=1,NY
          DO 10 I=1,NX
            EXS(I,J,K)=0.0
            EYS(I,J,K)=0.0
            EZS(I,J,K)=0.0
            HXS(I,J,K)=0.0
            HYS(I,J,K)=0.0
            HZS(I,J,K)=0.0
            IDONE(I,J,K)=0
            IDTWO(I,J,K)=0
            IDTHRE(I,J,K)=0
 10       CONTINUE
 20     CONTINUE
 30   CONTINUE
      DO 60 K=1,NZ1
        DO 50 J=1,NY1
          DO 40 I=1,4
            EYSX1(I,J,K)=0.0
            EYSX2(I,J,K)=0.0
            EZSX1(I,J,K)=0.0
            EZSX2(I,J,K)=0.0
 40       CONTINUE
 50     CONTINUE
 60   CONTINUE
      DO 90 K=1,NZ1
        DO 80 J=1,4
          DO 70 I=1,NX1
            EXSY1(I,J,K)=0.0
            EXSY2(I,J,K)=0.0
            EZSY1(I,J,K)=0.0
            EZSY2(I,J,K)=0.0
 70       CONTINUE
```

```
80    CONTINUE
90  CONTINUE
    DO 120 K=1,4
     DO 110 J=1,NY1
      DO 100 I=1,NX1
       EXSZ1(I,J,K)=0.0
       EXSZ2(I,J,K)=0.0
       EYSZ1(I,J,K)=0.0
       EYSZ2(I,J,K)=0.0
100     CONTINUE
110    CONTINUE
120  CONTINUE
    DO 130 L=1,9
     ESCTC(L)=0.0
     EINCC(L)=0.0
     EDEVCN(L)=0.0
     ECRLX(L)=0.0
     ECRLY(L)=0.0
     ECRLZ(L)=0.0
130  CONTINUE
    RETURN
    END
```

COMMONA.FOR

```
C
C   SPECIFY THE PROBLEM SPACE SIZE HERE.  NX, NY AND NZ
C   SPECIFY THE SIZE OF THE PROBLEM SPACE IN THE X, Y
C   AND Z DIRECTIONS.
C
    PARAMETER (NX=34,NY=34,NZ=34,NX1=NX-1,NY1=NY-1,
     NZ1=NZ-1)
C
C   DEFINE THE FIELD OUTPUT QUANTITIES HERE.  NTEST
C   SPECIFIES THE NUMBER OF FIELD SAMPLE LOCATIONS FOR
C   NEAR ZONE FIELDS.
C
    PARAMETER (NTEST=4)
C
C   DEFINE NSTOP, (MAXIMUM NUMBER OF TIME STEPS) HERE.
C
    PARAMETER (NSTOP=256)
C
C   DEFINE CELL SIZE (DELX, DELY, DELZ, IN METERS) HERE.
```

```
C
      PARAMETER (DELX=1./17.,DELY=1./17.,DELZ=1./17.)
C
C     SET INCIDENCE ANGLE OF INCIDENT PLANE WAVE HERE
C
C     SINCE THIS IS A SCATTERING CODE, ONE MUST BE CARE
C     FUL IN SPECIFYING THE INCIDENCE ANGLES, THINC AND
C     PHINC. THE  INCIDENCE ANGLES (ACCORDING TO SCAT-
C     TERING CONVENTION)  ARE TAKEN AS THE ANGLES THE
C     INCIDENT WAVE COMES FROM AND NOT THE ANGLES THE
C     PROPAGATION VECTOR MAKES WITH THE X AND Z AXES.
C
C     THINC AND PHINC ARE SPECIFIED AS FOLLOWS:
C     THINC IS SPECIFIED FROM THE +Z AXIS
C     PHINC IS SPECIFIED FROM THE +X AXIS
C     (ALL ANGLES ARE SPECIFIED IN DEGREES)
C
      PARAMETER (THINC=180.0,PHINC=180.0)
C
C     SPECIFY POLARIZATION OF INCIDENT PLANE WAVE HERE
C
C     SET INCIDENT WAVE POLARIZATION — ETHINC = 1 FOR
C     THETA POLARIZED ELECTRIC FIELD, EPHINC = 1 FOR PHI
C     POLARIZED
C
      PARAMETER (ETHINC=1.0,EPHINC=0.0)
C
C     SET INCIDENT WAVEFORM PARAMETERS
C
C     PARAMETER AMP IS THE MAXIMUM AMPLITUDE OF THE
C     INCIDENT PLANE WAVE AND BETA IS THE TEMPORAL
C     WIDTH OF A GAUSSIAN PULSE SPECIFIED IN TIME STEPS.
C
C     PARAMETER (AMP=1000.0,BETA=32.0)
C
C    CHANGE BETA TO 64 SINCE SPHERE HAS RELATIVE PERMIT-
C     TIVITY 4
      PARAMETER (AMP=1000.0,BETA=64.0)
C
      PARAMETER (EPS0=8.854E-12,XMU0=1.2566306E-
       6,ETA0=376.733341)
C
      COMMON/IDS/IDONE (NX, NY, NZ), IDTWO (NX, NY, NZ),
       IDTHRE (NX, NY, NZ)
```

```
      COMMON/ESCAT/EXS (NX, NY, NZ), EYS (NX, NY, NZ), EZS
      (NX, NY, NZ)
     COMMON/HSCAT/HXS (NX, NY, NZ), HYS (NX, NY, NZ), HZS
      (NX, NY, NZ)
C
     COMMON /MUR/  CXD,CXU,CYD,CYU,CZD,CZU
     COMMON /MUR2/ CXX, CYY,CZZ,CXFYD,CXFZD,CYFXD,
     CYFZD,CZFXD,CZFYD
C
     COMMON/RADSAV/EYSX1 (4,NY1, NZ1), EZSX1 (4,NY1, NZ1),
    $           EZSY1 (NX1, 4, NZ1), EXSY1 (NX1, 4, NZ1),
    $           EXSZ1 (NX1, NY1, 4), EYSZ1 (NX1, NY1, 4)
     COMMON/RADSV2/EYSX2 (4,NY1,NZ1),EZSX2(4,NY1,NZ1),
    $           EZSY2 (NX1, 4, NZ1),EXSY2 (NX1, 4, NZ1),
    $           EXSZ2 (NX1, NY1, 4),EYSZ2 (NX1, NY1, 4)
C
     COMMON /INCPW/ AMPX, AMPY, AMPZ, XDISP, YDISP, ZDISP,
      DELAY,
    $            TAU,OFF
     COMMON/EXTRAS/
N,DT,T,NPTS,C,PI,ALPHA,PERIOD,BETADT,W1,W2,W3
C
     COMMON/TERMS/
ESCTC(9),EINCC(9),EDEVCN(9),ECRLX(9),ECRLY(9),
    $ECRLZ(9),DTEDX,DTEDY,DTEDZ,DTMDX,DTMDY,DTMDZ
C
     COMMON/CONSTI/EPS(9),SIGMA(9)
C
     COMMON/SAVE/
STORE(NTEST),IOBS(NTEST),JOBS(NTEST),KOBS(NTEST),
    $NTYPE(NTEST)
```

DIAGS3D.DAT

```
THE PROBLEM SPACE SIZE IS  34 BY  34 BY  34 CELLS IN THE X,
Y, Z  DIRECTIONS

CELL SIZE: DELX=  0.058824 DELY=  0.058824, DELZ=  0.058824
METERS

TIME STEP IS 0.113283E-09 SECONDS, MAXIMUM OF   256 TIME
STEPS

INCIDENT GAUSSIAN PULSE AMPLITUDE= 1000. V/M
```

DECAY FACTOR ALPHA= 0.304E+18
WIDTH BETA= 64.

INCIDENT PLANE WAVE POLARIZATION:
RELATIVE ELECTRIC FIELD THETA COMPONENT= 1.0
RELATIVE ELECTRIC FIELD PHI COMPONENT= 0.0

PLANE WAVE INCIDENT FROM THETA =180.00 PHI =180.00
DEGREES

THE INCIDENT EX AMPLITUDE = 1000.00000 V/M
THE INCIDENT EY AMPLITUDE = 0.00009 V/M
THE INCIDENT EZ AMPLITUDE = 0.00009 V/M

RELATIVE SPATIAL DELAY = 0.0000002

QUANTITIES SAMPLED AND SAVED AT 4 LOCATIONS

SAMPLE: 1, NTYPE= 1, SAMPLED AT CELL I= 17, J= 18, K= 25

SAMPLE: 2, NTYPE= 1, SAMPLED AT CELL I= 17, J= 18, K= 18

SAMPLE: 3, NTYPE= 5, SAMPLED AT CELL I= 17, J= 18, K= 24

SAMPLE: 4, NTYPE= 5, SAMPLED AT CELL I= 17, J= 18, K= 17

EXIT TIME= 0.2900040E-07 SECONDS, AT TIME STEP 256

NZOUT3D.DAT

0.58824E-01	0.58824E-01	0.58824E-01	0.1132828E-09	256	4
0.000000	0.000000	0.000000	0.000000		
0.000000	0.000000	0.000000	0.000000		
0.000000	0.000000	0.000000	0.000000		
0.000000	0.000000	0.000000	0.000000		
0.000000	0.000000	0.000000	0.000000		
0.000000	0.000000	0.000000	0.000000		
0.000000	0.000000	0.000000	0.000000		
0.000000	0.000000	0.000000	0.000000		
0.000000	0.000000	0.000000	0.000000		
0.000000	0.000000	0.000000	0.000000		
0.000000	0.000000	0.000000	0.000000		
0.000000	0.000000	0.000000	0.000000		
0.000000	0.000000	0.000000	0.000000		

0.000000	0.000000	0.000000	0.000000
0.000000	0.000000	0.000000	0.000000
0.000000	0.000000	0.000000	0.000000
0.000000	0.000000	0.000000	0.000000
0.000000	0.000000	0.000000	0.000000
0.000000	0.000000	0.000000	0.000000
0.000000	0.000000	0.000000	0.000000
0.000000	0.000000	0.000000	0.000000
0.000000	0.000000	0.000000	0.000000
0.000000	0.000000	0.000000	0.000000
0.000000	0.000000	0.000000	-0.202235E-14
0.000000	-0.108238E-12	0.000000	-0.605891E-13
0.000000	-0.334043E-11	0.000000	-0.439806E-11
0.000000	-0.238393E-09	0.000000	-0.911907E-10
0.000000	-0.509543E-08	0.000000	-0.680289E-08
0.000000	-0.368675E-06	0.000000	-0.112241E-06
0.000000	-0.633948E-05	0.000000	-0.371248E-06
0.000000	-0.821796E-04	-0.505071E-22	-0.769339E-06
-0.270318E-20	-0.206003E-03	-0.215671E-20	-0.137985E-05
-0.117868E-18	-0.398780E-03	-0.133459E-18	-0.232803E-05
-0.724897E-17	-0.699520E-03	-0.382751E-17	-0.380618E-05
-0.211383E-15	-0.116874E-02	-0.209046E-15	-0.610326E-05
-0.113785E-13	-0.189847E-02	-0.521084E-14	-0.964751E-05
-0.289138E-12	-0.302657E-02	-0.871081E-13	-0.150680E-04
-0.492210E-11	-0.475712E-02	-0.433616E-11	-0.232825E-04
-0.236493E-09	-0.738953E-02	-0.868801E-10	-0.356201E-04
-0.486287E-08	-0.113593E-01	-0.123657E-08	-0.539920E-04
-0.705515E-07	-0.172950E-01	-0.873090E-07	-0.811224E-04
-0.473612E-05	-0.260960E-01	-0.310801E-06	-0.120861E-03
-0.742324E-04	-0.390386E-01	-0.662690E-06	-0.178593E-03
-0.188055E-03	-0.579171E-01	-0.121921E-05	-0.261787E-03
-0.366518E-03	-0.852305E-01	-0.210613E-05	-0.380690E-03
-0.647417E-03	-0.124426	-0.351114E-05	-0.549233E-03
-0.108921E-02	-0.180215	-0.571056E-05	-0.786165E-03
-0.178014E-02	-0.258969	-0.911023E-05	-0.111647E-02
-0.285170E-02	-0.369225	-0.143061E-04	-0.157310E-02
-0.449825E-02	-0.522306	-0.221711E-04	-0.219910E-02
-0.700506E-02	-0.733077	-0.339773E-04	-0.305005E-02
-0.107880E-01	-1.02086	-0.515631E-04	-0.419703E-02
-0.164489E-01	-1.41050	-0.775613E-04	-0.572985E-02
-0.248508E-01	-1.93361	-0.115704E-03	-0.776077E-02
-0.372212E-01	-2.62999	-0.171228E-03	-0.104285E-01
-0.552889E-01	-3.54915	-0.251408E-03	-0.139021E-01
-0.814660E-01	-4.75200	-0.366250E-03	-0.183856E-01
-0.119086	-6.31261	-0.529384E-03	-0.241213E-01

-0.172710	-8.31988	-0.759195E-03	-0.313933E-01
-0.248522	-10.8792	-0.108025E-02	-0.405301E-01
-0.354818	-14.1138	-0.152505E-02	-0.519050E-01
-0.502625	-18.1657	-0.213616E-02	-0.659350E-01
-0.706451	-23.1962	-0.296879E-02	-0.830775E-01
-0.985192	-29.3855	-0.409376E-02	-0.103823
-1.36320	-36.9311	-0.560099E-02	-0.128684
-1.87156	-46.0456	-0.760345E-02	-0.158181
-2.54947	-56.9523	-0.102414E-01	-0.192822
-3.44590	-69.8798	-0.136870E-01	-0.233077
-4.62126	-85.0548	-0.181492E-01	-0.279350
-6.14930	-102.693	-0.238784E-01	-0.331944
-8.11890	-122.987	-0.311707E-01	-0.391023
-10.6359	-146.098	-0.403718E-01	-0.456569
-13.8249	-172.136	-0.518796E-01	-0.528346
-17.8300	-201.153	-0.661451E-01	-0.605851
-22.8165	-233.119	-0.836710E-01	-0.688284
-28.9701	-267.919	-0.105008	-0.774510
-36.4968	-305.331	-0.130748	-0.863037
-45.6207	-345.021	-0.161511	-0.952006
-56.5810	-386.532	-0.197930	-1.03919
-69.6271	-429.284	-0.240634	-1.12203
-85.0126	-472.574	-0.290216	-1.19763
-102.987	-515.581	-0.347208	-1.26289
-123.787	-557.386	-0.412042	-1.31450
-147.624	-596.983	-0.485016	-1.34912
-174.671	-633.311	-0.566247	-1.36341
-205.051	-665.284	-0.655630	-1.35425
-238.824	-691.823	-0.752797	-1.31881
-275.967	-711.901	-0.857074	-1.25472
-316.366	-724.583	-0.967446	-1.16023
-359.804	-729.064	-1.08253	-1.03429
-405.949	-724.715	-1.20054	-0.876707
-454.347	-711.116	-1.31931	-0.688208
-504.423	-688.084	-1.43629	-0.470511
-555.479	-655.700	-1.54853	-0.226339
-606.701	-614.322	-1.65277	0.405996E-01
-657.173	-564.584	-1.74548	0.325672
-705.889	-507.393	-1.82292	0.623424
-751.778	-443.904	-1.88124	0.927730
-793.727	-375.495	-1.91660	1.23196
-830.614	-303.723	-1.92525	1.52922
-861.333	-230.275	-1.90372	1.81253
-884.836	-156.914	-1.84888	2.07510
-900.158	-85.4195	-1.75814	2.31057

-906.454	-17.5259	-1.62956	2.51322
-903.032	45.1383	-1.46198	2.67819
-889.373	101.105	-1.25514	2.80162
-865.162	149.119	-1.00979	2.88084
-830.302	188.173	-0.727758	2.91437
-784.927	217.545	-0.411998	2.90205
-729.409	236.813	-0.666306E-01	2.84492
-664.362	245.862	0.303084	2.74524
-590.632	244.882	0.690795	2.60632
-509.289	234.344	1.08916	2.43236
-421.610	214.979	1.48996	2.22833
-329.055	187.737	1.88433	1.99969
-233.242	153.745	2.26292	1.75221
-135.914	114.258	2.61618	1.49176
-38.8988	70.6126	2.93466	1.22410
55.9240	24.1729	3.20930	0.954698
146.673	-23.7147	3.43174	0.688585
231.507	-71.7643	3.59471	0.430215
308.675	-118.788	3.69223	0.183387
376.562	-163.724	3.72000	-0.488046E-01
433.738	-205.661	3.67556	-0.263989
478.999	-243.851	3.55849	-0.460505
511.409	-277.713	3.37053	-0.637359
530.334	-306.836	3.11557	-0.794159
535.468	-330.972	2.79963	-0.931035
526.847	-350.020	2.43067	-1.04855
504.860	-364.013	2.01836	-1.14762
470.237	-373.097	1.57379	-1.22938
424.038	-377.516	1.10902	-1.29515
367.615	-377.585	0.636717	-1.34632
302.581	-373.677	0.169615	-1.38430
230.751	-366.200	-0.279914	-1.41046
154.087	-355.584	-0.700345	-1.42606
74.6329	-342.266	-1.08142	-1.43228
-5.55064	-326.680	-1.41451	-1.43015
-84.4545	-309.245	-1.69291	-1.42054
-160.184	-290.364	-1.91204	-1.40419
-231.015	-270.414	-2.06951	-1.38170
-295.444	-249.747	-2.16516	-1.35355
-352.230	-228.688	-2.20093	-1.32007
-400.414	-207.534	-2.18067	-1.28153
-439.343	-186.554	-2.10986	-1.23810
-468.667	-165.994	-1.99531	-1.18989
-488.330	-146.072	-1.84475	-1.13698
-498.557	-126.983	-1.66649	-1.07941

-499.816	-108.900	-1.46898	-1.01723
-492.794	-91.9720	-1.26050	-0.950483
-478.345	-76.3282	-1.04882	-0.879263
-457.458	-62.0737	-0.840897	-0.803692
-431.208	-49.2918	-0.642728	-0.723955
-400.712	-38.0426	-0.459164	-0.640308
-367.095	-28.3626	-0.293857	-0.553087
-331.450	-20.2636	-0.149248	-0.462719
-294.809	-13.7327	-0.266170E-01	-0.369729
-258.118	-8.73203	0.738141E-01	-0.274740
-222.217	-5.19875	0.152746	-0.178476
-187.830	-3.04608	0.211639	-0.817534E-01
-155.550	-2.16485	0.252527	0.145205E-01
-125.845	-2.42566	0.277846	0.109366
-99.0532	-3.68184	0.290251	0.201747
-75.3933	-5.77317	0.292458	0.290590
-54.9712	-8.53013	0.287110	0.374812
-37.7925	-11.7787	0.276658	0.453347
-23.7757	-15.3457	0.263285	0.525177
-12.7666	-19.0637	0.248839	0.589363
-4.55339	-22.7769	0.234813	0.645076
1.11892	-26.3454	0.222330	0.691625
4.53420	-29.6498	0.212163	0.728486
5.99248	-32.5948	0.204762	0.755318
5.79818	-35.1113	0.200292	0.771984
4.25028	-37.1583	0.198684	0.778557
1.63416	-38.7225	0.199686	0.775320
-1.78477	-39.8176	0.202907	0.762762
-5.76560	-40.4814	0.207874	0.741561
-10.0948	-40.7732	0.214063	0.712561
-14.5874	-40.7687	0.220944	0.676746
-19.0873	-40.5554	0.228005	0.635202
-23.4660	-40.2275	0.234775	0.589086
-27.6209	-39.8801	0.240838	0.539579
-31.4728	-39.6047	0.245844	0.487854
-34.9634	-39.4839	0.249513	0.435033
-38.0524	-39.5881	0.251633	0.382158
-40.7147	-39.9721	0.252061	0.330161
-42.9379	-40.6735	0.250715	0.279841
-44.7199	-41.7117	0.247568	0.231853
-46.0672	-43.0882	0.242641	0.186695
-46.9930	-44.7875	0.235996	0.144715
-47.5165	-46.7793	0.227730	0.106110
-47.6610	-49.0208	0.217964	0.709464E-01
-47.4540	-51.4600	0.206839	0.391696E-01

-46.9263	-54.0382	0.194511	0.106293E-01
-46.1119	-56.6936	0.181146	-0.149000E-01
-45.0470	-59.3639	0.166913	-0.376958E-01
-43.7707	-61.9888	0.151983	-0.580656E-01
-42.3236	-64.5120	0.136524	-0.763260E-01
-40.7482	-66.8828	0.120702	-0.927854E-01
-39.0880	-69.0570	0.104675	-0.107728
-37.3869	-70.9973	0.885945E-01	-0.121405
-35.6887	-72.6737	0.726028E-01	-0.134022
-34.0363	-74.0624	0.568346E-01	-0.145738
-32.4708	-75.1461	0.414149E-01	-0.156663
-31.0309	-75.9125	0.264591E-01	-0.166856
-29.7516	-76.3540	0.120731E-01	-0.176332
-28.6640	-76.4667	-0.164748E-02	-0.185063
-27.7942	-76.2495	-0.146177E-01	-0.192987
-27.1630	-75.7041	-0.267635E-01	-0.200011
-26.7854	-74.8342	-0.380227E-01	-0.206022
-26.6702	-73.6453	-0.483458E-01	-0.210891
-26.8203	-72.1448	-0.576967E-01	-0.214484
-27.2324	-70.3424	-0.660533E-01	-0.216669
-27.8978	-68.2499	-0.734080E-01	-0.217320
-28.8025	-65.8820	-0.797680E-01	-0.216332
-29.9280	-63.2566	-0.851551E-01	-0.213618
-31.2520	-60.3950	-0.896054E-01	-0.209124
-32.7491	-57.3226	-0.931677E-01	-0.202831
-34.3919	-54.0691	-0.959026E-01	-0.194757
-36.1514	-50.6685	-0.978800E-01	-0.184963
-37.9980	-47.1593	-0.991762E-01	-0.173557
-39.9025	-43.5842	-0.998712E-01	-0.160687
-41.8362	-39.9897	-0.100045	-0.146545
-43.7716	-36.4255	-0.997734E-01	-0.131361
-45.6829	-32.9436	-0.991265E-01	-0.115398
-47.5462	-29.5972	-0.981633E-01	-0.989434E-01
-49.3398	-26.4397	-0.969293E-01	-0.823027E-01
-51.0438	-23.5227	-0.954539E-01	-0.657866E-01
-52.6409	-20.8955	-0.937484E-01	-0.497023E-01
-54.1155	-18.6027	-0.918049E-01	-0.343415E-01
-55.4543	-16.6831	-0.895962E-01	-0.199701E-01
-56.6458	-15.1682	-0.870766E-01	-0.681855E-02
-57.6803	-14.0810	-0.841832E-01	0.492718E-02
-58.5499	-13.4351	-0.808394E-01	0.151318E-01
-59.2483	-13.2336	-0.769582E-01	0.237155E-01
-59.7706	-13.4693	-0.724465E-01	0.306555E-01
-60.1138	-14.1238	-0.672111E-01	0.359857E-01
-60.2760	-15.1690	-0.611643E-01	0.397936E-01

-60.2570	-16.5668	-0.542300E-01	0.422145E-01
-60.0577	-18.2709	-0.463503E-01	0.434241E-01
-59.6807	-20.2278	-0.374911E-01	0.436291E-01
-59.1294	-22.3790	-0.276480E-01	0.430565E-01
-58.4088	-24.6623	-0.168512E-01	0.419428E-01
-57.5242	-27.0142	-0.516886E-02	0.405225E-01
-56.4822	-29.3720	0.728990E-02	0.390181E-01
-55.2896	-31.6756	0.203747E-01	0.376304E-01
-53.9536	-33.8694	0.338945E-01	0.365309E-01
-52.4811	-35.9036	0.476207E-01	0.358559E-01
-50.8793	-37.7360	0.612917E-01	0.357025E-01
-49.1547	-39.3327	0.746193E-01	0.361269E-01
-47.3134	-40.6686	0.872974E-01	0.371446E-01
-45.3610	-41.7277	0.990113E-01	0.387330E-01
-43.3026	-42.5028	0.109449	0.408347E-01
-41.1432	-42.9951	0.118314	0.433633E-01
-38.8878	-43.2136	0.125334	0.462087E-01
-36.5420	-43.1737	0.130276	0.492444E-01

INDEX

A

D

E

F

T

Z

Milton Keynes UK
Ingram Content Group UK Ltd.
UKHW021855071024
449327UK00021B/1578